Plant Secondary Metabolites Chemistry and Role

For centuries the vast and versatile pharmacological effects of medicinal plants and their constituents have played vital roles in biological, economic, social, spiritual, cultural, and physiological well-being. This unique text establishes a groundwork in natural product chemistry and phytochemistry by considering the biosynthesis and mechanistic way. There is abundant evidence showing that medicinal plants and their secondary metabolites are useful in preventing different ailments and this book discusses this as well as the mechanisms, amelioration, and biosynthesis of these metabolites. It helps readers to understand the computational, toxicological, cosmetic, and nutraceutical aspects of plant secondary metabolites.

Plant Secondary Metabolites Chemistry and Role

A Biosynthetic and Mechanistic Approach

Edited by
Malik Saadullah

CRC Press
Taylor & Francis Group
Boca Raton London New York

CRC Press is an imprint of the
Taylor & Francis Group, an **informa** business

First edition published 2025
by CRC Press
2385 NW Executive Center Drive, Suite 320, Boca Raton FL 33431

and by CRC Press
4 Park Square, Milton Park, Abingdon, Oxon, OX14 4RN

CRC Press is an imprint of Taylor & Francis Group, LLC

ISBN: 978-1-032-70250-6 (hbk)
ISBN: 978-1-032-70726-6 (pbk)
ISBN: 978-1-032-70727-3 (ebk)

DOI 10.1201/9781032707273

Typeset in Palatino
by KnowledgeWorks Global Ltd.

Table of Contents

About the Editor

Dr. Malik Saadullah earned his PhD in Pharmaceutical Chemistry and is an Assistant Professor in the Department of Pharmaceutical Chemistry, Faculty of Pharmaceutical Sciences, Government College University, Faisalabad, Pakistan. Dr. Malik has a robust background in mass spectrometry, chromatography, biomolecules, and the development of anticancer and anti-inflammatory models both in vitro and in vivo. He is experienced in, and passionate about advancing the field of chemistry.

Over the course of his career, he has accumulated more than 1300 citations, reflecting the significant impact of his research. His work, which spans a cumulative impact factor of more than 400, has been recognized and respected by peers worldwide. As the editor of two influential books and author of more than 30 chapters, he has contributed extensively to the academic and scientific community, shaping the discourse and direction of research. He had the privilege of supervising the research work of 26 MPhil students and 3 PhD students. This role allowed him to mentor the next generation of scientists, fostering their growth, and guiding their research projects to successful completion.

His research has focused on the application of mass spectrometry and chromatography techniques to identify and characterize bioactive compounds with potential therapeutic applications. He has optimized numerous methodologies for the analysis of complex biological matrices, contributing to the understanding of disease mechanisms and the discovery of novel therapeutic agents.

In the realm of anticancer and anti-inflammatory research, he has worked on several in-vitro and in-vivo models that have provided critical insights into the efficacy and mechanisms of action of potential drug candidates. His work in this area has led to the identification of promising compounds that can be advanced to preclinical and clinical development stages.

Contributors

Ghulam Abbas
Department of Pharmaceutics
Government College University
Faisalabad, Pakistan

Abeesha
Department of Pharmaceutical Chemistry
Government College University
Faisalabad, Pakistan

Khurram Afzal
Department of Pharmaceutical Chemistry
Government College University
Faisalabad, Pakistan

Samina Afzal
Department of Pharmaceutical Chemistry
Bahauddin Zakariya University
Multan, Pakistan

Mohsin Ali
Government College University
Faisalabad, Pakistan

Nasir Ali
Department of Pharmaceutical Chemistry
Government College University
Faisalabad, Pakistan

Asra Shanzeh Shabbir
Department of Pharmaceutical Chemistry
Government College University
Faisalabad, Pakistan

Muhammad Yasir Ali
Department of Pharmaceutics
Government College University
Faisalabad, Pakistan

Areej Asghar
Department of Pharmaceutical Chemistry
Government College University
Faisalabad, Pakistan

Muhammad Asif
Department of Pharmacology
The Islamia University
Bahawalpur, Pakistan

Rizwan Rasheed Bazmi
Department of Pharmaceutical Chemistry
Government College University
Faisalabad, Pakistan

Duryab
Department of Pharmaceutical Chemistry
Government College University
Faisalabad, Pakistan

Syeda Abida Ejaz
Department of Pharmaceutical Chemistry
The Islamia University
Bahawalpur, Pakistan

Muhammad Fakhar-e-Alam
Department of Physics
Government College University
Faisalabad, Pakistan

Maryam Farrukh
Government College University
Faisalabad, Pakistan

Minahil Fatima
Department of Pharmacology
Government College University
Faisalabad, Pakistan

Humaira Gul
Department of Pharmacology
Government College University
Faisalabad, Pakistan

Muhammad Hanif
Department of Pharmaceutics
Bahauddin Zakariya University
Multan, Pakistan

Iqbal Hussain
Department of Botany
Government College University
Faisalabad, Pakistan

Muhammad Iqbal
Government College University
Faisalabad, Pakistan

Kanz-ul-Eman
Department of Pharmaceutical Chemistry
Government College University
Faisalabad, Pakistan

Saher Madni
Government College University
Faisalabad, Pakistan

Abid Mahmood
Department of Pharmaceutical Chemistry
Government College University
Faisalabad, Pakistan

Maria Manan
Department of Pharmacology
Government College University
Faisalabad, Pakistan

Shahid Manzoor
Department of Pharmaceutical Chemistry
Government College University
Faisalabad, Pakistan

Noor ul Amin Mohsin
Department of Pharmaceutical Chemistry
The Islamia University
Bahawalpur, Pakistan

Muhammad Usama Munir
Department of Pharmaceutical Chemistry
Government College University
Faisalabad, Pakistan

Hassan Murtaza Naqvi
Department of Pharmaceutical Chemistry
Government College University
Faisalabad, Pakistan

Shahab Nasir
Department of Zoology
Government College University
Faisalabad, Pakistan

Hassan Nawaz
Department of Pharmaceutical Chemistry
Government College University
Faisalabad, Pakistan

Sameen Qamar
Department of Pharmaceutical Chemistry
Government College University
Faisalabad, Pakistan

Qurat-ul-Ain
Department of Pharmaceutical Chemistry
Government College University
Faisalabad, Pakistan

Ayesha Rafiq
Department of Pharmaceutical Chemistry
Government College University
Faisalabad, Pakistan

Afifa Rafique
Government College University
Faisalabad, Pakistan

Azhar Rasool
Department of Zoology
Baba Guru Nanak University
Nankana Sahib, Pakistan

M. Rabi Raza
Department of Pharmaceutical Chemistry
Government College University
Faisalabad, Pakistan

Muneeb ur Rehman
Department of Pharmaceutical Chemistry
Government College University
Faisalabad, Pakistan

Zia Ur Rehman
Institute of Biological Sciences,
Gomal University
Dera Ismail Khan, Pakistan

Farhan Sadeeq
Department of Pharmaceutical Chemistry
Bahauddin Zakariya University
Multan, Pakistan

Malik Saadullah
Department of Pharmaceutical Chemistry
Government College University
Faisalabad, Pakistan

Shahid Shah
Department of Pharmacy Practice
Government College University
Faisalabad, Pakistan

M. Zain Sharif
Department of Pharmaceutical Chemistry
Government College University
Faisalabad, Pakistan

Ayesha Shehzadi
Government College University
Faisalabad, Pakistan

Farwa Tahir
Department of Pharmaceutical Chemistry
Government College University
Faisalabad, Pakistan

Muhammad Umer
Department of Pharmaceutical Chemistry
Government College University
Faisalabad, Pakistan

Preface

Plant and natural products are still turning out to be an undiscovered miracle of nature to man. The plant secondary metabolites are very important in our everyday life, as they contribute to our overall health. A medicinal chemist must understand secondary metabolites to grasp the chemistry of natural products as these compounds generate the basis of drug discovery. This knowledge highlights the significance of chemical diversity in all aspects of drug development. It has always been important to discover and develop novel medicinal chemicals to keep pace with the world's need as the world will not stop needing new and better medicinal chemicals that could be used as drugs. This book is a small contribution toward the understanding of secondary metabolites, the roles they may perform and the ways they can be used. It also discusses how they can be beneficial and how they can be harmful to humans; how their chemistry is linked to nature; and their roles.

Dr. Malik Saadullah

1 The Building Blocks

Secondary Metabolites and Chemistry

Malik Saadullah, Farhan Sadeeq, Muhammad Asif, Kanz-ul-Eman, Qurat-ul-Ain, Minahil Fatima, and Afifa Rafique*

1.1 INTRODUCTION

All organisms need energy for growth, reproduction, structural upkeep, and reacting to their surroundings. Metabolism entails the essential chemical reactions that enable organisms to convert stored chemical energy into usable energy for cellular activities. Animals derive energy from food consumption, breaking down carbohydrates, lipids, proteins, and nucleic acids through metabolism. Conversely, plants utilize sunlight to convert it into chemical energy stored within molecules via the process of photosynthesis.

The sum of all cellular chemical processes crucial for sustaining life, including the synthesis and decomposition of complex macromolecules, is called metabolism. There are further two types of metabolism: primary and secondary metabolism. Primary metabolites are chief plants constituents, such as sugars, amino acids and lipids, broadly distributed in large quantities in plants which are inevitable for plant growth, development, and reproduction. Basic complex molecules are formed as a result of primary metabolism (Verpoorte 2000). Plants possess the ability to produce various organic compounds known as secondary metabolites, having unique carbon structures.

These molecules, while not essential for cellular survival, play vital roles in interactions with the environment, ensuring the organism's sustainability within ecosystems. Variability in secondary metabolites exists within plant populations, offering protection against various biotic and abiotic stresses such as pathogens, temperature fluctuations, and heavy metals. These compounds hold significant economic value for humans, serving as pharmaceuticals, flavorings, fragrances, insecticides, and dyes. Secondary metabolites are vital in molding ecological interactions and ecosystem dynamics, significantly influencing the survival, reproduction, and interactions among organisms within their environments. Secondary metabolites may be used as (Tiwari and Rana 2015):

1. Competitive weapons against other bacteria, fungi, amoebae, plants, insects, and large animals
2. Metal transporting agents
3. Agents of symbiosis between microbes and plants, nematodes, insects, and higher animals
4. Sexual hormones
5. Differentiation effectors

Some secondary metabolites stimulate spore formation and inhibit or stimulate germination. Secondary metabolites can be classified into three groups on the basis of their biosynthesis origin:

1. Terpenoids
2. Polyketides
3. Phenylpropanoids

*Corresponding author

DOI: 10.1201/9781032707273-1

1.2 THE BUILDING BLOCKS

Building blocks are basic units or modules that serve as the base for constructing more complex structures or systems. They resemble physical blocks that are used by children to build analogous extensive structures and are made up of primary metabolites such as sugar, amino acids, and fatty acids.

1.2.1 Acetyl CoA

It is produced by pyruvate, a product of glycolysis, which is processed by hydroxyethyl-thiamine pyrophosphate (TPP) intermediate within the enzyme pyruvate dehydrogenase complex, and it reacts with coenzyme A (CoA). The acetyl-CoA is then converted into acetate through the removal of CoA and TPP is regenerated. Acetate can also be generated by beta oxidation of fatty acids, which occurs in mitochondria of eukaryotes and cytoplasm of prokaryotes. Before transporting to mitochondria, fatty acids must be activated by attaching CoA to their carboxyl end in presence of fatty acyl-CoA synthetase (Krivoruchko et al. 2015). They yield acetate by several metabolic pathways such as catabolism of ketogenic amino acids, branched chain amino acids, aromatic amino acids through deamination and transamination. Certain amino acids can also be converted into pyruvate through transamination or metabolic reactions yielding acetate or acetyl-CoA. Acetyl-CoA is utilized in biosynthesis of various secondary metabolites including phenols and flavonoids.

1.2.2 Malonyl-CoA

It is produced by acetyl-CoA carboxylation using acetyl-CoA carboxylase, which controls the synthesis of malonyl-CoA. Acetyl-CoA serves as a precursor of malonyl-CoA which is produced by oxidation of carbohydrates, fatty acids, or amino acids. This factor requires adenosine triphosphate (ATP) and biotin as a cofactor. In some metabolic conditions, glucose can also convert into malonyl-CoA indirectly. Malonyl-CoA serves as a critical building block for synthesis of polyketides, terpenoids, etc.

1.2.3 Shikimic Acid

It is formed from the amino acid pathway which is also known as chorismate pathway. In this pathway, arogenate is formed by a series of reactions which is then converted to shikimic acid. In this process, arogenate is converted to phenylpyruvate in the presence of the enzyme arogenate dehydratase. Shikimic acid is also produced by the shikimate pathway in the plants. In this pathway, 3-D-arabino-heptulosonate-7-phosphate (DAHP) is produced by the combination of phosphoenolpyruvate (PEP) and erythrose 4-phosphate (E4P) in the presence of enzyme DAHP synthase. DAHP undergoes a series of enzymatic reactions and shikimic acid is formed. Shikimic acid is utilized in shikimate pathway which leads to the formation of phenols and various alkaloids (Figure 1.1).

1.2.4 Mevalonic Acid

Mevalonic acid is important for the production of isoprenoids, which are very important for various physiological processes. The formation of mevalonic acid starts with the conversion of the acetyl-CoA to acetoacetyl-CoA which occurs in the presence of the enzyme acetyltransferase. Then acetoacetyl-CoA is combined with other acetyl-CoA molecule to produce

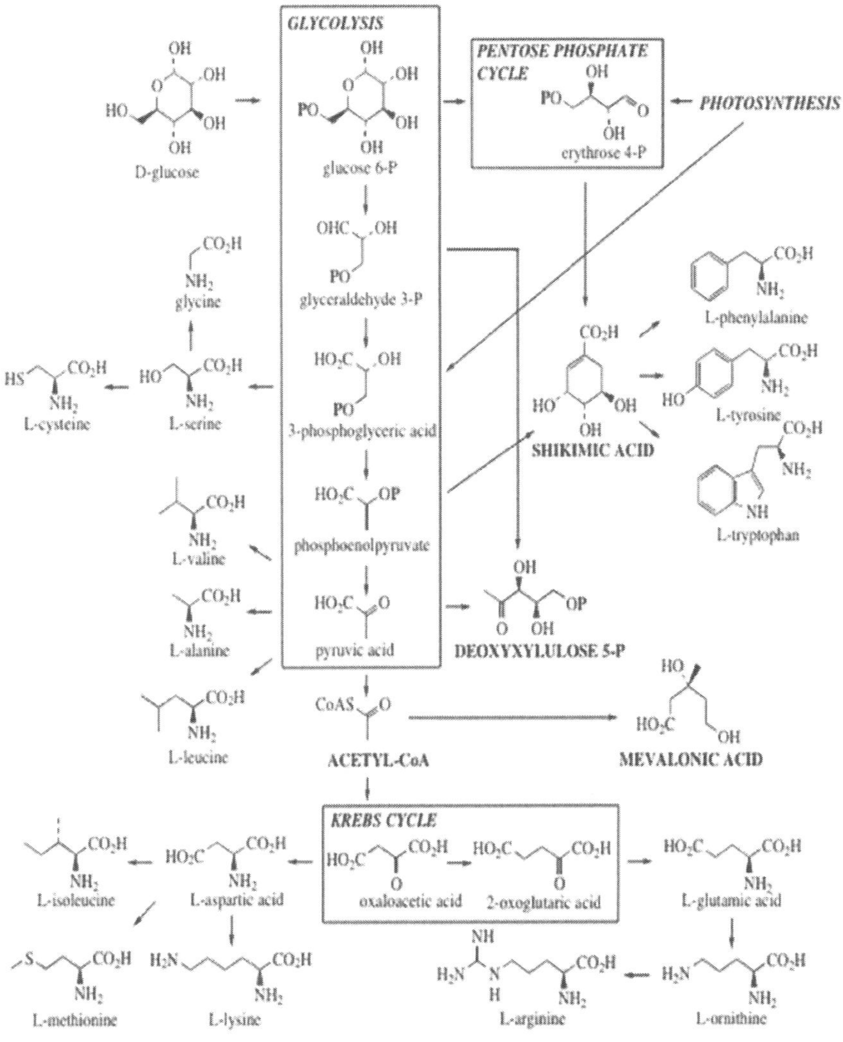

Figure 1.1 Shikimic acid pathway.

3-hydroxy-3-methylglutaryl-CoA (HMG-CoA) in the presence of enzyme HMG-CoA synthase (Brodie et al. 1963).

1.2.5 Deoxy Xylulose

Formation of deoxy xylulose occurs through the methylerythritol phosphate (MEP) pathway. The MEP pathway starts with the condensation of the glyceraldehyde 3-phosphate (GAP) and pyruvate in the presence of the enzyme 1-deoxy-D-xylulose 5-phosphate synthase (DXS). This condensation results in the production of 1-deoxy-D-xylulose 5-phosphate (DXP). These molecules are responsible for the formation of a diverse variety of steroid and terpenoid metabolites.

1.2.6 Isopentyl Diphosphate and Dimethyl Allyl Diphosphate

These are important precursors for biosynthesis of terpenoids, isoprenoids, and other secondary metabolites. They are formed in plant body from mevalonate which is phosphorylated to give mevalonate diphosphate. This mevalonate phosphate undergoes decarboxylation to yield isopentyl diphosphate (IPP) which isomerizes to dimethyl allyl diphosphate (DMAPP) in presence of isomerase. IPP and DMAPP can also be generated from MEP pathway that occurs inside plastid, in which deoxy xylulose undergoes series of enzymatic reactions yielding IPP and DMAPP.

Aromatic amino acids such as tryptophan, phenylalanine, and tyrosine are also essential building blocks for various secondary metabolites in plants including flavonoids, alkaloids, and indole derivatives (Alamgir 2018). The product of shikimate pathway, DHAP, undergoes enzymatic reactions and as a result prephenate is formed. Prephenate is converted to aromatic amino acids through processes such as transamination or hydroxylation.

Some *nitrogenous compounds, fatty acids,* and *sugars* also serve as precursors for biosynthesis of secondary metabolites including polyketides, fatty acid-derived alkaloids.

1.3 STRUCTURAL FEATURES OF BUILDING BLOCKS

The sources of building blocks of secondary metabolites are shown in Table 1.1, while Figure 1.2 mentions some building blocks.

Table 1.1: Sources of Building Blocks of Secondary Metabolites

Building Blocks	Source	Transformation	Example
C1	Derived from S-methyl of L-methionine	Methyl group attached to oxygen, nitrogen, or carbon	Methyl group, Methylenedioxy group
C2	Supplied by acetyl-CoA	Converted to malonyl-CoA before incorporation	Fatty acids, phenols
C5	Formed from mevalonate or deoxy xylulose phosphate	Three acetyl-CoA molecules from mevalonate, then converted to isoprene unit	Isoprenoids
C6C3	Obtained from C skeleton of L-phenylalanine or L-tyrosine	Loss of amino group, side chain modifications	Phenolic compounds, Aromatic derivatives
C6C2N	Formed from L-phenylalanine or L-tyrosine (more common)	Removal of carboxyl carbon	Aromatic derivatives
Indole.C2N	L-tryptophan	Decarboxylation	Indole derivatives
C4N	Produced by non-protein amino acid L-ornithine	Loss of carboxylic acid function and α-amino nitrogen	Pyrrolidine derivatives
C5N	Produced from L-lysine	Retains ε-amino nitrogen, forms piperidine system	Piperidine derivatives

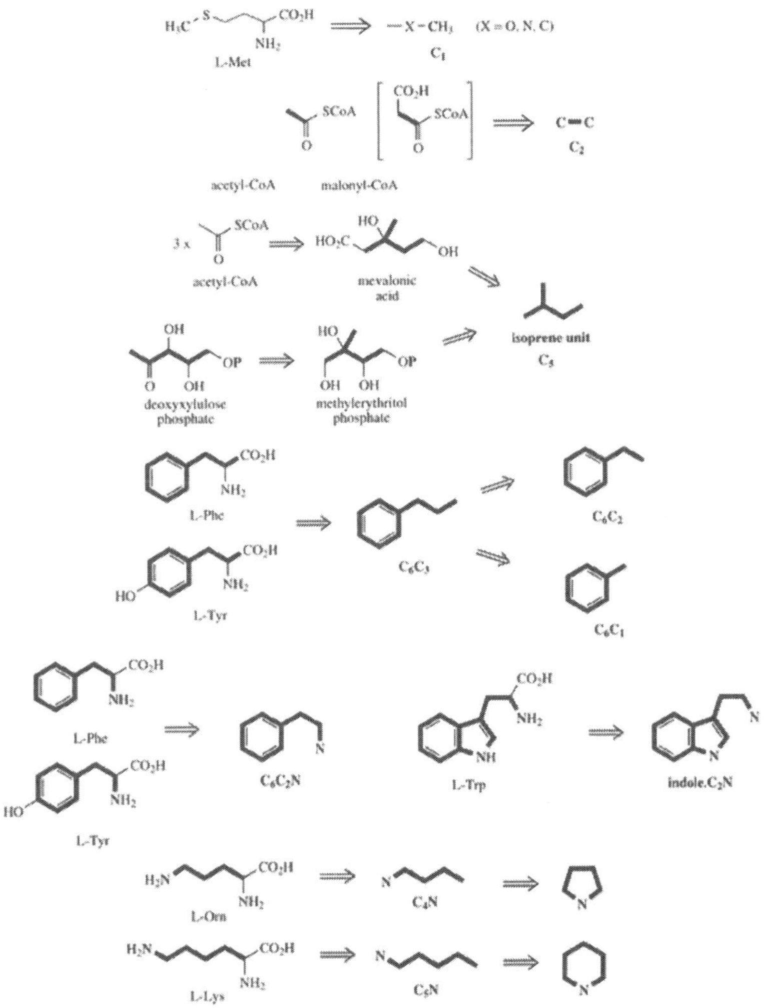

Figure 1.2 Building blocks.

1.4 CONSTRUCTION MECHANISMS

1.4.1 Nucleophilic Substitution Reaction

L-methionine serves as the source of the C1 methyl building unit, which is introduced through a nucleophilic substitution reaction. This process is naturally facilitated by the conversion of L-methionine to S-adenosylmethionine (SAM), resulting in positively charged sulfur that enhances the leaving group and enables a nucleophilic substitution reaction (SN2 mechanism). This reaction allows for the formation of O-methyl and N-methyl linkages with hydroxyl and amino groups as nucleophiles as shown in Figure 1.3. Furthermore, C-methyl linkages can be formed through the participation of nucleophilic carbon, typically at specific positions adjacent to phenol or carbonyl groups. DMAPP, a C5 isoprene unit, can also function as an alkylating agent. This process involves a nucleophilic displacement reaction, where the diphosphate group

THE CONSTRUCTION MECHANISMS

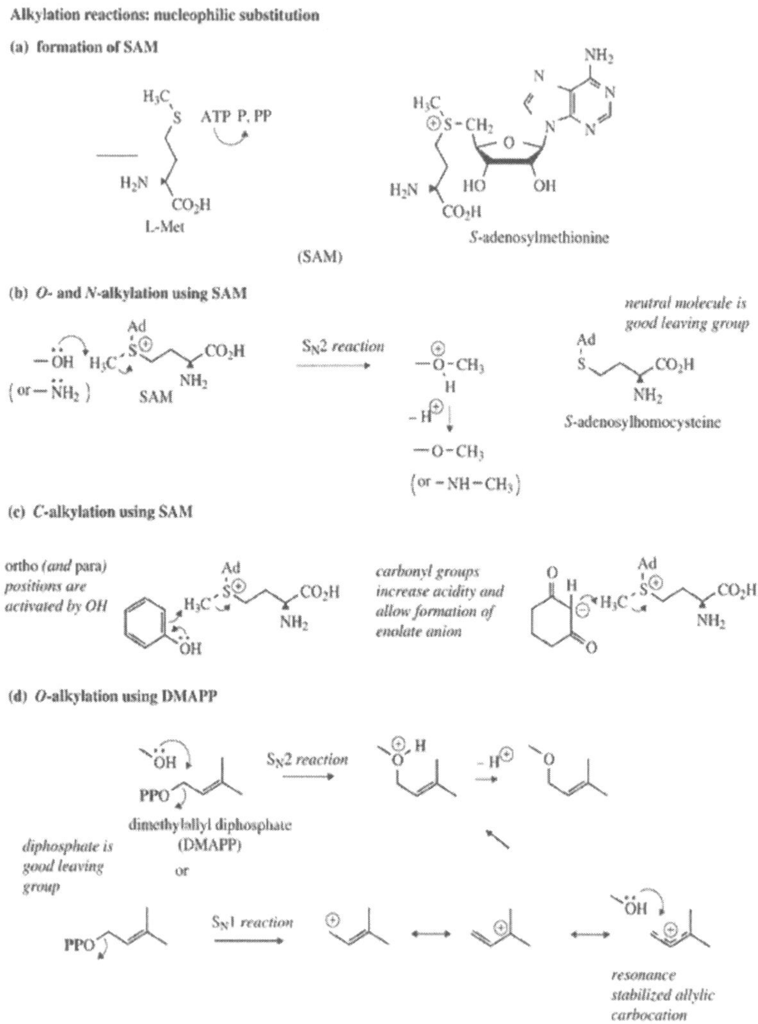

Figure 1.3 Nucleophilic substitution reactions.

acts as a good leaving group. In some instances, DMAPP may first ionize to form a resonance stabilized allylic carbocation, leading to an SN1 process. This C-alkylation reaction, facilitated by DMAPP, is similar to the C-methylation process, where activated positions are targeted for alkylation.

1.4.2 Alkylation Reactions, Electrophilic Addition

The C5 isoprene component found in DMAPP can be utilized for nucleophile alkylation. In the synthesis of terpenoids and steroids, multiple C5 units are combined, and the processes are explained through carbocation chemistry, including the electrophilic addition of carbocation onto alkenes. DMAPP might ionize, creating a resonance-stabilized allylic carbocation, which can then react with an alkene (e.g., IPP). The resulting carbocation could then shed a proton, yielding the neutral product geranyl diphosphate (GPP). When the alkene and carbocation functionalities are within the same molecule, this mechanism can lead to cyclization reactions. Various mechanisms can generate the initial carbocation, including leaving group expulsion, particularly diphosphate (SN1-type ionization), alkene protonation, and epoxide protonation/ring opening. S-adenosylmethionine can also alkylate alkenes through an electrophilic addition method, adding a C1 unit and forming an intermediate carbocation. The final carbocation can be eliminated by proton release (resulting in an alkene or occasionally forming a cyclopropane ring) or by being neutralized with a suitable nucleophile, especially water as shown in Figure 1.4.

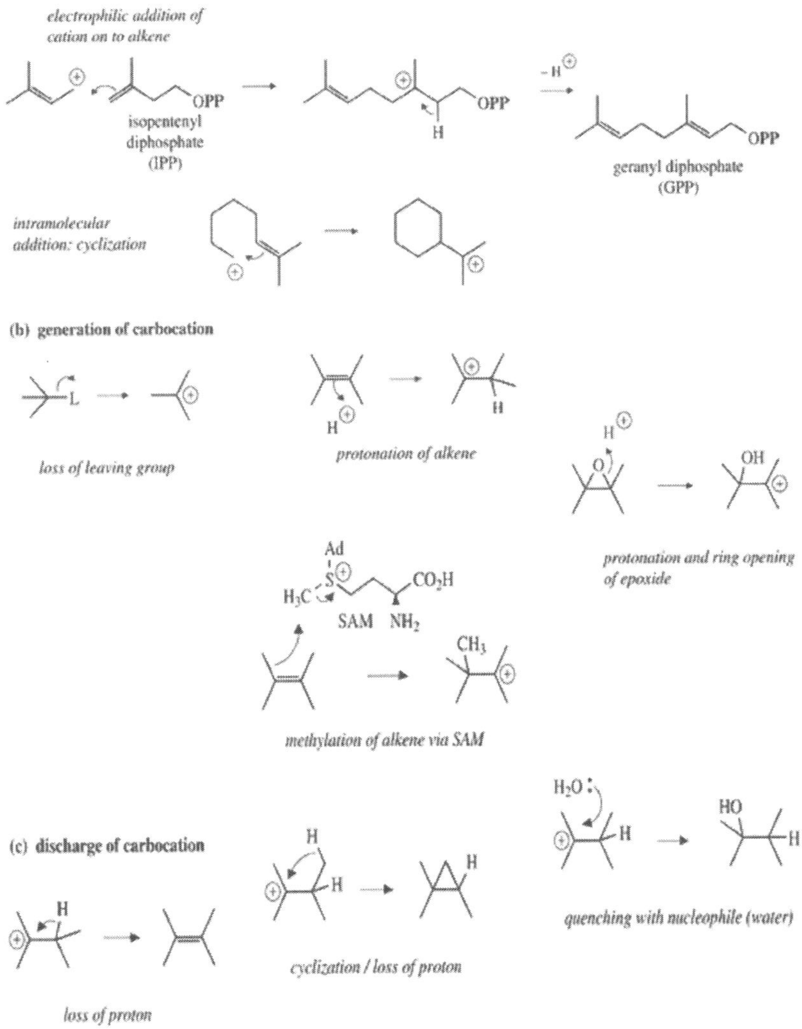

Figure 1.4 Electrophilic addition.

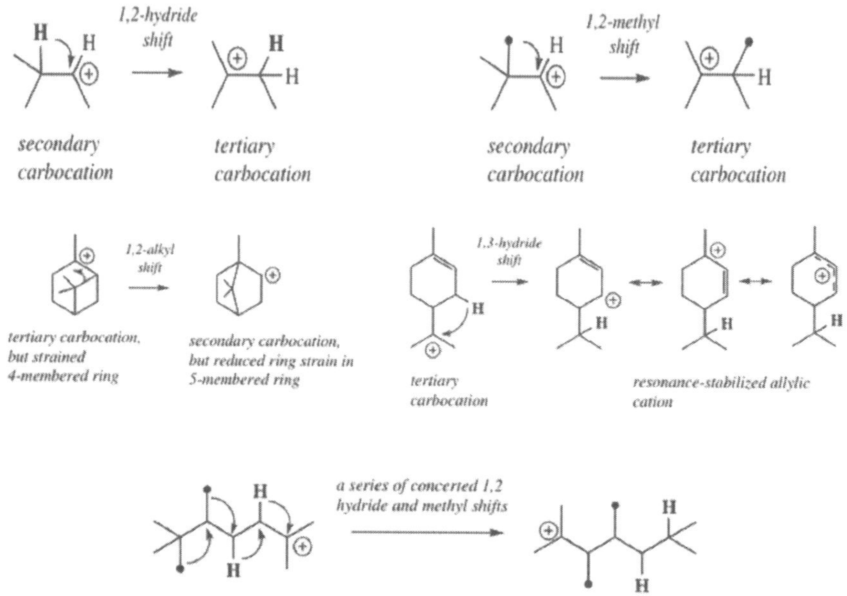

Figure 1.5 Wagner–Meerwien rearrangements.

1.4.3 Wagner–Meerwien Rearrangement

The Wagner–Meerwien rearrangement is an important tool in organic chemistry, providing chemists with means to create diverse compounds with specific characteristics and enabling construction of secondary metabolites. Various secondary metabolites can be formed by building blocks after some rearrangements during their biosynthesis such as terpenoids and steroids produced from C_5 isoprene units. This rearrangement involves the shift of alkyl or aryl group from one carbon atom to other carbon atom forming a carbocation. These shifts are called as **Wagner–Meerwien rearrangement**, as shown in Figure 1.5. Both secondary and tertiary carbocations are formed, but tertiary carbocations are chosen over secondary. First, a carbocation intermediate is formed which may be stabilized by resonance effects following shifting of alkyl or aryl group forming a more stable carbocation intermediate. These shifts involve usually 1, 2 shifts of alkyl or aryl groups but longer shifts are also encountered. The rearranged carbocations undergo elimination, addition, or nucleophilic reaction depending on nature or substrate.

1.4.4 Aldol Claisen Reaction

In typical base-catalyzed chemical processes, the aldol and Claisen reactions both result in the creation of carbon–carbon bonds and rely on the generation of a resonance-stabilized enolate anion from an appropriate carbonyl system. Whether an aldol-type or Claisen-type product is formed depends on the nature of X and its potential as a leaving group. Thus, chemically, two molecules of acetaldehyde produce aldol, but two ethyl acetate molecules can produce ethyl acetoacetate (Dewick 2002). These procedures are essential to biochemistry because these facilitate the development of primary and secondary metabolites, although the enzyme's catalysis eliminates the requirement for strong bases and most likely indicates that the enolate anion is very briefly present, as shown in Figure 1.6. However, the reactions seem to mimic the chemistry of enolate anion and are often involved in the combining of C2 acetate groups.

Coenzyme A esters, such as acetyl-CoA, are typically involved in biological reactions. This thioester of acetic acid offers a number of significant advantages to oxygen esters, such as ethyl acetate, because it increases the possibility of producing the enolate anion as the α-methylene hydrogens are now more acidic—they are really comparable to those in the corresponding ketone. The ester function's electron delocalization is used to explain why this is more pronounced in the oxygen ester than in the sulfur ester; it is because of the smaller size of oxygen and the lone pair's closer closeness to the orbitals of carbon. The thioester also has a significantly more favorable leaving group than the oxygen ester, which together have the effect of making the aldol- and Claisen-type reactions more reactive (Figure 1.7).

Acetyl-CoA-related Claisen reactions are further enhanced by first turning acetyl-CoA into malonyl-CoA via a carboxylation reaction with CO_2 that uses ATP and the coenzyme biotin. The coenzyme in a biotin–enzyme complex is carboxylate by the mixed anhydride, which is formed by ATP and CO_2 (as bicarbonate, HCO_3–). Not only does acetyl-CoA contain biotin–enzyme complexes that fix CO_2, but the synthesis of oxaloacetate from pyruvate during the synthesis of glucose from non-carbohydrate sources is another significant example as shown in Figure 1.6.

The α-hydrogens now have two carbonyl groups on either side and are more acidic due to the conversion of acetyl-CoA into malonyl-CoA. This gives the Claisen reaction a more favorable nucleophile. The carboxyl group added to malonyl-CoA is simultaneously removed by a decarboxylation process during the Claisen condensation, and no acylated malonic acid derivatives are generated in this process (Figure 1.6).

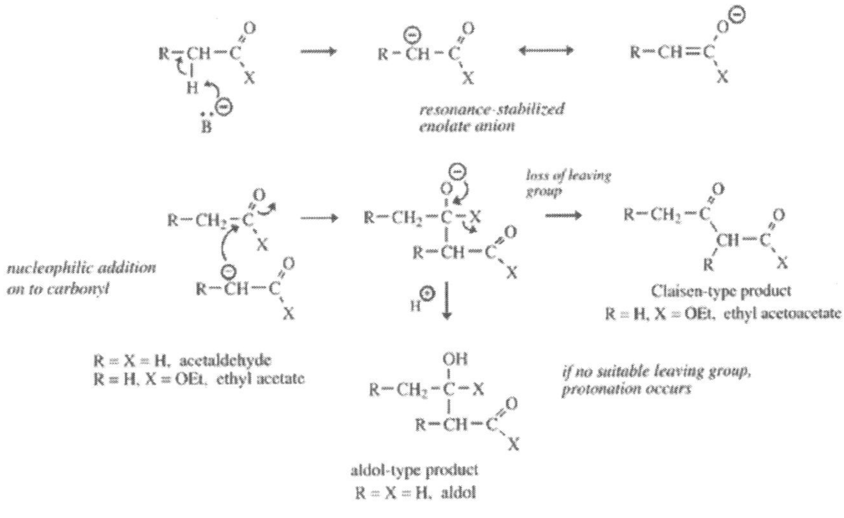

Figure 1.6 Aldol Claisen reactions.

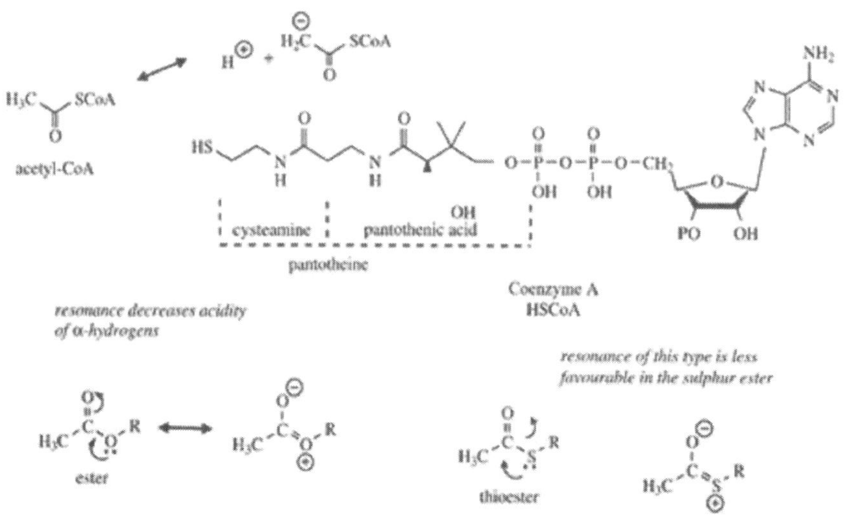

Figure 1.7 Biochemical pathway illustrating the synthesis of pantothenic acid derivatives and coenzyme A, with resonance effects on ester and thioester structures.

An alternative explanation is that the acetyl enolate anion is produced via decarboxylation of the malonyl ester which does not require a strong base. The result of acetyl-CoA acting as an electrophile and malonyl-CoA acting as a nucleophile is acetoacetyl-CoA, which is essentially the same as what happens when two acetyl-CoA molecules condense. As a result, the function of the carboxylation phase is obvious: after activating the α carbon

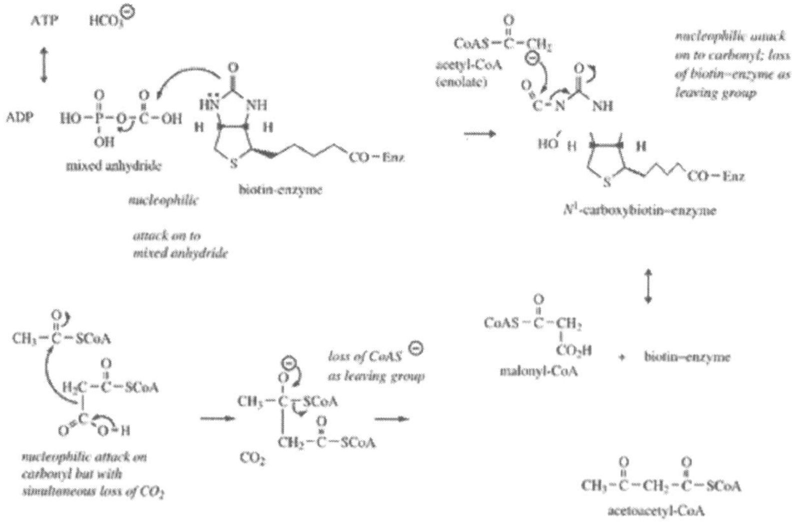

Figure 1.8 Biochemical pathway illustrating the enzymatic carboxylation of acetyl-CoA to form malonyl-CoA and acetoacetyl-CoA via nucleophilic attack and loss of CO_2.

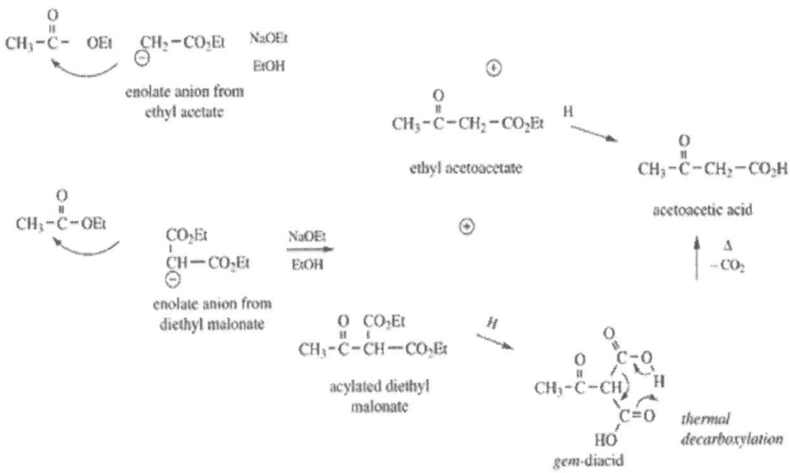

Figure 1.9 Enolate formation and decarboxylation in secondary metabolism leading to acetoacetic acid.

to aid in the Claisen condensation, the carboxyl is quickly eliminated (Figure 1.8).

By comparison, the chemical Claisen condensation that employs the enolate anion from diethyl malonate progresses far more favorably than the one that employs the enolate from ethyl acetate. Through decarboxylation of the gem-diacid and hydrolysis of the acylated malonate intermediate, the identical acetoacetic acid product can be produced in the malonate condensation (Figure 1.9).

Reverse Claisen and reverse aldol reactions can both occur when natural product molecules are modified. Such reactions may increase the diversity of structures, but they also remove pieces from the basic skeleton that has already been produced. One prominent characteristic of the β-oxidation sequence for the catabolic breakdown of fatty acids is the reverse Claisen reaction, whereby a fatty acid chain's C2 unit, acetyl-CoA, is broken off, leaving the chain two carbons shorter (Figure 1.10).

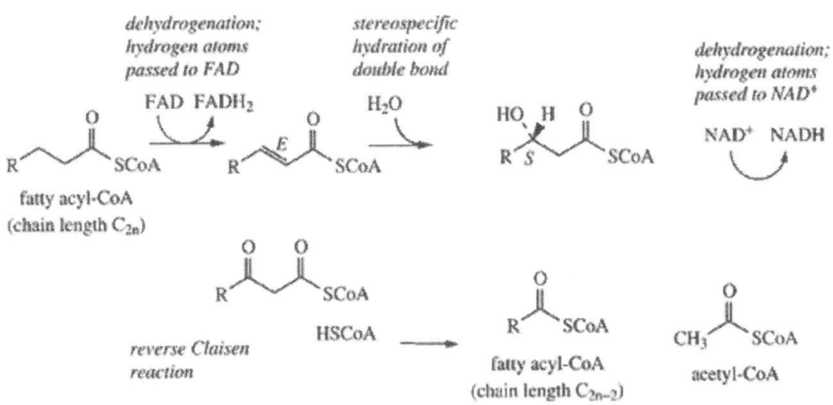

Figure 1.10 β-oxidation of fatty acids.

1.4.5 Schiff Base Generation and Mannich Reaction

The establishment of C–N bonds often occurs through condensation reactions, pairing amines with aldehydes or ketones. This begins with a typical nucleophilic addition and progresses to water removal, resulting in the formation of an imine or Schiff base (Figure 1.11(a)). Equally crucial is the reverse mechanism: hydrolyzing imines to regenerate amines and aldehydes/ketones.

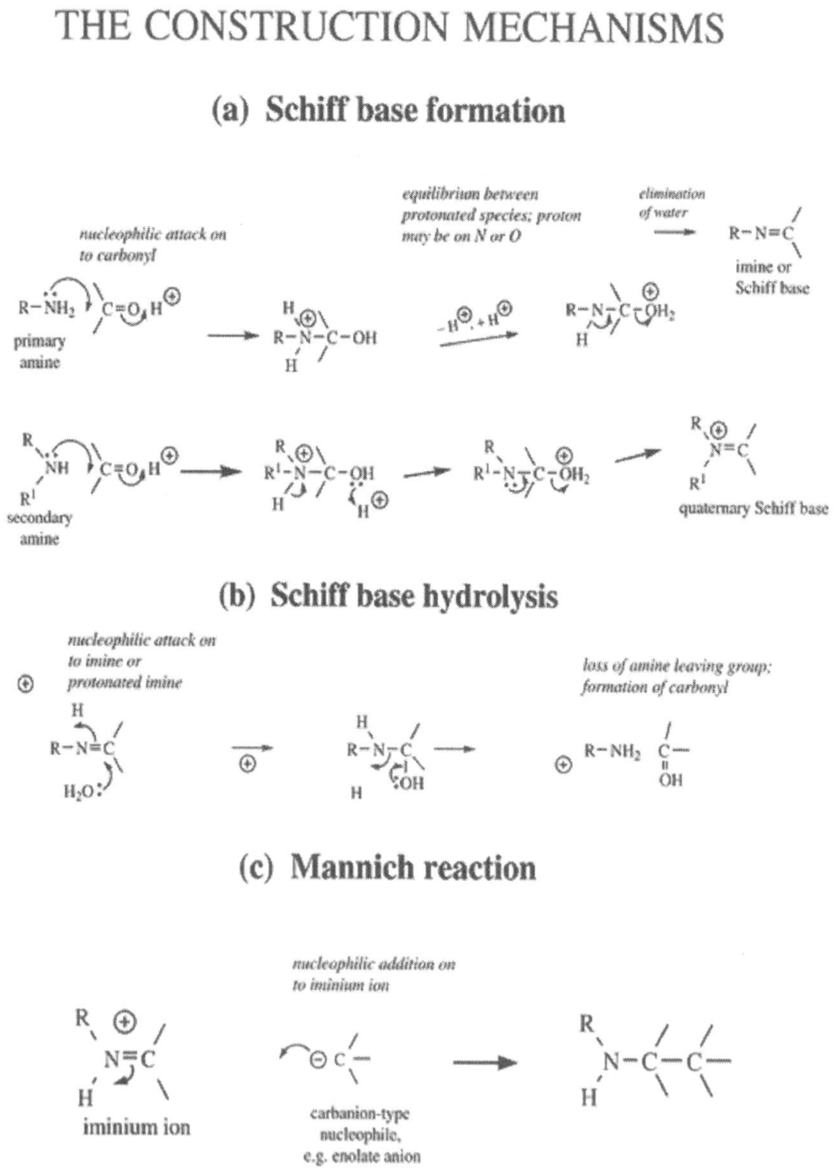

Figure 1.11 Schiff bases.

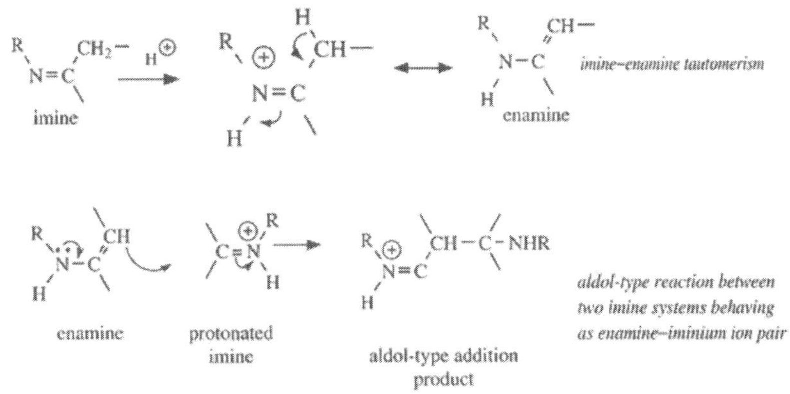

Figure 1.12 Secondary metabolism of Schiff bases.

The resulting imines or more commonly their protonated form, the iminium ion, can then function as an electrophile in a Mannich reaction (Figure 1.11(c)). The nucleophile may be provided by an enolate anion or, in many instances, by a suitably activated site within an aromatic ring system. The Mannich reaction is prevalent in alkaloid biosynthesis and, in its broadest context, involves the combination of an amine (primary or secondary), an aldehyde or ketone, and a nucleophilic carbon. Secondary amines will directly react with the carbonyl compound to form an iminium ion (quaternary Schiff base), eliminating the need for an additional protonation step.

Mannich-like addition reaction, depicted in Figure 1.11(c), is essentially similar to nucleophilic addition to a carbonyl group. In fact, the imine/iminium ion acts as the nitrogen equivalent of a carbonyl/protonated carbonyl. Expanding on this comparison, protons adjacent to an imine group, similar to those adjacent to a carbonyl group, exhibit acidity. The isomerization to the enamine, illustrated in Figure 1.12, parallels keto–enol tautomerism. Just as two carbonyl compounds can undergo an aldol reaction, two imine systems can also react, as shown in Figure 1.12. In enzymatic reactions, aldehyde/ketone substrates often form covalent bonds with enzymes through imine linkages, retaining the carbonyl activation despite the altered bonding configuration.

1.4.6 Transamination

The most common biochemical reaction for the transfer of amino group from one amino acid to another ketoacid to form a new amino acid and ketoacid is called transamination.

The usual donor for this reaction is glutamic acid (amino acid) and acceptor is 2-oxaglutaric acid (ketoacid). Glutamic acid is synthesized by the body (nonessential amino acid) and serves as a precursor for the synthesis of other amino acids while 2-oxoglutaric acid is derived from glutamic acid by transamination and also serves as a precursor for the synthesis of amino acids, including glutamic acid.

The transamination process typically involves two substrates: one amino acid which serves as a donor and other keto acid which serves as an acceptor. The reaction is catalyzed by different enzymes known as transaminases or

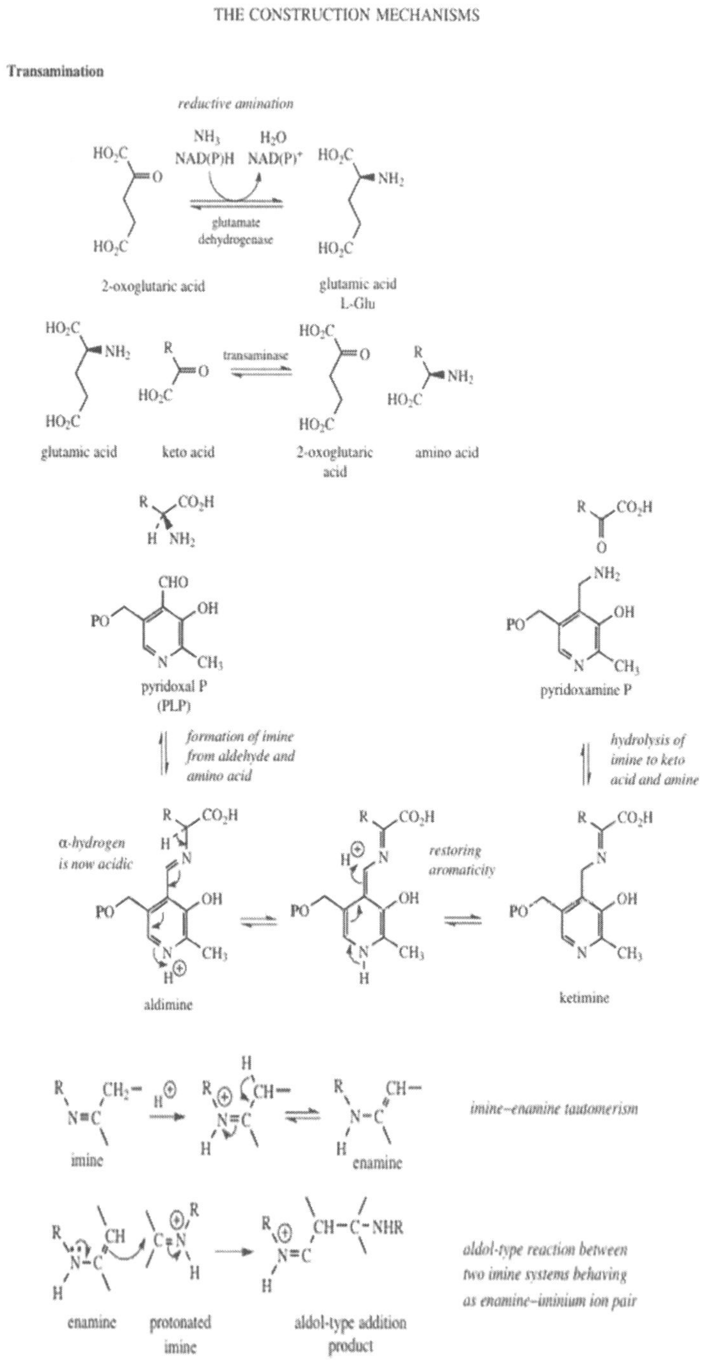

Figure 1.13 Biochemical transformations involved in constructing secondary metabolites.

aminotransferases which are specific in their actions (Guo and Berglund 2017). The enzymes also need some coenzymes, e.g., pyridoxal phosphate (PLP). PLP serves as a covalent intermediate in transamination process and is derived from vitamin B6. First of all, the amino group from the donor amino acid, i.e., glutamic acid is transferred to the PLP and there is a formation of Schiff base intermediate. The alpha hydrogen of the original amino acid plays an important role in the formation of the Schiff base linkage with the PLP coenzyme. The alpha hydrogen of the amino acid facilitates the nucleophilic attack on the coenzyme leading to the formation of Schiff base/imine intermediate (aldimine). Then the Schiff base intermediate transfers the amino group to the keto acid and forms a new amino acid and keto acid. The coenzyme (PLP) is regenerated in the reaction for the next reaction (Figure 1.13).

1.4.7 Decarboxylation Reactions

Decarboxylation reactions play important roles in various natural product pathways, involving the removal of carbon atoms from the molecular skeleton. *Amino acids*, in particular, undergo decarboxylation to yield fundamental building blocks such as C_6C_2N and indole. C_2N, which are important components of various biomolecules. This decarboxylation process depends upon *PLP* and involves the formation of *Schiff bases*, and it has mechanistic similarities with transamination reactions to some extent. The decarboxylation of amino acids consists of three crucial steps:

- Protonation of the α-carbon
- Hydrolysis of the Schiff base function
- Release of the amine from the coenzyme

1.4.8 Importance of Decarboxylation Reaction

This process is a crucial aspect of the biosynthetic utilization of amino acids, leading to the formation of various biomolecules. Furthermore, β-keto acids and ortho-phenolic acids also undergo decarboxylation, potentially through cyclic mechanisms involving enol intermediates. The acetate pathway frequently yields structures containing phenol and carboxylic acid functions, which may undergo further decarboxylation modifications, showing the importance of decarboxylation reactions in the biosynthesis of natural products. Despite the fact that carboxyl group may originate from hydrolysis of the thioester portion of acetyl-CoA, methyl groups can also be oxidized to carboxyl groups, which then undergo decarboxylation, revealing the versatility and complexity of decarboxylation reactions. Decarboxylation of α-keto acids is a vital feature of primary metabolism which plays a crucial role in various biochemical pathways, including glycolysis (pyruvic acid → acetaldehyde) and the Krebs cycle (pyruvic acid → acetyl-CoA). Thiamine diphosphate (TPP), a coenzyme containing a thiazole ring, plays a vital role in these reactions, acting as a nucleophile toward carbonyl groups, and its involvement is essential for the decarboxylation process. The decarboxylation of pyruvic acid to acetaldehyde regenerates the carbanion, while the oxidation step involves the enzyme-bound coenzyme lipoic acid, resulting in the reduction of lipoic acid and the release of acetyl-CoA. A similar reaction takes place in the Krebs cycle, converting 2-oxoglutaric acid into succinyl-CoA, highlighting the significance of decarboxylation reactions in the regulation of metabolic pathways (Dewick 2002) (Figures 1.14 and 1.15).

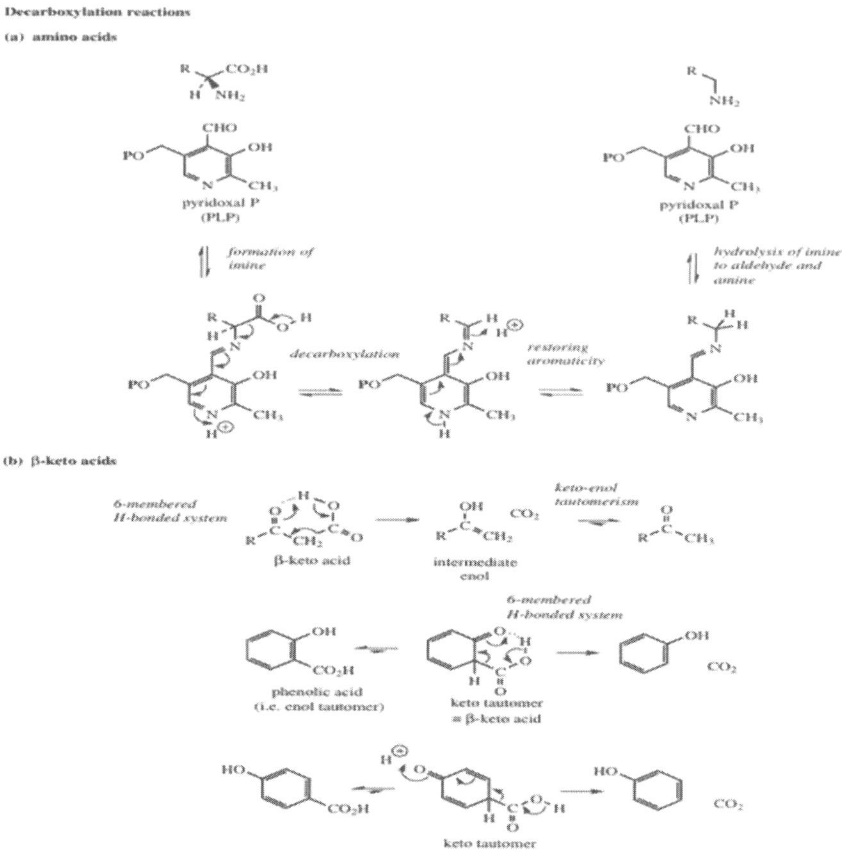

Figure 1.14 The construction mechanisms.

1.4.9 Oxidation and Reduction Reactions

Oxidation and reduction reactions are essential for the synthesis of many secondary metabolites. Oxidation reduction reactions form diversity of structures, are involved in activation of precursors in synthesis of secondary metabolites, modify functional groups, and regulate the biosynthetic pathways of secondary metabolites.

1.4.10 Dehydrogenase

Dehydrogenases facilitate transfer of hydrogen atoms (protons and electrons) between substrate molecule, catalyze the oxidation and reduction reactions, and play vital role in metabolic pathways, such as fatty acid oxidation, and synthesis of other secondary metabolites. For example, in GAP dehydrogenase reaction, GAP is converted into 1, 2-bisphosphoglycerate. In this reaction NAD+ accepts hydride ion from substrate producing NADH (Figure 1.16). Reduced forms NADH and NADPH are conveniently regarded as hydride-donating reducing agents. Generally, NADPH is employed in reductive processes, while NAD+ is used in oxidation. In tricarboxylic acid (TCA) cycle, dehydrogenases also transfer hydride ion to form $FADH_2$. Flavin adenine dinucleotide (FAD)

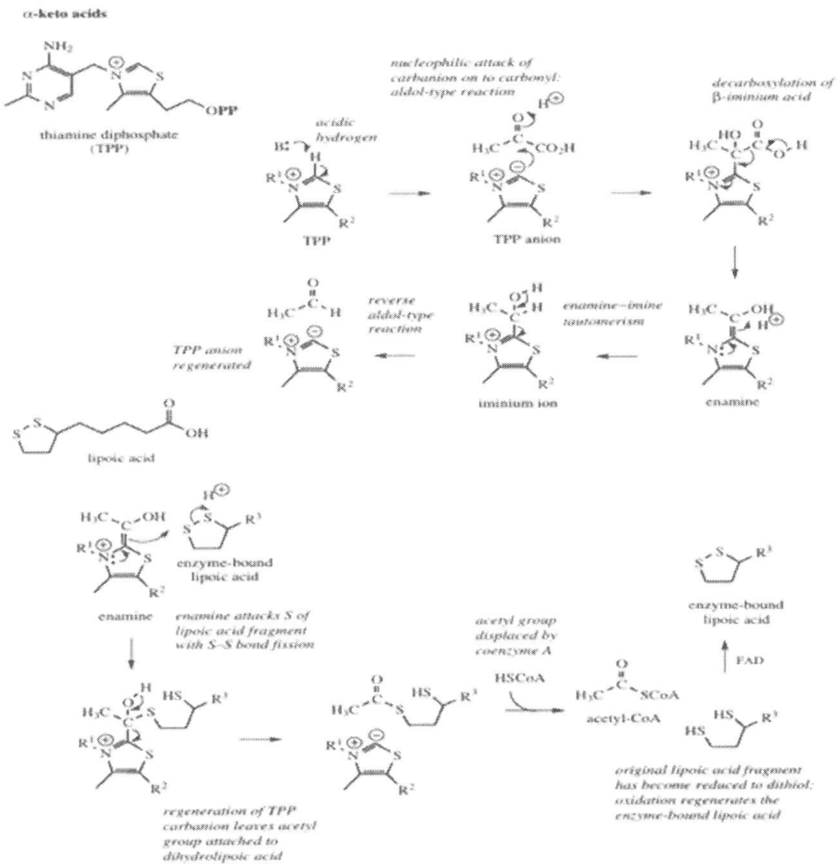

Figure 1.15 Biochemical transformations of α-keto acids involving decarboxylation, enamine formation, and enzyme-bound lipoic acid regeneration.

and flavin mononucleotide (FMN) take up two hydrogen atoms, as shown in Figures 1.16 and 1.17. Alternative mechanisms have also been proposed. They are essential in maintaining redox balance within cells and for regulating metabolic pathways. These oxidation processes employing pyridine nucleotides and flavoproteins are especially important in liberating energy in the primary metabolism.

1.4.11 Dehydrogenases: NAD and NADP

Figure 1.16 illustrates the structures and biochemical transformations involving nicotinamide adenine dinucleotide (NAD+) and nicotinamide adenine dinucleotide phosphate (NADP+), which are critical cofactors in redox reactions. The primary distinction between NAD+ and NADP+ is the additional phosphate group at the 2' position of the ribose in NADP+, represented by "R = P," while NAD+ has a hydrogen atom at this position ("R = H"). Both cofactors consist of an adenine group, a nicotinamide ring, and a ribose sugar, with the nicotinamide ring involved in electron transfer during oxidation-reduction reactions. The figure further depicts the reduction

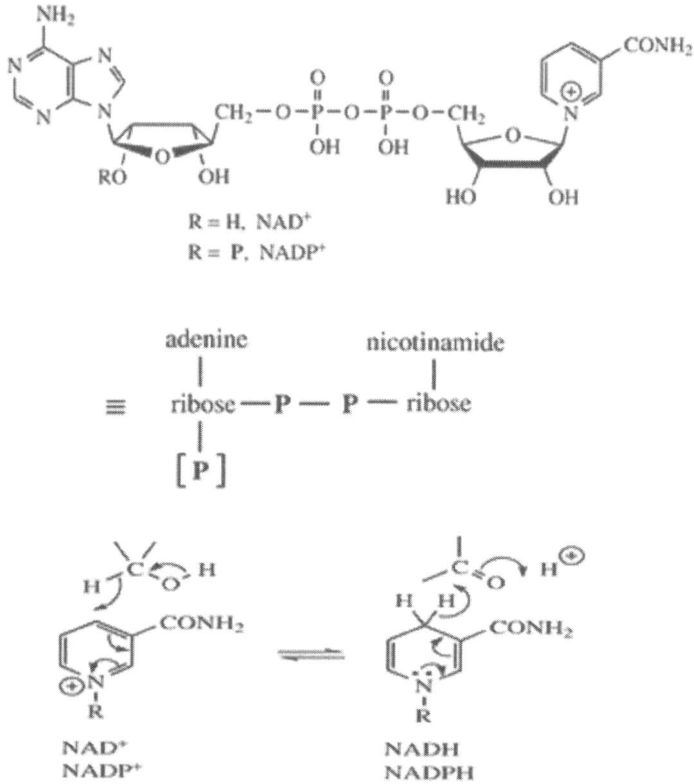

Figure 1.16 Dehydrogenases: NAD and NADP.

of NAD+ to NADH and NADP+ to NADPH, highlighting the conversion between the oxidized and reduced forms of these cofactors. NADH and NADPH play essential roles in cellular energy transfer and metabolic processes, participating in reactions where they either accept or donate electrons.

1.4.12 Dehydrogenases: FAD and FMN

Flavin adenine dinucleotide (FAD) and flavin mononucleotide (FMN), both are crucial cofactors in redox reactions. Figure 1.17 (top section) shows the structure of FAD, which consists of an adenine group, ribose sugar, and the FMN component. FAD plays an essential role in electron transfer during oxidation-reduction reactions. The middle section of the figure highlights the linkage between the adenine and flavin groups through ribose and ribitol, demonstrating the structural connection. In the bottom section, the figure depicts the conversion between the oxidized form of FAD and the reduced form, $FMNH_2$, emphasizing the role of these cofactors in biochemical redox processes.

1.4.13 Oxidases

Oxidases take hydrogen from a substrate, but transfer these atoms to either hydrogen peroxide or molecules of oxygen, which together produce water.

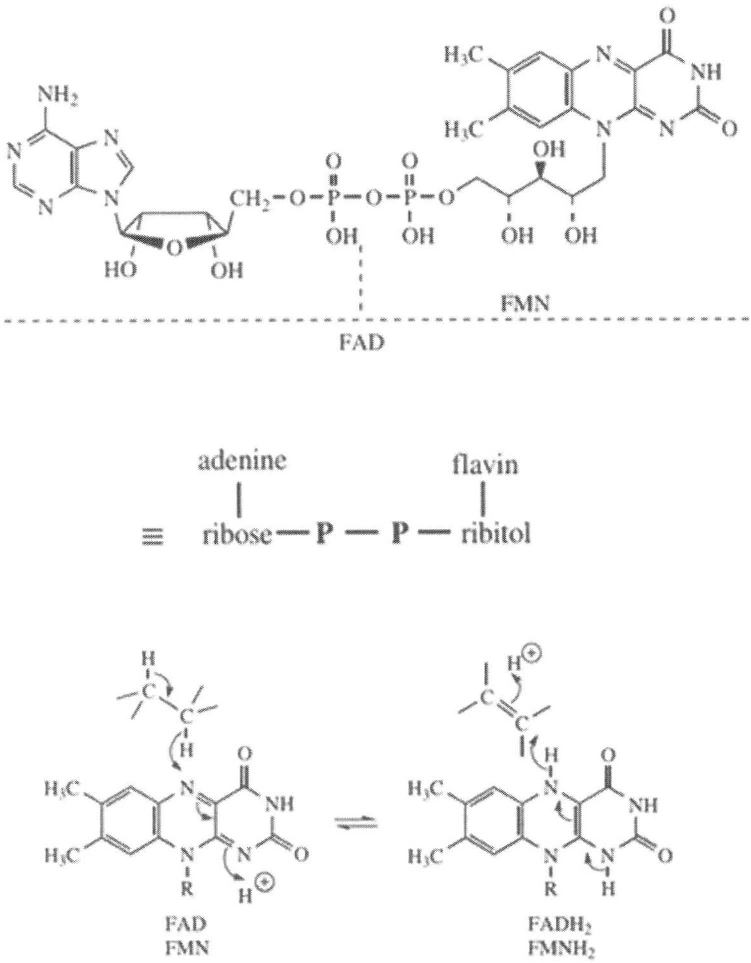

Figure 1.17 Dehydrogenases: FAD and FMN.

Oxidases that utilize hydrogen peroxide are referred to as peroxidases. Different mechanisms of action of oxidases are not necessary to be discussed here. The oxidation of ortho- and para-quinols to quinones as well as the peroxidase-induced phenolic oxidative coupling activities are significant transformations in secondary metabolism (Figure 1.18).

1.4.14 Monooxygenases

The direct addition of oxygen from molecular agents to the substrate is catalyzed by oxygenases. These are divided into mono- and dioxygenases, respectively, depending upon whether one or both of the substrate is filled with oxygen atoms. With monooxygenases, a suitable hydrogen donor, such as NADH, NADPH, or ascorbic acid (vitamin C), reduces the second oxygen atom from O_2. This makes them potentially oxidase-like in behavior; in fact, these enzymes are also referred to as "mixed-function oxidases." A notable example of these enzymes is the monooxygenases that are dependent on cytochrome P-450. These

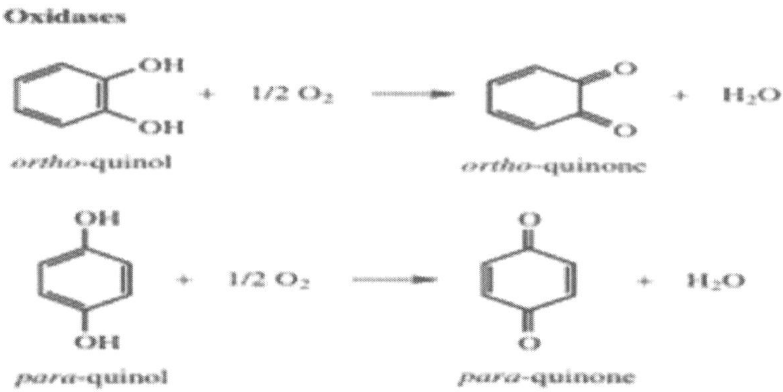

Figure 1.18 Oxidases.

enzymes are known as "hydroxylases" and are commonly involved in biological hydroxylation, either in biosynthesis or in the detoxification and metabolism of foreign substances like medicines by mammals.

Because cytochrome P-450 is a potent inhibitor of these enzymes, it gets its name from its strong absorption band at 450 nm when exposed to CO. It features a redox change involving the Fe atom that permits binding and the presence of an iron–porphyrin complex (hemes), which is attached to the enzyme, splitting apart an oxygen atom. Numerous such systems can hydroxylase aromatic or aliphatic systems in addition to converting alkenes into epoxides. NADPH is typically present as a hydrogen donor (Wedde et al. 2019). Arene oxide (epoxide) intermediates are likely involved in the aromatic hydroxylation process, which is catalyzed by monooxygenases, including cytochrome P-450 systems. An intriguing byproduct of this process is that the hydrogen atom that was initially bound to the location that becomes hydroxylated can move to the nearby carbon on the ring when the epoxide opens up. Even though part of this hydrogen is lost during enolization, a significant percentage of these hydrogen atoms are eventually retained in the final product. This migration was first noticed at the National Institute of Health in Bethesda, Maryland, USA, and is referred to as the "NIH shift" (Figure 1.19).

1.4.15 Amine Oxidases

Aside from being oxidizing enzymes, amine oxidases play a crucial role in metabolic processes by converting amines to aldehydes. These include monoamine and diamine oxidases. Monoamine oxidases use FAD and molecular oxygen to dehydrogenate to an imine, then these hydrolyze to aldehyde and ammonia (Figure 1.20). Diamine oxidases use molecular oxygen to oxidize one amino group at a time, resulting in the corresponding aldehyde. The process also produces hydrogen peroxide and ammonia. The resulting amino aldehyde can be converted into a cyclic imine through Schiff base production as shown in Figure 1.20.

1.4.16 Baeyer–Villiger Oxidations

The Baeyer–Villiger oxidation of ketones by peracids produces an ester by causing an alkyl group to migrate from the ketone (see Figure 1.21). Similar ketone-to-ester transformations in biochemistry include cytochrome-P-450- or

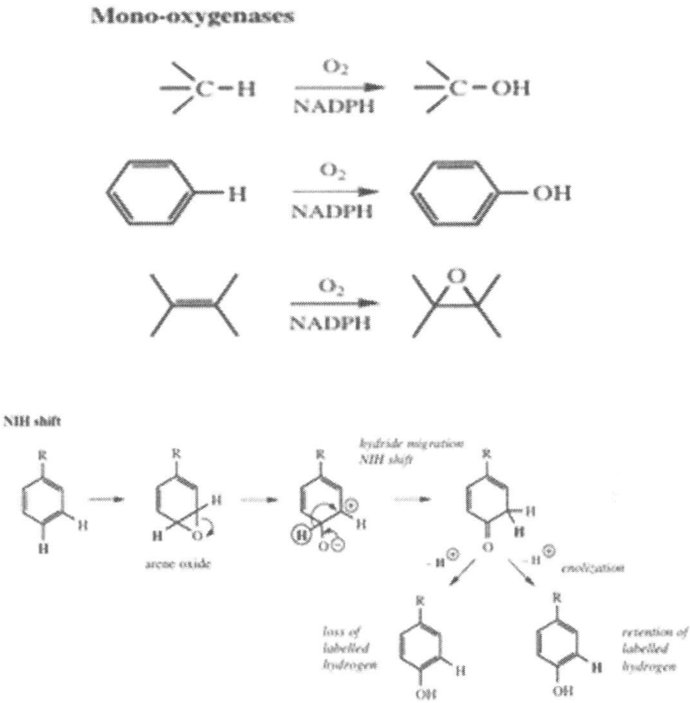

Figure 1.19 Monooxygenases.

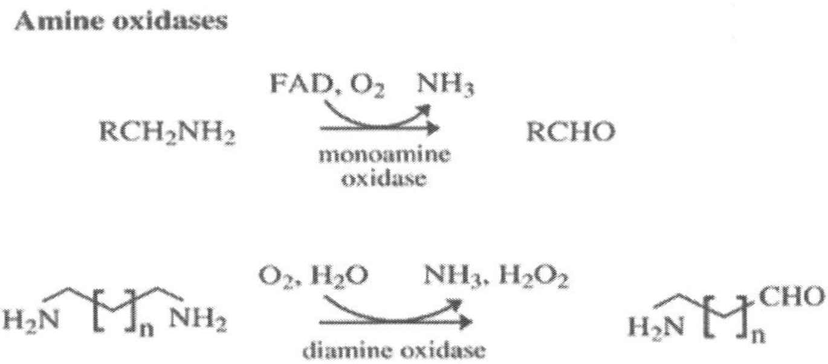

Figure 1.20 Amine oxidases.

FAD-dependent enzymes that require NADPH and O_2. This process creates a peroxy–enzyme complex, comparable to the chemical Baeyer–Villiger oxidation mechanism. The oxygen atom inserted is derived from O_2.

1.4.17 Phenolic Oxidative Coupling

Natural products are created by combining several phenolic systems through free radical reactions. The reactions that can be brought by oxidase enzymes, such as peroxidase and laccase systems, are known to generate radicals. Enzymes that catalyze phenolic oxidative coupling are

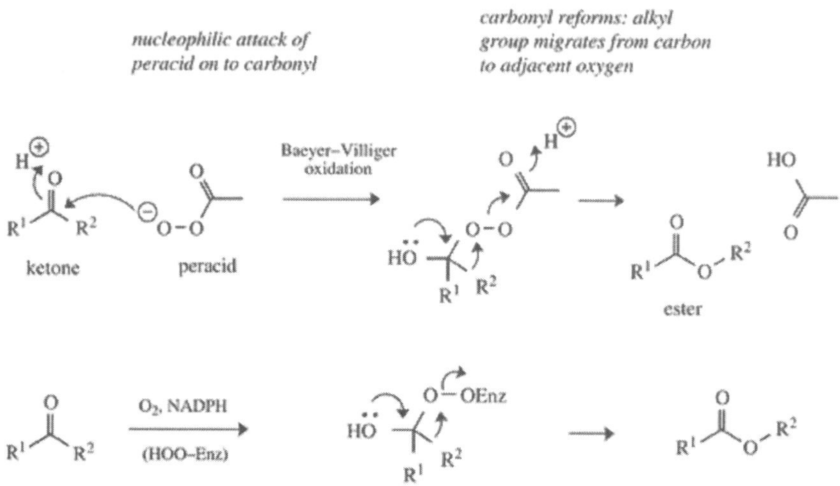

Figure 1.21 Baeyer–Villiger oxidation.

cytochrome-P-450-dependent and require NADPH and O_2 cofactors. However, the substrate does not contain any oxygen. After one-electron oxidation of a phenol, the unpaired electron can be delocalized by resonance forms, which disperse the free electron ortho and para to the initial oxygen function (Lewis 2017).

Combining two mesomeric structures results in a variety of dimeric systems, as illustrated in Figure 1.22, enolization restores aromaticity to the rings, resulting in the indicated end products.

Carbon–carbon bonds with ortho or para locations to the initial phenols, also known as ether linkages, can occur. The reactive dienone complexes generated as intermediates can be attacked by various nucleophilic groups, resulting in a wider range of structures from the initial reaction sequence.

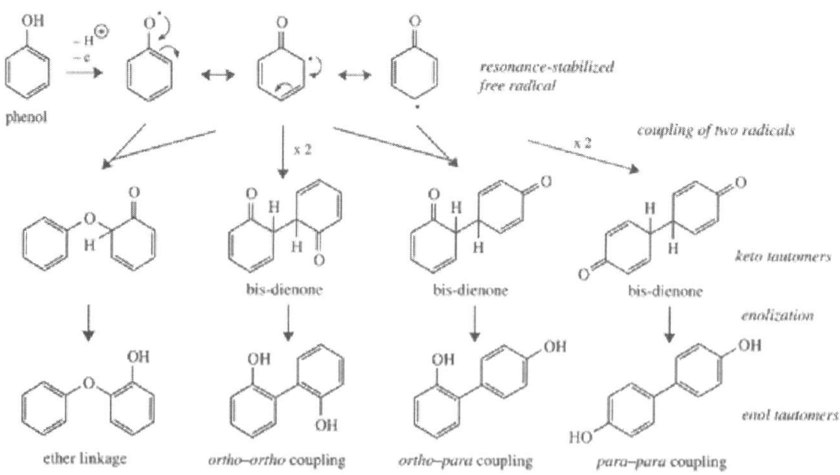

Figure 1.22 Phenolic oxidative coupling.

1.4.18 Glycosylation Reaction

The abundance of glycosides and polysaccharides necessitates attachment mechanisms.

Sugar units can be attached to an aglycone atom to form glycosides or polysaccharides. Oxygen is often used for linkages, but S-, N-, and C-glycosides are also commonly used. Glycosylation uses a uridine diphosphate (UDP) sugar, such as UDP glucose (Crich 2010). The process of glycosylation involves the synthesis of glucose 1-phosphate and UTP, followed by a simple SN2 nucleophilic displacement reaction (Figure 1.23(a)). UDP glucose's leaving group is in the configuration, resulting in a product with the β-configuration, similar to natural glucosides.

Many essential carbohydrates, such as sucrose and starch, include α-linkages that appear to arise from double SN2 processes. Other UDP sugars, such as UDP galactose or UDP xylose, are used to synthesize glycosides with various sugar units. Hydrolysis of glycosides requires specific enzymes such as β-glucosidase for β-glucosides and β-galactosidase for β-galactosides. These enzyme activities resemble acid-catalyzed activities (Figure 1.23(b)). Under acidic conditions, the α- and β-anomeric hemiacetal forms can equilibrate through an open-chain sugar. While acid can hydrolyze O-, N-, and S-glycosides, C-glycosides remain stable. C-glycosides can be synthesized using a nucleophilic carbon, such as aromatic systems triggered by phenol

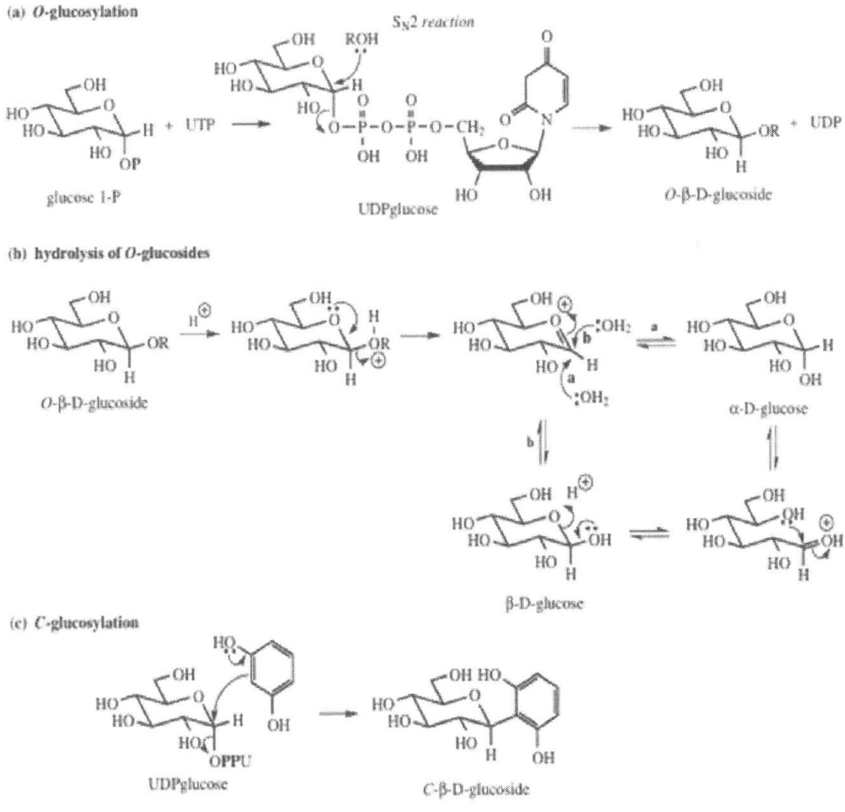

Figure 1.23 Glycosylation reaction.

groups, similar to the C-alkylation process (Figure 1.23(c)). The resulting C-glycoside comprises a novel carbon–carbon bond, requiring oxidation rather than hydrolysis for cleavage.

REFERENCES

Alamgir, A. (2018). "Secondary metabolites: Secondary metabolic products consisting of C and H; C, H, and O; N, S, and P elements; and O/N heterocycles." *Therapeutic use of medicinal plants and their extracts: Volume 2: Phytochemistry and bioactive compounds*: 165–309.

Brodie, J. D., et al. (1963). "The participation of malonyl coenzyme A in the biosynthesis of mevalonic acid." *Journal of Biological Chemistry* **238**(4): 1294–1301.

Crich, D. (2010). "Mechanism of a chemical glycosylation reaction." *Accounts of Chemical Research* **43**(8): 1144–1153.

Dewick, P. M. (2002). *Medicinal natural products: A biosynthetic approach.*

Guo, F. and P. Berglund (2017). "Transaminase biocatalysis: Optimization and application." *Green Chemistry* **19**(2): 333–360.

Krivoruchko, A., et al. (2015). "Microbial acetyl-CoA metabolism and metabolic engineering." *Metabolic Engineering* **28**: 28–42.

Lewis, N. G. (2017). "Plant phenolics." *Antioxidants in higher plants*: 135–169.

Tiwari, R. and C. Rana (2015). "Plant secondary metabolites: A review." *International Journal of Engineering Research and General Science* **3**(5): 661–670.

Verpoorte, R. (2000). "Secondary metabolism." *Metabolic engineering of plant secondary metabolism*: 1–29.

Wedde, S., et al. (2019). "The recent developments of enzymatic oxidation." *Green oxidation in organic synthesis*: 439–496.

2 Isolation and Characterization Techniques of Plant Secondary Metabolites

Malik Saadullah, Muhammad Umer, M. Zain Sharif, Shahid Manzoor, and Farhan Sadeeq*

2.1 INTRODUCTION

Plants produce a vast variety of secondary metabolites which are, apparently, not involved directly in the growth and development of the plants. In recent years, the importance of the plant secondary metabolites has evolved commercially. These are the remarkable sources for pharmaceuticals, food additives, flavors, and industrial materials. The secondary metabolites have also been widely used to prevent and cure chronic diseases. This chapter focuses on the analytical methodologies such as extraction, isolation, and characterization of plant secondary metabolites. Extraction of plant materials is the most important step for further separation and characterization of plant secondary metabolites (UNIDO/ICS, 2005). The plant extract is a complex mixture of various types of secondary metabolites. Isolation of these metabolites becomes a big challenge for further characterization and identification (Bajpai et al., 2016). Various chromatographic techniques such as thin-layer chromatography, high-performance liquid chromatography (HPLC), vacuum liquid chromatography (VLC), column chromatography (CC), ion exchange chromatography (IEC), and size exclusion chromatography (SEC) and spectroscopic techniques such as ultraviolet (UV) spectroscopy, infrared (IR) spectroscopy, nuclear magnetic resonance (NMR) spectroscopy, and mass spectrometry (MS) are used for analysis isolation and characterization of plant secondary metabolites in the plant extracts.

2.2 PLANT SECONDARY METABOLITES: AN OVERVIEW

From times long ago, individuals have been studying the natural world, focusing on plants, in the pursuit of discovering novel medications. Phytochemicals are compounds produced by plants, and they exhibit biological effects (Mendoza & Silva, 2018). These compounds are commonly referred to as natural products, secondary metabolites, or phyto-constituents. Secondary metabolites are organic compounds that do not participate in the regular growth and development of an organism. Lack of secondary metabolites does not lead to instant death, but these compounds are important for plants' biological functions, such as surviving harsh conditions and protecting against enemies (Das and Gezici, 2018). Plant extracts, whether in the form of pure compounds or standardized extracts, offer endless possibilities for discovering new drugs due to the vast range of chemical diversity they provide. As per the World Health Organization (WHO), more than 80% of the world population depends on traditional medicine for their main healthcare requirements. In addition to their healing properties, these secondary compounds are also utilized for making different dyes, producing agricultural insecticides, providing ingredients for industries, creating perfumes and cosmetics, and acting as coloring and flavoring agents. Examples of secondary metabolites include glycosides, alkaloids, carbohydrates, proteins, lipids, tannins, flavonoids, terpenoids, steroids, polyphenols, phytosterols, resins, glycoalkaloids, and more (Savithramma et al., 2011). In the past, plants or microbial extracts in crude or partially purified forms were the only sources of medication available for the

*Corresponding author

DOI: 10.1201/9781032707273-2

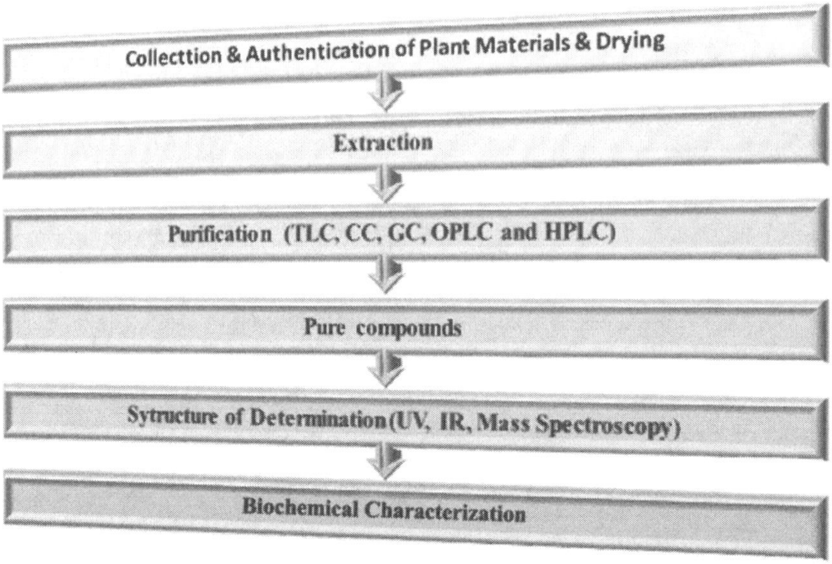

Figure 2.1 Flowchart of plant material processing, including authentication, extraction, purification, structural determination, and biochemical characterization.

treatment of human and animal diseases (Savithramma et al., 2011). This gave the idea that the effect of a drug in human body is due to an interaction of drug with biological molecules. This opened new doors in pharmacology, as pure, isolated chemicals, instead of extracts, were now the standard for the treatment of diseases (Figure 2.1).

2.3 EXTRACTION

Extraction is the initial stage in studying medicinal plants and a vital one, as it is essential to isolate the specific chemical compounds from the plant materials for subsequent separation and identification (Handa, 2008). The fundamental process involves various steps such as pre-washing, drying or freeze drying plant materials; grinding to achieve a uniform sample; and frequently enhancing analytic extraction kinetics while encouraging more contact between the sample surface and the solvent system (Abubakar & Haque, 2020). It is necessary to take proper measures to ensure that potential active components are preserved, undamaged, or unharmed during the process of extracting plant samples (Fonmboh et al., 2020). Different extraction methods are applied for the separation of secondary metabolites from plants (Sasidharan et al., 2011). Below are some of these extraction processes are discussed in subsequent subsections.

2.3.1 Maceration

It is a traditional technique utilized for creating medicine. Obtaining natural products from plant material is a popular and cost-effective method. Maceration is a technique used for extracting compounds from solids into liquids (Rasul et al., 2018). In this procedure, the solid powdered substances are put into a sealed container (Figure 2.2) and then the solvent is added. The mixture is

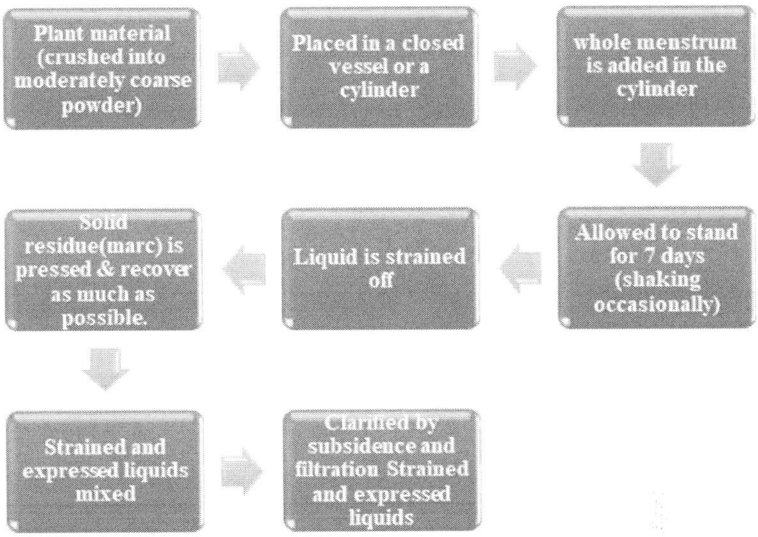

Figure 2.2 Maceration process for extraction.

allowed to stand for an extended period of time, ranging from several hours to several days, with intermittent shaking. Adequate time is given for the solvent to pass through the cell membrane and dissolve the components found in the plant. The process occurs solely through molecular diffusion (Rasul et al., 2018). Once the desired time passes, the liquid is strained out and the leftover solid is squeezed to extract as much solvent as can be recovered. When water is used as the solvent and the maceration period is extended, a small amount of alcohol can be included to hinder the growth of microbes.

In maceration, heat is transferred by convection and conduction, and the choice of the solvent is crucial to determine the type of phytochemical extracted from the samples.

2.3.2 Infusion

Fresh infusions are made by soaking the raw material in cold or boiling water for a brief period of time. These are weak solutions containing easily soluble components of raw drugs. Aromatic tea infusions are made from tea bushes: *Camellia sinensis* var. *assamica*, whose leaves are gathered at different growth stages to make dried tea. The method of tea leaf processing is determined by the type of tea and the desired quality of the end product. For premium tea, manual techniques are mainly used, with care taken to pick young shoots, underdeveloped buds, delicate stems, and the earliest, small leaves from tea plants during harvesting.

2.3.3 Decoction

Decoction is a method of using water to extract beneficial compounds from medicinal plant materials. In this procedure, the liquid mixture is created by boiling the plant material in water (Figure 2.3). Decoction is distinct from infusion because the latter is not boiled actively. Decoction is preferred for handling hard and fibrous plants, barks, and roots, as well as plants containing

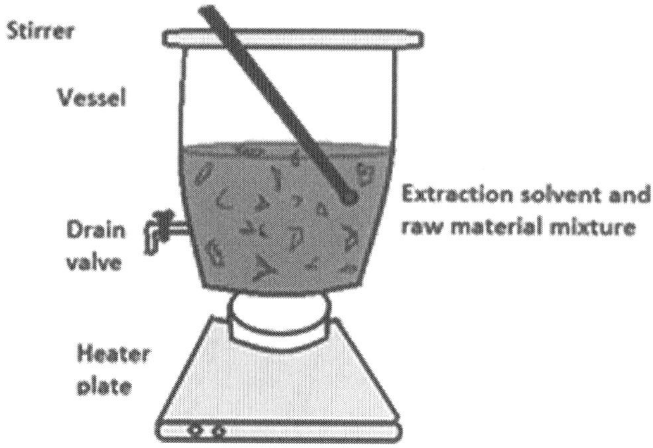

Figure 2.3 Decoction method for extraction.

water-soluble chemicals. Plant material is typically crushed into small fragments or ground into powder. This method is commonly employed when making Ayurvedic extracts known as "quath" or "kawath" (Sowa et al., 2016). The initial ratio of raw medicine to water is set at, for example, 1:4 or 1:16; the volume is reduced to one-fourth of its original volume through boiling in this extraction process. Next, the filtered concentrated extract is either used in its current form or undergoes additional processing.

2.3.4 Digestion

This is a method of extracting compounds which uses moderate heat in the extraction procedure. The extraction solvent is poured into a fresh container along with the powdered drug material. During this extraction process, the mixture is heated either over a water bath or in an oven at approximately 50°C to reduce the viscosity of the extraction solvent and improve the extraction of secondary metabolites.

It is employed when slightly raised temperature is tolerable. The method can be changed by blending the substance with the liquid using a magnetic stirrer, mechanical stirrer, or by intermittently shaking by hand. After 8 to 12 hours, the extract is strained, new solvent is introduced, and the procedure is repeated until all the needed substances are extracted (Figure 2.4).

2.3.5 Percolation

This method is commonly used to extract the phytochemicals for creating tinctures and fluid extracts. Typically, a slender cone-shaped vessel, open at both ends, is commonly utilized as a percolator (Figure 2.5). The dry components are mixed with a specific quantity of the designated liquid and left to sit for around 4 hours in a sealed vessel. Then, the mass is compressed and the lid of the percolator is shut. More liquid is added as necessary, until the percolate fills approximately three-fourth of the final product's volume. Then the marc is squeezed and the resulting liquid is combined with the percolate. Enough solvent is added to reach the necessary volume, and the blended solution is clarified through either filtration or allowing it to settle before pouring off the top layer.

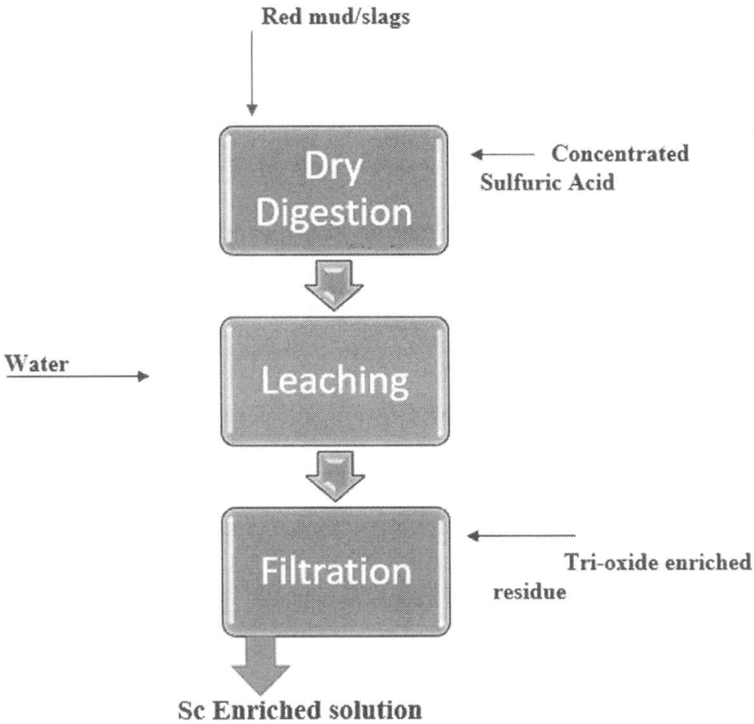

Figure 2.4 Flow chart of digestion.

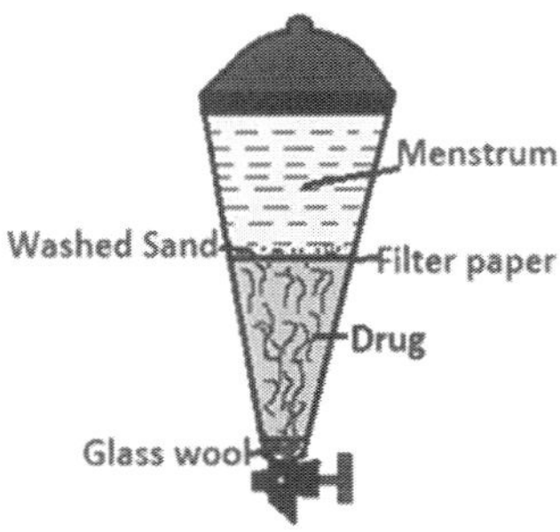

Figure 2.5 Simple percolator.

2.3.6 Hot Continuous Soxhlet Extraction

In this process, finely ground crude drug is placed in a porous bag or "thimble" made of strong filter paper, which is then put in the chamber of the Soxhlet apparatus (Figure 2.6). The solvent in the flask is warmed up, and its fumes are cooled down so that they condense into liquid in the condenser. The concentrated extract drips onto the thimble with the raw drug and pulls it out through contact. When the liquid in the chamber reaches the top of the siphon

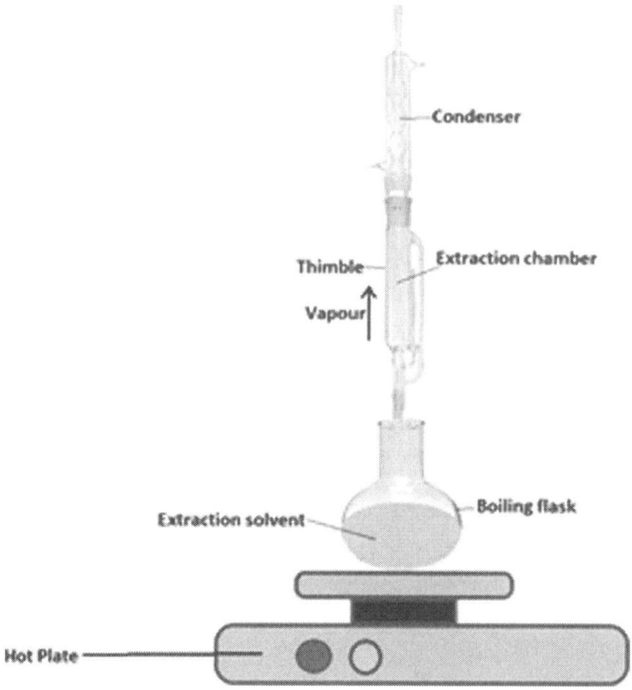

Figure 2.6 Soxhlet extractor.

tube, it siphons into the flask. This activity is ongoing and continues until the solvent dropping from the siphon tube no longer leaves any residue upon evaporation.

The advantage of this method over others is its ability to extract large quantities of bioactive metabolites using minimal solvent, which conserves time, energy, and money. On a small level, it is primarily carried out in batches but can be more cost-effective and feasible if transformed into a continuous extraction method in medium- or large-scale extractions.

2.3.7 Countercurrent Extraction

In countercurrent extraction (CCE), moist raw material is crushed with toothed disc disintegrators to create a fine slurry. During this procedure, the substance to be removed is transported in a specific direction (usually as a thin slurry) inside a cylindrical extractor where it meets the extraction solvent. The extract becomes more concentrated as the starting material moves further. Full extraction can be achieved by maximizing the amounts of solvent and material used, as well as optimizing their flow rates. The procedure is very effective, taking up minimal time and presenting no danger due to elevated temperatures. In the end, a highly concentrated extract emerges from one side of the extractor, while the marc, which is mostly free of visible solvent, is released from the opposite end. In contrast to co-current extraction, countercurrent extraction is more effective as it has a greater average driving force, making it easier to conduct multiple extraction stages in one device. Nevertheless, conducting countercurrent extraction on a small scale is challenging due to the prevalence

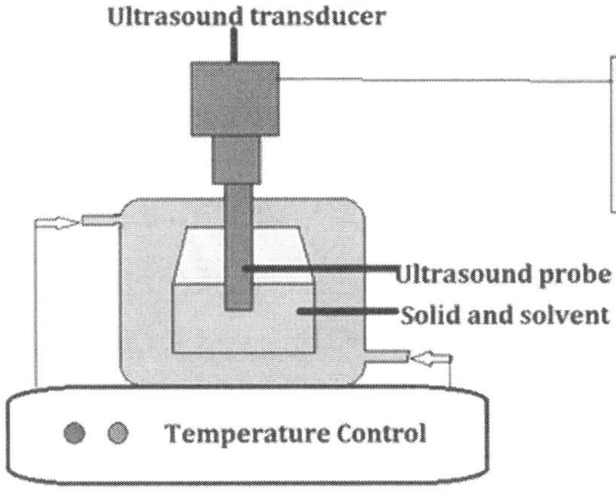

Figure 2.7 Ultrasound extraction.

of viscous and interfacial forces compared to inertial force and gravity, which are crucial for countercurrent extraction on a larger scale.

2.3.8 Ultrasound Extraction

It is an efficient technique that eliminates the need for excessive solvent (Chemat et al., 2017) and bulky extraction equipment such as Soxhlet and maceration (Figure 2.7). The decreased impact on the environment due to ultrasound extraction (UAE) is clearly beneficial in terms of both energy and time.

The method utilizes ultrasound frequencies between 20 kHz and 2000 kHz, which enhance cell wall permeability and create cavitation. While the technique has benefits for certain situations, such as obtaining *Rauwolfia* root, its widespread usage is constrained by its expensive nature. Ultrasound energy (> 20 kHz) can sometimes harm the active components of medicinal plants by creating free radicals which can lead to unwanted alterations in the drug molecules.

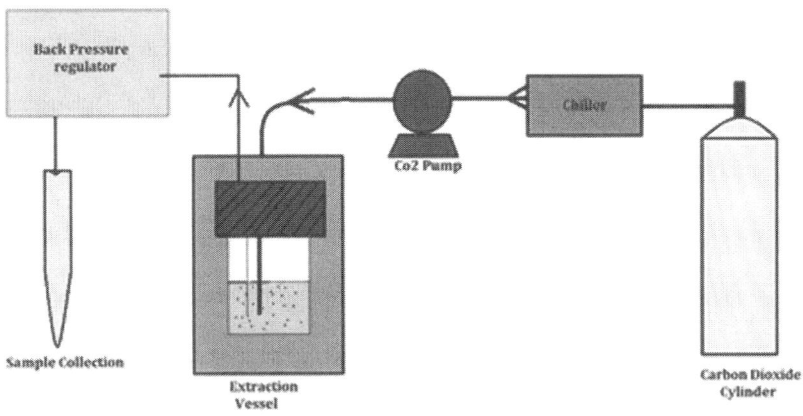

Figure 2.8 Supercritical fluid extraction.

2.3.9 Supercritical Fluid Extraction

This extraction system is most technologically advanced (Sapkale et al., 2010). Supercritical fluid extraction (SFE) uses gases, typically carbon dioxide (CO_2), that are compressed to form a dense liquid. This fluid is subsequently pushed through a tube that holds the substance intended for extraction. Next, the liquid containing the extract is transferred to a separation chamber to separate the extract from the gas and recover the gas for reuse. CO_2 is commonly utilized as a supercritical fluid, occasionally altered with co-solvents like ethanol or methanol. Supercritical CO_2 extraction requires temperatures above 31°C and pressures above 74 bar. Supercritical extraction (Figure 2.8) primarily utilizes high pressure CO_2 to extract valuable products from natural materials. In contrast to other methods, SFE does not result in any leftover solvent residue. Furthermore, CO_2 is not poisonous, nonflammable, odorless, flavorless, inert, and low-cost gas.

2.3.10 Microwave-Assisted Extraction

This method is simply termed microwave extraction and it combines microwave and traditional solvent extraction (Mandal et al., 2007) (Figure 2.9). Revolution in organic compound synthesis has been promoted by microwave-assisted organic syntheses (MAOS) by which small molecules are built up into large polymers in a fraction of time.

In microwave-assisted extraction (MAE), heat is applied selectively and efficiently within a closed system, minimizing heat loss to the surrounding environment. This special heating process can drastically cut down the extraction time (typically less than 30 minutes) in comparison to Soxhlet extraction. Heating with microwaves relies on interacting with polar materials or solvents through ionic conduction and dipole rotation, usually happening at the same time. Ionic conduction is the movement of ions caused by the

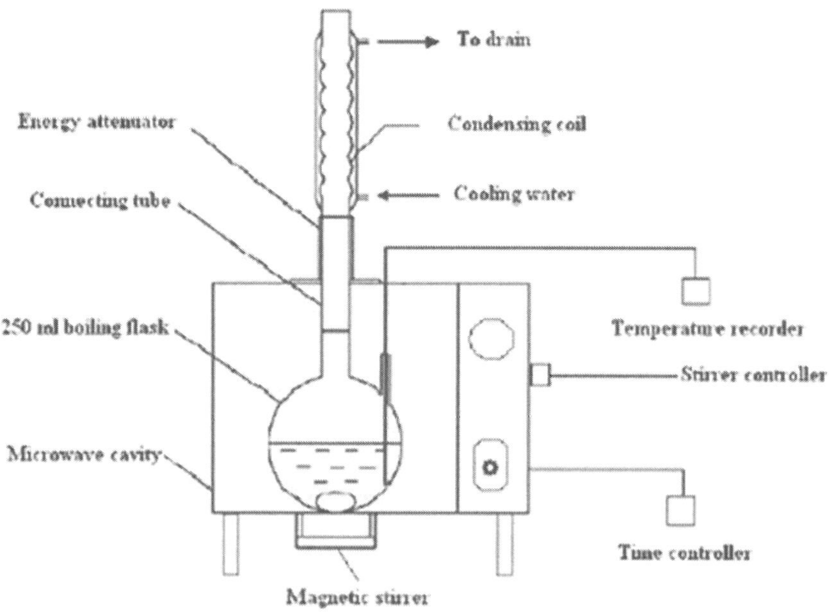

Figure 2.9 Microwave-assisted extraction.

fluctuating electric field. Dipole rotation involves the reorientation of the molecule's dipoles in response to the fluctuating electric field.

2.4 CHROMATOGRAPHIC TECHNIQUES

Chromatography is a method where molecules are separated mainly based on their form, duration, and charge. There are a number of unknown ingredients present in each extract and some of them are in very small amounts. During chromatography, mixture is dissolved in a solvent and passed through a strong phase that acts as a sieve. As the mixture progresses further through the molecular sieve, it gets separated. Thus, there are different chromatographic strategies likely to be used for the isolation of a plant's secondary metabolite (Agostini-Costa et al., 2012). These are discussed in subsequent subsections.

2.4.1 Thin-Layer Chromatography

This is an easy, short, and mostly inexpensive approach to determine the number of components in a combination. Thin-layer chromatography (TLC) is adsorption chromatography where samples are separated entirely by interaction between a thin layer of adsorbent associated with a plate. This approach is well equipped and widely used for the separation of low-molecular-weight compounds. Various adsorbents may be used such as silica gel, aluminum, cellulose powder, starch, and other materials. This process can be used to break down many compounds consisting of alkaloids, phenols, steroids, and so on.

TLC is used to manually identify a compound in an aggregate, while the retention factor (Rf) of a compound is evaluated with the Rf of said compound. Rate of modification of Rf is calculated using the following equation:

$$Rf = \text{distance traveled by the compound/distance traveled by solvent front}$$

The Rf charge shows difficulty in maintaining, i.e., the higher the molecular weight, the greater the distance that can be covered using isolates.

Basically, TLC is used for the following themes:

- Enables rapid assessment of natural extracts with minimal requirement for sample smoothing
- Provides qualitative and semi-quantitative data on separated compounds
- Enables the quantification of chemical elements.

Controls for the isolation and identification of plant secondary metabolites include performing phytochemical screening, which involves the spraying with vanillin in phosphoric or sulfuric acid or 5% sulfuric acid in ethanol. Moreover, a plate may be subjected to UV light to observe daily changes in shade, depicting the presence of phytochemicals in the plant extract. For example, vegetation with secondary metabolites along with lignans absorb moderate UV radiation; chemical-free plates can be located below a wavelength of 254 nm (Patyra et al., 2019). In addition, several authors use the 366-nm wavelength to check the impurities in samples. Other detection techniques include spraying dye dealers that contain vanillin in phosphoric or sulfuric acid and 5% sulfuric acid in ethanol. This proved to be further useful to verify the purity of distant compounds.

Although TLC was previously also used in quantitative analysis (densitometry) as well as for isolation and purification procedures (preparative TLC), it has been nowadays replaced with more advanced methods and its use is rather limited (Table 2.1).

Table 2.1: TLC Mobile Phase for Important Classes of Plant Secondary Metabolites

Plant Constituents	Stationary Phase	Mobile Phase	Detection
Alkaloids/ phenanthrenes	Silica gel	Toluene:ethyl acetate:diethylamine (7:2:1)	Dragendroff reagent
Flavonoids	Silica gel	Ethyl acetate:formic acid:glacial acetic acid:water (10:1.1:1.1:2.6)	UV 254 nm or 366 nm
Tannins	Silica gel	Ethyl acetate:formic acid:glacial acetic acid:water (7.5:0.3:0.2:2)	Vanillin sulfuric acid reagent
Saponin	Silica gel	Chloroform:glacial acetic acid:methanol:water (6.4:3.2:1.2:0.8)	Vanillin sulfuric acid reagent

2.4.2 High-end General Performance Liquid Chromatography (HPLC)

It is a flexible, robust, and widely used approach to the isolation of natural compounds. High performance liquid chromatography (HPLC) is an analytical approach for the separation and identification of backbone of natural and inorganic solutes in any samples. Currently, this approach is gaining recognition among various analytical techniques as well as it is the basic preference for fingerprint investigation. The amount or degree of separation is through use and is largely determined by the method of use of table-bound phase selection and cell section. The biologically active entity is often present at best as a minor matter in the extract, and the resolving power of HPLC is perfectly suited to the rapid processing of such multicomponent samples at any analytical and preparative scale. In order to discover any compound with a useful HPLC resource, a detector should first be chosen, HPLC gadgets are now modular and include a solvent transport pump, a patterning device that includes an automotive sampler or manual injection valve, an analytical column, a protective column, and recorder or printer.

Chemical separations can be completed by means of HPLC, taking advantage of the fact that fancy compounds have particular migration costs given the specific column and cell phase. In general, the identification and separation of phytochemicals can be completed using an isocratic tool (using unmarried invariant cell section device). Gradient elution, in which the proportion of natural solvent to water varies with time, may be appropriate if a multi-sample problem is being studied and each exact varies greatly in retention under the employed conditions. Purification of a hobby compound using HPLC is the process of setting aside or extracting the target compound from other (most likely structurally related) compounds or contaminants. Each compound must have a signature at the top under positive chromatographic conditions. Depending on what needs to be separated and how closely related the samples are, the chromatographer can also select prime situations to achieve advanced separations.

Identification of compounds by HPLC is an essential part of any HPLC assay. In order to capture any compound by HPLC, the detector must first be decided upon. Once the detector is selected and prepared for the most effective detection settings, a better separation test must be performed. The parameters of this test should be such that the height of the specified sample can be easily determined from the chromatograph. The detection peak must have a less expensive retention time and must be well separated from extraneous peaks at detection levels at which the assay is likely to be terminated. UV detectors are famous among all detectors (Rocha et al., 2018) because they actually offer extraordinary sensitivity and also because most of the real-life compounds we encounter have little UV absorbance at low wavelengths (190–210 nm). The extreme sensitivity of UV detection is a bonus if the compound of interest is found in high quality in small portions within the sample. In addition to UV, other detection techniques are also used for phytochemicals, including a diode array detector (DAD) coupled to a mass spectrometer (MS). HPLC is useful for compounds that cannot be volatilized or those that decompose at excessive temperature, and it provides an excellent complement to gas chromatography for the detection of compounds.

2.4.3 Vacuum Liquid Chromatography

VLC is used as a separation technique in plant product research (Paranagama, 2011). It was first used for the separation of diterpenes (Coll & Bowden, 2007), a plant secondary metabolite, in Australia. In 1997, a brief description of a method for diterpene isolation from Australian pure coral was published, but no facts about the equipment and technique were suggested. The name vacuum liquid chromatography (VLC) was first coined by Targett (Targett et al., 1983). VLC is considered preparative TLC (PTLC) because the separation is completed on silica gel or TLC grade alumina and the column is dried after each fraction as in PTLC, the plates are dried and rerun to garnish the separation. Filled VLC columns can be reused for the same or comparable separation by washing the columns thoroughly with methanol and scraping the decomposed polar substance or band from the top of the adsorption column. Gradient elution can be very powerful and can be used for the separation of small to massive forms of combinations There is no specific requirement for VLC equipment. This may carried out with glassware in a laboratory (Maurya et al., 2008). It requires a sintered glass funnel with a quality disc frit. A laboratory water pump or a water aspirator that offers a vacuum of 20–70 mm Hg is needed for suction. TLC grade binder, unfixed silica gel (silica gel H) or alumina is generally used in VLC; however, amazing adsorbents can also be tested based on their TLC profile. Packing VLC columns is easy and simple. The dry adsorbent is filled into a sintered funnel and allowed to settle primarily under gravity while gently tapping the column with a rubber tube as in CC. The layer can first be compressed with a glass or rubber stopper. Once the absorbent has settled, a vacuum should be applied and the adsorbent layer should be similarly compressed with a plug to finally achieve packed-uniform and tight column. To check the uniformity of the column, it is necessary to release the vacuum and quickly pour a nonpolar solvent (which includes hexane) on top of the adsorbent and apply the vacuum. The solvent should pass through the column evenly indicating a uniform column, otherwise it needs to be packed again. In VLC, samples are preferably loaded as liquids for simplicity, but solids can also be loaded as slurries if needed. After loading, the vacuum is completed to remove polar solvents and the column is fed with appropriate solvents under vacuum to elute the compound, with fractions collected in round bottom flasks to avoid channeling. Although gradient column elution is preferred, isocratic elution can also be completed. Volatile solvent loss can be minimized by using a sintered glass funnel with a large pore length or by using a drift enhancement source.

2.4.4 Column Chromatography

CC is the most effective technique used to separate components in their natural form. It consists of ion exchange, molecular sieves, and adsorption phenomena. This is a preparative chromatography technique (Becker et al., 2010). In this method, a long cylindrical suspension glass column containing a table-top solid phase (silica gel) is slowly collided from the top with a liquid solvent (cellular phase) that flows down the column due to gravity or external pressure. This approach is used to purify compounds from a mixture. Once the column is ready, the sample is loaded at the top of the column. The mobile solvent is then allowed to flow through the column. The compounds in the aggregate have a superior ability to interact with the table-bound phase (silica gel) and cell segment so that they will slide along the cell section at unique time periods or layers. In this way, the separation of the compounds from the aggregate is achieved. Individual compounds are collected as fractions and similarly analyzed for structure elucidation.

2.4.5 Ion Exchange Chromatography

Like other chromatography methods we have discussed, IEC also involves the use of a mobile and stationary phase to separate molecules depending on their charge. The setup involves ion exchange resins in a column as the stationary phase, with the mixture to be separated added to an aqueous buffer system as the mobile phase. Counter ions in the system are evenly distributed between the two phases, resulting in two possible IEC configurations: anion exchange and cation exchange. Resins that exchange positively charged ions are referred to as catex, whereas those that exchange negatively charged ions are known as annex. Different kinds of matrix counter ions available for exchange include protons; hydroxyl groups; monoatomic single-charge ions such as sodium, potassium, and chlorine; monoatomic ions with a double charge such as calcium and magnesium; inorganic polyatomic ions such as sulfate and phosphate, as well as organic bases and acids. Ion exchange is a commonly utilized technique for the chromatographic isolation and cleansing of charged biological molecules such as polypeptides, proteins, polynucleotides, and nucleic acids. Some applications this method in industry involve separating and purifying blood components such as albumin and recombinant growth factors, examining different varieties of wheat, linking proteinuria to kidney diseases, and overseeing fermentation processes for β-galactosidase production with cation exchange resins.

A pH gradient can be used to separate different proteins based on their isoelectric points (pI), which is the pH at which a protein carries no net charge and does not move in an electric field. Because amino acids are zwitterionic compounds, they have groups with both positive and negative charges. Proteins have a positive, negative, or neutral charge depending on the pH of their surroundings. At their pI, they will not bind to the charged groups on the column resin and will therefore be washed out. Proteins can be eluted using a decreasing pH gradient with an anion exchange resin and using an increasing pH gradient with cation exchange resins. This happens because when the buffer pH of the mobile phase is raised, the protein becomes less protonated (less positively charged), preventing it from forming an ionic interaction with the negatively charged resin and allowing it to be eluted. On the other hand, if the pH of the mobile phase is reduced, the molecule will become more protonated and less negatively charged, which facilitates its elution. The wide applicability of IEC, along with its large handling capacity, high capacity, simplicity, and high resolution (including high performance and high-throughput formats), moderate cost, powerful resolution, and ease of scale-up and automation, have made it a successful separation method and one of the most versatile and commonly used liquid chromatography techniques.

2.4.6 Size Exclusion Chromatography

SEC was developed to separate amino acids and protein fragments on the basis of size, using starch in a process known as "partition chromatography." Chromatography includes utilizing a gel as the stationary phase in a column to enable a liquid mobile phase to flow through. This holds the analyte in a water-based buffer solution which exits the column at a rate that is directly related to its size, specifically the hydrodynamic volume. The researchers at Queen Charlotte's Hospital in London first conducted experiments with potato and maize starch, which showed variations in size and behavior, resulting in different elution volumes in each column. The development of dextran and crosslinked agarose further enhanced this, both being derived from seaweed as polysaccharide polymer materials. Currently, gel matrices are identified by trade names such as Sephadex (Cytiva, Marlborough, MA, USA) for dextran polymers, Sepharose (Cytiva, Marlborough, MA, USA) for agarose, and Sephacryl (Cytiva, Marlborough, MA, USA) or BioGel P (Bio-Rad, Hercules, CA, USA) for polyacrylamide, designed as beads and tailored for experimental accuracy. The duration it takes for an analyte to travel through a substance must be determined through experimentation and is generally influenced by the size of the pores in the gel matrix. Furthermore, various gels require a specific molecular weight range for optimal linear separation, and it is advised to calibrate this using a standard curve with known weights. Additional variables that will impact the procedure include the size of the column, the implementation of column frits, and the uniformity of the gel formation process within the column, all of which can be controlled by utilizing commercially available columns that adhere to industry standards. Larger bed volumes can lead to an improved resolution of larger particles, but this must be balanced by a longer time taken to elute each fraction.

The amount of literature discussing SEC as a technique for isolating extracellular vesicles (EVs) has increased steadily, as shown by the growing number of papers in this current review. SEC has advantages such as being quick, with a processing time of 10 to 20 minutes per sample. In addition, it is cost-effective and can effectively eliminate more than 95% of blood proteins, making it highly suitable for clinical use compared to density gradient ultracentrifugation or other methods. Challenges in the procedure involve initial tasks of eliminating big impurities, potential simultaneous occurrence of lipoproteins of comparable sizes, and any necessary concentration steps before or after. Lipoproteins like very-low-density lipoprotein (VLDL) are comparable in size to EVs and are known to coprecipitate from blood plasma or serum samples, distinguished by the existence of apolipoproteins and a lipid monolayer. Considering lipoproteins as carriers of potentially functional miRNAs is an important factor to take into account, as it could complicate the discovery of potential biomarkers. It is possible to combine SEC with other methods like immunoprecipitation, as shown in various papers reviewed. Additional research may be required to understand how EVs interact with frequently utilized gels. EVs possess a significant surface area to volume ratio for potential exploration in future studies. SEC in EV research is also utilized for isolating vesicles from unbound dye, like cleaning out PKH or Carboxyfluorescein diacetate succinimidyl ester (CFSE), from EVs before further examination, and for separating EVs from iodixanol in density gradient method.

This use of orthogonal approaches, i.e., both size and density purification in EV isolation and purification, is increasing as protocols are shared and familiarity with different methods improves.

2.5 CHARACTERIZATION TECHNIQUES OF PLANT SECONDARY METABOLITES

Characterization of secondary metabolites extracted using numerous techniques is important in order to improve the knowledge about the considered compounds and to find new compounds as well. Currently, no single analytical technique or combination of analytical strategies (i.e., chromatography combined with MS or NMR spectroscopy) can capture all metabolites within a given natural sample. The chemical composition of an aggregate can be determined through a combination of various analytical techniques, which can profile all the chemical components present in a mixture. Spectroscopic instruments such as UV/visible and IR absorption spectroscopy, NMR, and MS currently serve as the foundation for modern shape assessment.

2.6 SPECTROSCOPIC TECHNIQUES

Spectroscopy studies investigate how light or electromagnetic radiation interacts with matter, resulting in either the absorption or emission of the energy by the matter in discrete amounts termed as quanta. Important spectroscopic techniques are described in subsequent subsections.

2.6.1 UV-Visible Spectroscopy

UV spectroscopy is emerging as a powerful analytical tool that utilizes UV radiation or visible light, with ranges of wavelengths from 200 to 800 nm (Guo et al., 2013). This method is flexible and can analyze all colorless compounds in the UV region (400–200 nm) and colored compounds in the visible region (800–400 nm) (Aleixandre-Tudo & Du Toit, 2018). When a monochromatic beam of light passes through a substance, some amount of light may be absorbed and the remaining is transmitted through the sample.

There are three types of ground state orbitals that may be of concern:

i. σ (bonding) molecular orbital

ii. π (bonding) molecular orbital

iii. n (bonding) atomic orbital

There are two forms of anti-bonding orbitals that can be of concern:

i. σ * orbital

ii. π * orbital.

There is always no n* anti-bonding orbital because n electrons do not form bonds.

The virtual transition occurring via the UV absorption pathway and visible light rays observed are as follows:

- σ to σ*
- π to π*
- n to σ*
- n to π*

The transition from σ to σ* and n to σ* requires an exorbitant energy that approaches at far UV area in the range of 180 to 240 nm.

2.6.1.1 Absorption Laws

Different molecules absorb a unique amount of energy, and this can be shown as a plot of absorbance on the y axis and wavelength on the x axis (Antonov

& Stoyanov, 2013). This plot is commonly called an **absorption spectrum** (Sudharshan & Swetha, 2016). We can determine the **λmax** (the wavelength at which the pattern shows the greatest absorbance) of a sample and thus its concentration (Mandru et al., 2021). It is entirely based on the **principle of the Beer–Lambert law**, which states that "the absorbance of a sample is simultaneously proportional to the concentration of the absorbing species and the path length" (Upstone, 1994).

$$\mathbf{A = \varepsilon cl}$$

where, A is absorbance, ε is molar extinction coefficient, c is concentration of absorbing substances, and l is path length. For a fixed path duration, it can be used to decide the concentration of the absorber in the sample. So, it is important to understand how fast absorbance changes with concentration.

2.6.2 IR Spectroscopy

The secondary metabolites interact with IR light and make IR spectroscopy a useful tool for quantitative and qualitative evaluation (Huck, 2015).

Principle of IR spectroscopy: IR radiation interacts with a sample and results in an increase in molecular energy that would trigger three incredible transitions: electronic excitation, vibrational change, and rotational change, depending on the wavelength of the incident radiation.

IR regions: IR regions are described as near-IR (13,500–4,000 cm^{-1}; 780–2,500 nm), mid-IR (4,000–400 cm^{-1}; 2,500–25,000 nm), and extended-IR (far IR, 400–10 cm^{-1}; 25,000–one million nm).

Absorption in mid-IR spectroscopy: Absorption in mid-IR spectroscopy corresponds to important vibrations of chemical bonds associated with atoms of molecules (Theodoridou et al., 2014). The chemical bonds will vibrate extra vigorously while the molecule interacts with IR, causing vibrational and rotational modifications in the molecule. Vibration will be seen as a stretching when only two atoms are involved. In case of three or more atoms, bending of the bond can also occur. Typical mid-IR spectra form several absorbance peaks due to fundamental transitions and are divided into four regions generalized as X–H stretching region (4000–2500 cm^{-1}), around triple bonds region (2500–2000 cm^{-1}), double bond region (2000–1500 cm^{-1}), and fingerprint region (1500–600 cm^{-1}). The frequency shift of the band is related to the structural change in the molecules under evaluation (Wetzel, 2002).

Quantitative information can be obtained from mid-IR spectra because of the proportionality of the absorbance to the number of functional groups through following the Lambert–Beer law. This has been demonstrated for some herbal ingredients (Huck-Pezzei et al., 2013).

Absorption in near-IR spectroscopy: The spectral range here is narrower than the mid-IR variety; therefore, the near-IR molar absorptivity is typically quite small, but there is an extra increased penetration depth of the samples (Manley, 2014). Two processes generate prominent absorptions in this region: undertones and combinations of fundamental CH, –NH, –OH (and –SH) functional groups vibrations. All hydrogen-containing molecules have a measurable near-IR spectrum, implying that a greater range of natural substances in plant samples are suitable for near-IR evaluation than for mid-IR evaluation (Chen et al., 2020).

IR imaging/micro-spectroscopy: In conventional IR spectroscopy it is difficult to resolve the distinction between bulk and localized components within the plant samples, as it does not give information on spatial localization of molecular structures. IR imaging/microspectroscopy exhibits great advantages for the broad analysis on the spatial distribution patterns of plant constituents (Kumar et al., 2015).

2.6.3 NMR Spectroscopy

NMR spectroscopy uses a physical phenomenon where a magnetic nucleus in a fixed external magnetic field absorbs and reemits measurable electromagnetic radiation (Macomber, 1998).

NMR techniques offer metabolic fingerprints with authentic chemical specificity for compounds containing elements with nonzero magnetic moments, which include ^{1}H, ^{13}C, ^{15}N, and ^{32}P, which can be generally found in most organic metabolites (Bligny & Douce, 2001).

Most tested nuclei: The two most examined nuclei are ^{1}H and ^{13}C.

Chemical shift: A parallel alignment of previously randomly oriented nuclei will occur when exposed to an external magnetic field. The nuclei align themselves along or against the direction of the magnetic field; the latter alignment requires least energy. The energy difference between the two spin states will increase with the increased external magnetic field. The sample is now exposed to radiant energy in the radio frequency (rf) range. When nuclei are exposed to electromagnetic radiation with a frequency equal to its Larmour frequency, transition from low-energy state to high-energy state occurs. After turning off the rf radiations, the nuclei come back to lower-energy state by the emission of the absorbed rf energy, depending upon the external magnetic field, and this produces a measurable signal. This signal, known as the resonance frequency, creates a way to obtain better structural information about the analyzed molecule. The phenomenon is known as chemical shift.

Overview of NMR experiments, specifically focusing on 1D and 2D NMR techniques:

- *1D experiments:* 1D experiments are spectroscopic evaluations of a single nucleus.
- The most widely used NMR control is ^{1}H-NMR, providing records of chemical shifts, multiplet systems, homonuclear coupling constants, and integrations of all protons donated in the sample.
- *2D NMR experiments:* The nuclei themselves behave like small magnets. Therefore, they can influence and change the energy of adjacent nuclei as they resonate, spin–spin coupling. This forms the basis for 2D NMR experiments.

2.6.4 Mass Spectrometry

MS, in simple words, is the ability to encounter the mass-to-charge (m/z) ratio and abundance of numerous analytes generated by ionization of sample extract or chromatographic fraction (Gross, 2004). High selectivity mass analyzers have the ability to distinguish ions of comparable masses.

MS instrument consists of three modules:

- A supply of ions that could convert gas phase molecules into ions
- A mass analyzer that sorts ions according to their masses using electromagnetic fields
- A detector that measures the value of indicator quantity to calculate the amount of each ion present (Kind & Fiehn, 2010).

2.6.4.1 Ion Source

It is the part of the MS that ionizes the sample. The ions are then transported using electromagnetic fields to the mass analyzer.

Depending on the probe beam used to interrogate tissue samples, a variety of ionization techniques are available. Electrospray ionization and matrix-assisted laser desorption/ionization (MALDI) techniques are commonly used for liquid and solid natural samples.

MALDI: In MALDI, there is the mixing of matrix and analyte powder together in a mortar (Tanaka et al., 1988). Matrix is a crystalline substance, most commonly used matrices are 3,5 dimethoxy-4-hydroxy cinnamic acid (sinapinic acid), alpha–cyano-4-cinnamic acid (alpha-matrix), and 2,5 dihydroxy benzoic acid (DHB). Matrix protects analyte from destruction by direct laser beam (Zhigilei et al., 1999).

Electrospray ionization: In this technique, the sample solution is pumped constantly from a fine, charged stainless steel capillary needle (which has maintained at high voltage of about 4 KV), which produces the mist or aerosol of fine charged droplets from the liquid (Fenn et al., 1989). The charged droplets then travel through desolvating capillary where there is attachment of charge by the evaporation of solvent under vacuum (Bruins, 1987). After the evaporation process of droplets, analyte molecules become more closely spaced. As the molecules of similar charge come nearer to each other, they become unstable and explode again (Banerjee & Mazumdar, 2012). This continues until the analyte is devoid of solvent resulting in an isolated ion. This ion then starts moving toward the mass analyzer (Patel et al., 2015).

2.6.4.2 Mass Analyzer

There are many mass analyzers using either static or dynamic field and magnetic or electrically driven field.

There are many important characteristics of analyzers, one of them being the **mass resolving power** which is the ability to differentiate two peaks of slightly distinct m/z ratios (Xian et al., 2012).

2.6.4.3 Detector

The last element of an MS is the detector. The detector records either the induced charge or the current created as the ions hit the surface of the detector. It will result in the formation of mass spectrum, a record of ions as a function of m/Q (Park et al., 2007).

2.7 CONCLUSION

It can be very challenging to separate natural products from the plant extracts, as varying component mixtures with different polarities are present in plant extracts. So, the process of separation and characterization becomes difficult. Extraction process is crucial for the isolation and identification of various natural substances. Various chromatographic and non-chromatographic techniques are used to isolate and purify different plant secondary metabolites.

REFERENCES

Abubakar, A.R. and M. Haque, Preparation of medicinal plants: Basic extraction and fractionation procedures for experimental purposes. *Journal of Pharmacy and Bioallied Sciences*, 2020. **12**(1): p. 1–10.

Aleixandre-Tudo, J.L. and W. Du Toit, The role of UV-visible spectroscopy for phenolic compounds quantification in winemaking. *Frontiers and new trends in the science of fermented food and beverages*. 2018. p. 200–204.

Antonov, L. and S. Stoyanov, Analysis of the overlapping bands in UV-vis absorption spectroscopy. *Applied Spectroscopy*, 1993. **47**(7): p. 1030–1035.

Bajpai, V.K., R. Majumder, and J.G. Park, Isolation and purification of plant secondary metabolites using column-chromatographic technique. *Bangladesh Journal of Pharmacology*, 2016. **11**(4): p. 844–848.

Banerjee, S. and S. Mazumdar, Electrospray ionization mass spectrometry: a technique to access the information beyond the molecular weight of the analyte. *International Journal of Analytical Chemistry*, 2012. **2012**: p. 282574.

Becker, H., K. Hostettmann, A. Marston, M. Hostettmann, *Preparative Chromatography Techniques: Applications in Natural Product Isolation*, 1998, 244 pages, (Hardcover). Price: DM 198.00/öS 1.446/sFr. *179.00*. 1999, Urban & Fischer.

Bligny, R. and R. Douce, NMR and plant metabolism. *Current Opinion in Plant Biology*, 2001. **4**(3): p. 191–196.

Bruins, A., Mass spectrometry with ion sources operating at atmospheric pressure. *Mass Spectrometry Reviews*, 1991. **10**(1): p. 53–77.

Chemat, F., V. Tomao, and M. Virot, Ultrasound-assisted extraction in food analysis. *Handbook of food analysis instruments*. 2008. p. 85–103.

Chen, J.-B., S.-Q. Sun, and Q. Zhou, Chemical morphology of *Areca* nut characterized directly by Fourier transform near-infrared and mid-infrared microspectroscopic imaging in reflection modes. *Food Chemistry*, 2016. **212**: p. 469–475.

Coll, J.C. and B.F., Bowden, The application of vacuum liquid chromatography to the separation of terpene mixtures. *Journal of Natural Products*, 1986. **49**(5): p. 934–936.

da S. Agostini-Costa, T., R.F. Vieira, H.R. Bizzo, D. Silveira, and M.A. Gimenes. Secondary metabolites. *Chromatography and its applications*. 2012. p. 131.

Das, K. and S. Gezici, Secondary plant metabolites, their separation and identification, and role in human disease prevention. *Annals of Phytomedicine An International Journal*, 2018. **7**: p. 13–24.

Fenn, J.B., et al., Electrospray ionization for mass spectrometry of large biomolecules. *Science*, 1989. **246**(4926): p. 64–71.

Fonmboh, D.J., et al., An overview of methods of extraction, isolation and characterization of natural medicinal plant products in improved traditional medicine research. *Asian Journal of Research in Medical and Pharmaceutical Sciences*, 2020. **9**(2): p. 31–57.

Gross, J.H. and J.H., Gross, *Electron ionization*. Mass spectrometry: A textbook. 2004. p. 193–222.

Guo, Y., et al., Advances on water quality detection by UV-vis spectroscopy. *Applied Sciences*, 2020. **10**(19): p. 6874.

Handa, S., An overview of extraction techniques for medicinal and aromatic plants. *Extraction Technologies for Medicinal and Aromatic Plants*, 2008. **1**(1): p. 21–40.

Huck, C.W., Advances of infrared spectroscopy in natural product research. *Phytochemistry Letters*, 2015. **11**: p. 384–393.

Huck-Pezzei, V., et al., Fourier transform infrared imaging analysis in discrimination studies of St. John's wort (*Hypericum perforatum*). *Analytical and Bioanalytical Chemistry*, 2012. **404**: p. 1771–1778.

Kind, T. and O. Fiehn, Advances in structure elucidation of small molecules using mass spectrometry. *Bioanalytical Reviews*, 2010. **2**: p. 23–60.

Kumar, S., et al., Infrared spectroscopy combined with imaging: a new developing analytical tool in health and plant science. *Applied Spectroscopy Reviews*, 2016. **51**(6): p. 466–483.

Macomber, R.S., A complete introduction to modern NMR spectroscopy. 1997.

Mandal, V., Y. Mohan, and S. Hemalatha, Microwave assisted extraction—an innovative and promising extraction tool for medicinal plant research. *Pharmacognosy Reviews*, 2007. **1**(1): p. 7–18.

Mandru, A., J. Mane, and R. Mandapati, A review on UV-visible spectroscopy. *Journal of Pharma Insights and Research*, 2023. **1**(2): p. 091–096.

Manley, M., Near-infrared spectroscopy and hyperspectral imaging: non-destructive analysis of biological materials. *Chemical Society Reviews*, 2014. **43**(24): p. 8200–8214.

Maurya, A., et al., Vacuum liquid chromatography: simple, efficient and versatile separation technique for natural products. *Organic & Medicinal Chemistry International Journal*, 2018. **7**: p. 1–3.

Mendoza, N. and E.M.E. Silva, Introduction to phytochemicals: secondary metabolites from plants with active principles for pharmacological importance. *Phytochemicals: Source of Antioxidants and Role in Disease Prevention*, 2018. **25**: p. 1–5.

Paranagama, P.A., *Vacuum liquid chromatography (VLC) and gel permeation chromatography in natural product research*. 2016.

Park, M.A., J.H. Callahan, and A. Vertes, An inductive detector for time-of-flight mass spectrometry. *Rapid Communications in Mass Spectrometry*, 1994. **8**(4): p. 317–322.

Patel, R., M. Roy, and G. Dutta, Mass spectrometry-A review. *Veterinary World*, 2012. **5**(3): p. 185–192.

Patyra, A., et al., Extraction techniques and analytical methods for isolation and characterization of lignans. *Plants*, 2022. **11**(17): p. 2323.

Pauli, G.F., B.U. Jaki, and D.C. Lankin, Quantitative 1H NMR: development and potential of a method for natural products analysis. *Journal of Natural Products*, 2005. **68**(1): p. 133–149.

Rocha, F.S., et al., Experimental methods in chemical engineering: ultraviolet visible spectroscopy (UV-Vis). *The Canadian Journal of Chemical Engineering*, 2018. **96**(12): p. 2512–2517.

Sapkale, G., et al., Supercritical fluid extraction. *International Journal of Chemical Science*, 2010. **8**(2): p. 729–743.

Sasidharan, S., et al., Extraction, isolation and characterization of bioactive compounds from plants' extracts. *African Journal of Traditional, Complementary and Alternative Medicines*, 2011. **8**(1): p. 1–10.

Savithramma, N., M.L. Rao, and D. Suhrulatha, Screening of medicinal plants for secondary metabolites. *Middle-East Journal of Scientific Research*, 2011. **8**(3): p. 579–584.

Sowa, M.G., et al., The utility of near infrared imaging in intra-operative prediction of flap outcome: a reverse McFarlane skin flap model study. *Journal of Near Infrared Spectroscopy*, 2012. **20**(5): p. 601–615.

Stuart, B.H., *Infrared spectroscopy: fundamentals and applications*. 2004.

Sudharshan, N. and V. Swetha, UV-visible spectroscopy: a comprehensive review on instrumentation. *World Journal of Pharmaceutical Research*. 2023. **12**(19): p. 1342–1363.

Tanaka, K., et al., Protein and polymer analyses up to m/z 100 000 by laser ionization time-of-flight mass spectrometry. *Rapid Communications in Mass Spectrometry*, 1988. **2**(8): p. 151–153.

Targett, N., J. Kilcoyne, and B. Green, Vacuum liquid chromatography: an alternative to common chromatographic methods. *The Journal of Organic Chemistry*, 1979. **44**(26): p. 4962–4964.

Theodoridou, K., S. Vail, and P. Yu, Explore protein molecular structure in endosperm tissues in newly developed black and yellow type canola seeds by using synchrotron-based Fourier transform infrared microspectroscopy. *Spectrochimica Acta Part A: Molecular and Biomolecular Spectroscopy*, 2014. **120**: p. 421–427.

United Nations Industrial Development Organization and the International Centre for Science and High Technology (UNIDO/ICS), *Extraction technologies for medicinal and aromatic plants*. 2008, International Centre for Science High Technology.

Upstone, S.L., Ultraviolet/visible light absorption spectrophotometry in clinical chemistry. *Encyclopedia of analytical chemistry*. 2000. p. 1699–1714.

Wetzel, D.L., Mid-IR and near-IR chemical imaging: complementary for biological materials. *Vibrational Spectroscopy*, 2012. **60**: p. 29–33.

Xian, F., C.L. Hendrickson, and A.G. Marshall, High resolution mass spectrometry. *Analytical Chemistry*, 2012. **84**(2): p. 708–719.

Zhigilei, L.V., et al., Molecular dynamics simulations of matrix-assisted laser desorption—connections to experiment. *International Journal of Mass Spectrometry*, 2003. **226**(1): p. 85–106.

BIBLIOGRAPHY

Aenugu, H.P.R., et al., Near infra red spectroscopy: An overview. *International Journal of ChemTech Research*, 2011. **3**(2): p. 825–836.

Du, C., et al., Microdroplet-based continuous countercurrent extraction with high phase ratio. *Separation and Purification Technology*, 2022. **295**: p. 121269.

Dubois, F., R. Knochenmuss, and R. Zenobi, Optimization of an ionto-photon detector for large molecules in mass spectrometry. *Rapid Communications in Mass Spectrometry*, 1999. **13**(19): p. 1958–1967.

Dunn, W.B., N.J. Bailey, and H.E. Johnson, Measuring the metabolome: current analytical technologies. *Analyst*, 2005. **130**(5): p. 606–625.

Gendrin, C., Y. Roggo, and C. Collet, Pharmaceutical applications of vibrational chemical imaging and chemometrics: a review. *Journal of Pharmaceutical and Biomedical Analysis*, 2008. **48**(3): p. 533–553.

Koehn, F.E. and G.T. Carter, The evolving role of natural products in drug discovery. *Nature Reviews Drug Discovery*, 2005. **4**(3): p. 206–220.

Koparde, A.A., C. Magdum, and R. Doijad, Phyto active compounds from herbal plant extracts: its extraction, isolation and characterization. *World Journal of Pharmaceutical Research*, 2017. **6**(8): p. 1186–1205.

Li, L., Q. Zhang, and D. Huang, A review of imaging techniques for plant phenotyping. *Sensors*, 2014. **14**(11): p. 20078–20111.

Mangam, V.T., et al., Novel colorimetric approach for amikacin estimation in pure powder and its pharmaceutical formulations. *World Journal of Biology Pharmacy and Health Sciences*, 2023. **14**(1): p. 270–279.

Rasul, M.G., Conventional extraction methods use in medicinal plants, their advantages and disadvantages. *International Journal of Basic Sciences and Applied Computing*, 2018. **2**: p. 10–14.

Rasul, M.G., Extraction, isolation and characterization of natural products from medicinal plants. *The International Journal of Basic Sciences and Applied Computing*, 2018. **2**(6): p. 1–6.

Rehman, M.U., F. Khan, and K. Niaz, *Introduction to natural products analysis. Recent advances in natural products analysis*. 2020. p. 3–15.

Suganya, R. and M. Thangaraj, Isolation and characterization of leaf extract of Derris trifoliate. *International Journal of ChemTech Research*, 2014. **6**(9): p. 4115–4122.

Sumner, L.W., P. Mendes, and R.A. Dixon, Plant metabolomics: large-scale phytochemistry in the functional genomics era. *Phytochemistry*, 2003. **62**(6): p. 817–836.

Tarapatskyy, M., et al., Magnetic field extraction techniques in preparing high-quality tea infusions. *Applied Sciences*, 2018. **8**(10): p. 1876.

3 Biological Aspects of Plant Natural Products

Muhammad Umer, Azhar Rasool, M. Zain Sharif, Muneeb ur Rehman, Usama Munir, Farwa Tahir, and Malik Saadullah*

3.1 INTRODUCTION OF PLANT PRODUCTS

Since the beginning of time, the plant kingdom has provided an endless supply of beneficial medications, meals, gums, lubricants, flavoring agents, and coloring agents (Aiyegoro and Okoh 2009). Plant extracts are made up of several substances that are unique to the plant from which they are taken (Šernaitė 2017). Metabolic specialization is observable within individual members of the plant kingdom, and it is frequently seen that the capacity to synthesize particular chemicals is limited to specific plant lineages (Field et al. 2006).

Simple inorganic precursors can be used by plants to synthesize a wide variety of low organic molecules. This ability of synthesis aids in the colonization of difficult and diverse settings by plants. Low organic molecules are typically classified as primary metabolites, secondary metabolites (also known as specialized metabolites or natural products), and plant hormones based on their perspective function (Erb and Kliebenstein 2020).

3.1.1 Plant Secondary Metabolites

Organic substances that are not directly engaged in an organism's regular growth, development, or reproduction are referred to as secondary metabolites (Fraenkel 1959). In contrast to primary metabolites, the lack of secondary metabolites causes long-term impairment of the organism's capacity to survive, reproduce, or look well, or it may cause no discernible change at all (Tiwari and Rana 2015).

Alkaloids, glycosides, amines, tannins, steroids, flavonoids, and related metabolites are examples of secondary metabolites. Numerous secondary metabolites found in plants are constitutive, meaning they are present in healthy plants in their physiologically active forms. However, other secondary metabolites are found as inactive precursors and are activated in response to pathogen attack or tissue damage. Of all the secondary metabolites listed in the Dictionary of Natural Products, terpenoids make up 33,000; alkaloids make up 16,000; and flavonoids make up 8,182 of the total (Jain et al. 2019).

3.1.1.1 Alkaloids

Heterocyclic nitrogen molecules are known as alkaloids. The term "alkaline" is the source of the name "alkaloids," which was once used for any base that contained nitrogen (Roy 2017). It has been discovered that the bulk of alkaloids, including tyrosine, phenylalanine, anthranilic acid, tryptophan/tryptamine, ornithine/arginine, lysine, histidine, and nicotinic acid, are generated from amino acids (Roberts et al. 2010). Approximately 5,500 alkaloids are known to exist, making them the most abundant class of secondary plant metabolites. They are also extensively distributed (Jain et al. 2019). It was discovered that alkaloids are present in 20% of plant species. Overview of biosynthetic pathways of major groups of alkaloids are given in Figure 3.1.

* Corresponding author

 DOI: 10.1201/9781032707273-3

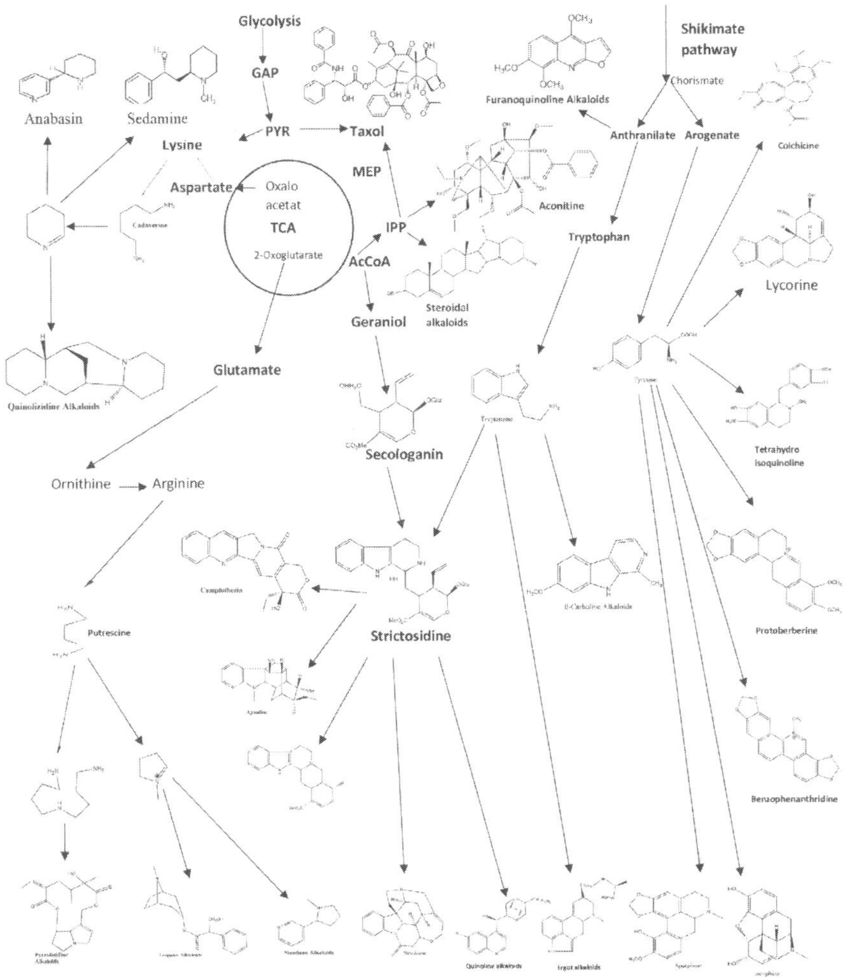

Figure 3.1 Overview of biosynthetic pathways of major groups of alkaloids.

Several scholars have suggested various classification schemes for alkaloids. One common classification scheme that splits the entire class of chemicals into three groups is discussed in subsequent text.

Chemicals such as atropine, nicotine, and other chemicals derived from amino acids and a heterocyclic-ring-containing nitrogen are known as true alkaloids. Proto-alkaloids, such as adrenaline, ephedrine, and other substances that do not belong in the heterocyclic ring, are molecules that have a nitrogen atom generated from an amino acid. Pseudo-alkaloids are substances such as coffee, theobromine, and others that do not come from amino acids (Eagleson et al. 1994).

It has been discovered that the bulk of alkaloids, including tyrosine, phenylalanine, anthranilic acid, tryptophan/tryptamine, ornithine/arginine, lysine, histidine, and nicotinic acid, are generated from amino acids. Though terpenoids, which become "aminated" after the primary skeleton has been synthesized, aconitine or the steroidal alkaloids are found in the *Solanaceae* and *Liliaceae* families; alkaloids can also be generated from other precursors such as purines in the case of caffeine. Alkaloids can also be produced from polyketides generated from acetate by adding amino nitrogen, as in the case of the hemlock alkaloid coniine (Roberts et al. 2010).

Alkaloids were once believed to be mostly plant products; however, microbes and animals can also contain these fundamental molecules. While most alkaloids, currently under research, are generated from amino acids, more and more alkaloids originating from terpenoid or polyketide sources are being found in insects and marine species (Roberts et al. 2010) (Figure 3.2).

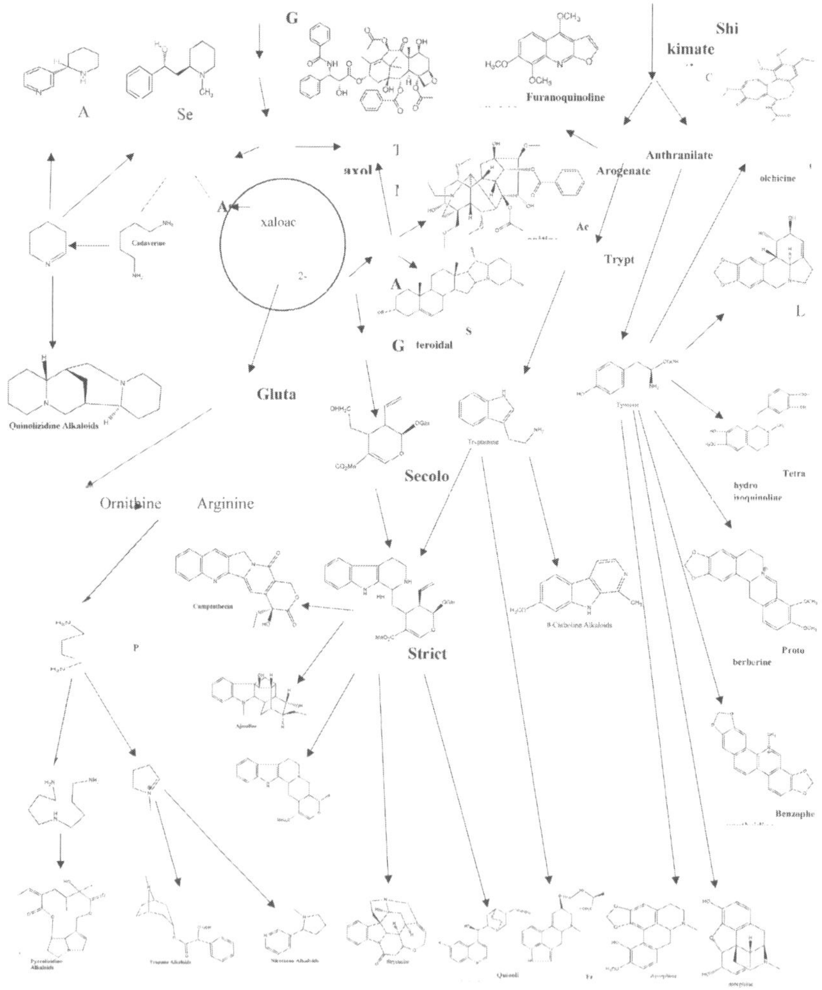

Figure 3.2 Overview of biosynthetic pathways of major groups of alkaloids.

It appears that alkaloids are critical to the health of the organisms that make them. Even one alkaloid can frequently have many biological functions. Alkaloids can perform a number of physiological activities concurrently. For example, they can occasionally act as hazardous nitrogen storage and transport molecules; providing chemical defense against herbivores and predators is one of the primary purposes. Certain alkaloids have antiviral, antibacterial, and antifungal characteristics that may also make them harmful to animals. Plants can employ alkaloids as herbicides to suppress rival plants (Wink 1993, 2000, 2008).

3.1.1.1.1 Mode of Action

Alkaloids must be able to interfere with key cellular and molecular targets in the species they are produced in order to prevent, repel, or suppress the wide range of possible enemies, which include infections, fungus, viruses, arthropods, and vertebrates, among other organisms. A molecular target's ability to communicate with other members of the cellular network—particularly proteins (cross-talk between proteins) or signal transduction components—will be adversely affected by modification of the target. This will ultimately have an impact on the metabolism and operation of all cells, tissues, organs, and ultimately the entire body, leading to a general physiological or toxic effect (Wink 1993, 2008, Van Wyk and Wink 2018).

There are two types of alkaloids that are classified as widely active: those that can establish covalent connections with proteins and nucleic acids, and those that use noncovalent bonding to modify the shape of these molecules (Wink 2008).

Plants have developed allelochemicals that have a wide range of activities as well as those that are specific target interferers. Neural signal transduction, the endocrine hormone system, and nerve cells are targets that are found in mammals but absent in plants (Wink 1993). Strong neurotoxins, such as alkaloids, were chosen as a defense against animals. Amino acids, the majority of which are amines that protonate under physiological conditions, are the building blocks of neurotransmitters. It should come as no surprise that some alkaloids share structural similarities with neurotransmitters, as alkaloids are also derived from amino acids, often the same ones as neurotransmitters (Wink 2000).

A lot of alkaloids are notorious for having a potent cytotoxic effect. The cytotoxic effect may result from inhibition of cytoskeleton components or from the permeabilization of cell membranes caused by substances such as steroidal alkaloids or saponins. Especially well-known are the spindle poisons vinblastine, vincristine, colchicine, and taxol. Alkaloids can also target DNA: planar and lipophilic alkaloids, such as sanguinarine and berberine, are chemicals that assemble between the stacks of paired nucleotides in the DNA double helix by an intercalating mechanism. It is known that certain alkaloids can form covalent adducts with DNA bases. Examples of these alkaloids are pyrrolizidine alkaloids, aristolochic acids, cycasin, and furoquinoline alkaloids. Apoptosis is typically the result of disruptions to the cytoskeleton, DNA replication, DNA topoisomerase, or DNA alkylation and intercalation. The cytotoxic effects typically don't just harm animals, they also have an impact on other plants, bacteria, fungi, and even viruses (Wink 2008).

3.1.1.2 Glycosides

These are chemical molecules consisting of two subunits: the non-sugar part (aglycone) and the sugar part (glycone). Hydrolysis can be used to separate these two components. β-D-glucose is typically the composition of the glycone. Rhamnose, digitoxose, and cymarose are constituents of glycone part of glycosides.

In glycosides, nitrogen, sulfur, or oxygen are bonded to the glycones via the C1 bond. An oxide ring is formed by the condensation of the secondary hydroxyl group. They are known as acetal or sugar ethers for this reason. Alcohol, phenol, or any other heterocyclic hydroxyl molecule can be the aglycone. Only β-glycosidic linkage is found in plants, although many other carbohydrates, including glycogen, sucrose, and starch, also include α-glycosidic linkage (Bartnik 2023) (Figure 3.3).

Classification of glycosides: Glycosides can be categorized according to two factors: the type of glycosidic bond and the type of aglycone moiety (Zhu and Yu 2018). This is how glycosides separate based on the type of glycosidic bond.

O-glycosides: These glycosides comprise sugar-coupled hydroxyl groups that are phenolic or alcoholic, like digitoxin.

N-glycosides: The nitrogen amino acid is one of the components of these glycosides. The group (—NH_2/—NH—) is unified by nucleosides, which are sugars.

S-glycosides: These glycosides, like isothiocyanate glycosides, are connected to the sulfur of an aglycone by a sugar moiety.

C-glycosides: A sugar undergoes direct condensation. C-glycosides, like aloin, are formed when a carbon atom is added to cascaroside. These glycosides do not undergo hydrolysis, using acids, alkalis, or enzymes.

Some glycosides, depending on the characteristics of the aglycone molecule, are as follows:

Steroidal glycosides: A sterol functions as an aglycone in these glycosides, like diosgenin.

Flavonoid glycosides: Some glycosides, such as rutin, have a flavonold aglycone.

Glycosides anthracene: These glycosides, such as frangulin, barbaloin, have a sugar moiety attached to an anthracene aglycone. They go by the name anthraquinone glycosides as well.

Glycosides cyanophorize: These glycosides possess cyanogen and aglycone. They hydrolyze to form hydrocyanic acid. For example, amygdalin and prunasin.

Triterpene glycosides: A triterpene molecule is condensed with a sugar component (e.g., glycyrrhizin).

Alcohol glycosides: One type of alcohol glycoside is salicin.

Lactone glycosides: Examples are hydroxycoumarin and glycosides.

Saponin glycosides: Higher plants have substantial amounts of highly complex glycosides known as saponins. When agitated, a colloidal solution containing saponins produces a foam that resembles soap.

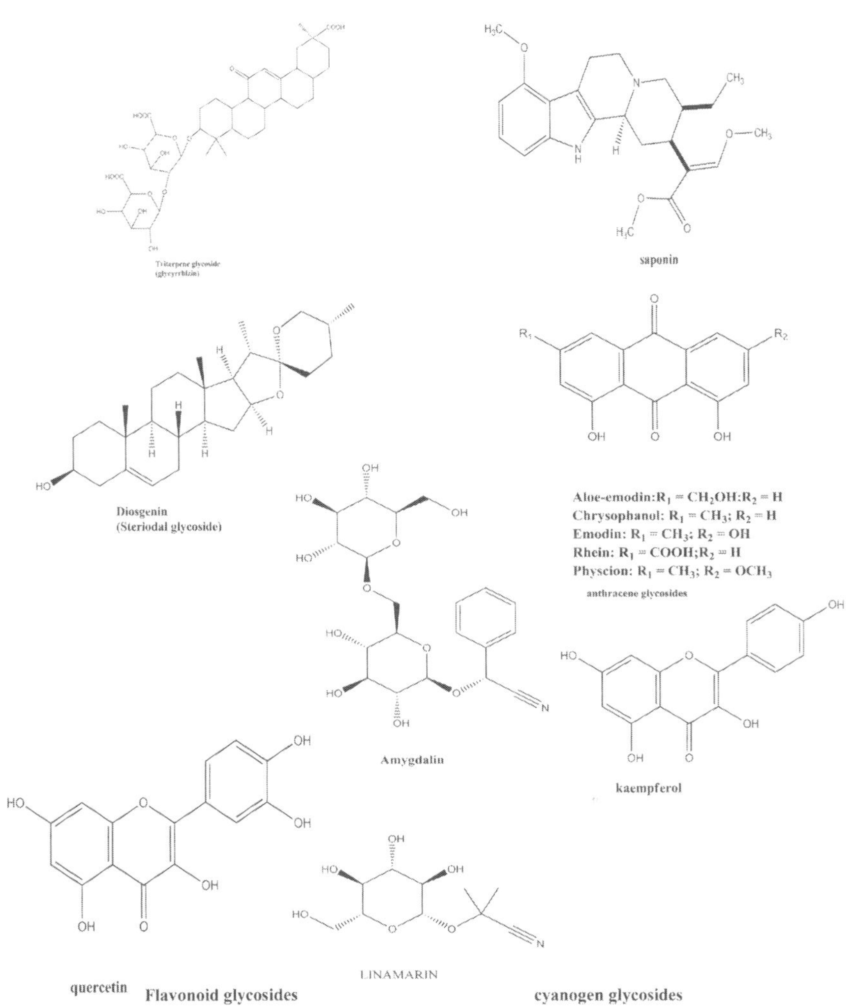

Figure 3.3 Chemical structure of different glycosides.

3.1.1.2.1 Role of Glycosides in Plants

Glycosides are among the most common secondary metabolites in plant tissues and have been shown to play an important role in many studies on the interactions between plants and herbivores, such as willows, poplar, and aspen (Boeckler et al. 2011). Numerous insect and mammalian herbivore species have been found to be poisoned by and deterred by these plants, according to various researches. In order to determine whether glycosides are truly protective, researchers experimented on salix leaves. The findings indicate that salix leaves containing phenol glycosides are generally less consumed by insects than salix leaves lacking phenol glycosides (Rowell-Rahier 1984).

3.1.1.2.2 Medicinal Uses

Pharmacological activity of glycosides is dependent upon their actions. For example, cardiac glycosides show their effects on the heart. Glycosides are employed in the management of certain diseases and disorders. *Digitalis* and *Strychnos*, which are used as heart stimulants, include cardiac glycosides. Aloe, senna, cascara, and rhubarb all contain anthraquinone glycosides, which are used as diuretics. The picroside glycosides found in *Picrorhiza kurroa* rhizomes and roots are used as a bitter tonic and these protect wounded liver tissue. Expectorant properties are found in Sdilla glycosides and the bark of wild cherries. When the tubers of the *Dioscorea* species are hydrolysed, a glycoside saponin called dioscin is produced that yields aglyconin diosgenin. Diosgenin synthesis has been used to manufacture a number of steroidal medicines. While some glycosides are less beneficial in their natural condition, when they hydrolyse, they produce active compounds. Take black glycoside, for example. Mustard is not irritating in its natural condition, but when it hydrolyses, allylisothiocyanate—a potent irritant—is created (Kytidou et al. 2020).

3.1.1.3 Phenolic Compounds

"Plant phenolics" and "polyphenols" are naturally occurring secondary metabolites that originate from the shikimate/phenylpropanoid pathway, which yields phenylpropanoids directly, or the "polyketide" acetate/malonate pathway, which can yield simple phenols or both. Consequently, monomeric and polymeric phenols as well as polyphenols are produced, which serve a wide variety of physiological functions in plants. Many thousand distinct phenolic chemicals are known to be synthesized by higher plants. Over the history of evolution, several plant lineages have chosen to retain the capacity to synthesize phenolic chemicals, which has allowed plants to adapt over time to the ever-changing environmental challenges.

When environmental stressors, including intense light, low temperatures, pathogen infection, herbivores, and nutrient deficiencies, cause an increased formation of free radicals and other oxidative species in plants, plant phenolics are thought to play a crucial role as defense chemicals. A change in the available resources is brought about by the stimulation of carbon fluxes from the primary to the secondary metabolic pathways by both biotic and abiotic stressors. This results in the synthesis of secondary products. The increase of the stress metabolite proline and the energy transfer toward phenylpropanoid production via the oxidative pentose phosphate pathway are linked in an intriguing fashion by primary and secondary metabolism. Phenolic molecules accumulate simultaneously due to the oxidative pentose phosphate pathway's two oxidative stages reducing NADP+ and oxidizing NADPH by proline synthesis (Lattanzio 2013).

In biosynthesis of phenolic compound there are two main processes at play: the malonic acid pathway and the shikimic acid pathway. Simple carbohydrate precursors from glycolysis and the pentose phosphate pathway are converted to aromatic amino acids via the shikimic acid pathway (Hartmann 1999).

Shikimic acid, which gives this entire series of reactions its name, is one of the intermediates. The well-known broad-spectrum herbicide glyphosphate kills plants by obstructing a step in this pathway. Cinnamic acid, one of the most prevalent groups of secondary phenolic compounds in plants, is formed when phenylalanine loses one ammonia molecule (Meland et al. 2024). Phenylalanine ammonia lyase (PHL), arguably the most researched enzyme in plant secondary metabolites, catalyzes this process.

3.1.1.4 Resins

Complex combinations of phenolic and/or volatile and nonvolatile terpenoid chemicals make up resins. They are created and stored in specialized structures (schizogenous or schlzolysigenous ducts) within or on the surface of plants. They are usually lipid-soluble (Murthy 2021).

Chemical substances or compounds derived from plants are known as resins. Their biosynthesis, extraction, identification, measurement, structural clarification, physical and chemical properties, and reactions are all included in the chemistry. Either the primary or secondary metabolic route produces them (Finar 1956). A primary metabolite plays a direct role in health, development, growth, and reproduction, such as alcohol, fats and oils, proteins, and carbohydrates. In contrast, resin is a secondary metabolite. Although they serve an ecological purpose, they are not directly engaged in the growth, development, or reproduction of an organism. The secondary metabolites produced by plants can be found in their leaves, stem, roots, or bark, depending on the type of metabolite (Hill 1937).

3.1.1.4.1 Classification of Resins

Resins are intricate blends of organic substances that can be categorized according to their characteristics and makeup. Few of these resins are as follows:

Oleoresin: Natural blends of essential oils and resins are called oleoresins. They are usually sticky, viscous materials that smell strongly. Resin canals or resin ducts are specialized plant structures that produce oleoresins. Plants that produce oleoresins include spruce, pine, and fir trees (Langenheim 2003).

Oleogum resin: Oleogum resins consist of gums, resins, and essential oils combined. Normally, they are solid at normal temperature, but when heated, they become sticky. These resins are used in incense, fragrances, and medications and frequently have a fragrant smell. Plants that produce oleogum resins include myrrh trees and frankincense (De Pooter et al. 1985).

Balsam: These are sticky, fragrant oleoresins with a high concentration of derivatives of cinnamic acid. They are frequently utilized in medicine, aromatherapy, and perfumery. Many plants, such as Peru balsam trees, Tolu balsam trees, and balsam firs, produce balsams (Rojas and Buitrago 2015).

Gum resins: It is a particular kind of resin that has a sizable concentration of gum, a polysaccharide that dissolves in water. Usually, they are solid at room temperature, but when heated, they get sticky. Gum resin-producing plants include asafoetida, frankincense, and myrrh.

Glycoresins: One kind of resin with a high carbohydrate content is called a glycosin. They are frequently discovered in plant latex. Examples of plants that generate glycoresins are milkweed plants and rubber trees.

3.1.1.4.2 Functions of Resins

Monoterpenes gather in resin ducts found in the needles, twigs, and trunks of conifers like pine and fir. Insects such as beetle are harmed due to these chemicals. Majorly, some plants use this defense mechanism to protect themselves from herbivorous plants, as terpenes are often harmful to them. In this situation, the plant may also cry out for assistance. The insect's feeding of the plant will cause it to produce secondary metabolites, which will draw in carnivores who will then eat the bug. Herbivore-induced plant volatile is the term used for this. For example, the diamond back moth parasitoid is seen. In a Y tube olfactometer, *Diadegma semiclausum* is drawn to volatiles from *Brassica oleracea* plants infested with diamond back moth (*Plutella xylostella*) larvae (Wink 1988). Terpene-containing resins are used to provide fragrances that attract insects and facilitate pollination. These volatile oils are extracted to produce products like citronella and eugenol, which are derived from *Osmium gratissimum*. Secondary metabolites generally serve as an attractant to animals for seed distribution or pollination (fragrance and color). Secondary metabolites, such as anthocyanin or volatile terpene, which can be both attractant and repellant in flowers, but also have insecticidal and antibacterial properties, are typically not aimed against a single creature but rather against a range of possible foes. The day flower, *Commelina communis*, attracts animals for pollination thanks to its blue hue. Resin can seep to the wound site of an injured plant, closing it off and halting additional harm or infection.

3.1.1.5 Terpenoids

Terpenoids are a class of chemical substances that are found in all living things. But compared to other living things, green plants—and flowering plants especially—show an exceptionally large quantity of terpenoids, both overall and per species (Chen et al. 2011). Both terpenes and compounds having terpene moieties connected to other moieties originating from various routes are referred to as terpenoid compounds. The tree resin known as terpentine (German: Terpentin; Latin: Balsamum Terebinthinae) is the source of the term "terpene." Terpentine is composed of hydrocarbons with a repeating isoprene unit as a monomer. The terms "terpene" and "terpenoid" are frequently used synonymously in the literature, with no clear distinction between them (Pichersky and Raguso 2018).

Many genera of plants, algae, and sponges produce terpenoids. Terpenoids are found in higher medicinal plants primarily in the following groups: Compositae, Ranunculaceae, Araliaceae, Oleaceae, Magnoliaceae, Lauraceae, Aristolochiaceae, Rutaceae, Labiatae, Pinaceae, Umbelliferae, Celastraceae, Acanthaceae, Taxaceae, etc. Terpenoids are mostly found in the form of volatile oils. The majority of bioactive terpenes have been identified and extracted from medicinal plants. Larger-molecular-weight terpenes, such triterpene, are primarily found in balsam and resin; monoterpenes and sesquiterpenes are primarily found in the essential oils of the medicinal plant (Yang et al. 2020). The genus *Eucalyptus*, whose fragrant leaves attract the attention of early European immigrants in Australia, is a major source of terpenoids. For example, the first chemical investigation of the components of eucalyptus oil resulted in the creation of eucalyptus-based cough medications, which are still quite popular and have positive effects on the bronchia. Following that the terpene content of a handful of the approximately 900 *Eucalyptus* species and their subspecies that are currently known was thoroughly examined (Cribb and Cribb 1981, Sebei et al. 2015). It has been discovered that the leaves of several species, in particular *Eucalyptus globulus* and *E. citriodora*, contain significant amounts of essential oils. Depending on the tree's environment and the time of year when the leaves are harvested, different species have different essential oil compositions and terpene content (Maciel et al. 2010).

Chemically, all terpenoids can be thought of as deriving from the basic branched C5 unit isoprene (2-methyl-1,3-butadiene (Figure 3.1). Terpenoids molecules include an integral number of C5 units. Terpenoids are categorized into hemi-, mono-, sesqui-, di-, sester-, tri-, and tetraterpenoids (carotenoids) based on the number of these C5 units present in the molecule; these terpenoids have 1, 2, 3, 4, 5, 6, and 8 isoprenoid C5 residues, respectively.(Jaeger and Cuny 2016).

Triterpenoids and carotenoids (e.g., steroidal alkaloids, cardenolides, and bixin), as well as monoterpenes, sesquiterpenes, diterpenes, and triterpenes found in floral and vegetative portions (e.g., menthol, artemisinin, taxol) are examples of terpenoids. All vascular plants produce gibberellins, which are diterpenoid hormones. Brassinosteroids are a type of hormones present in all plants, while sterols are abundant membrane components in plants. The

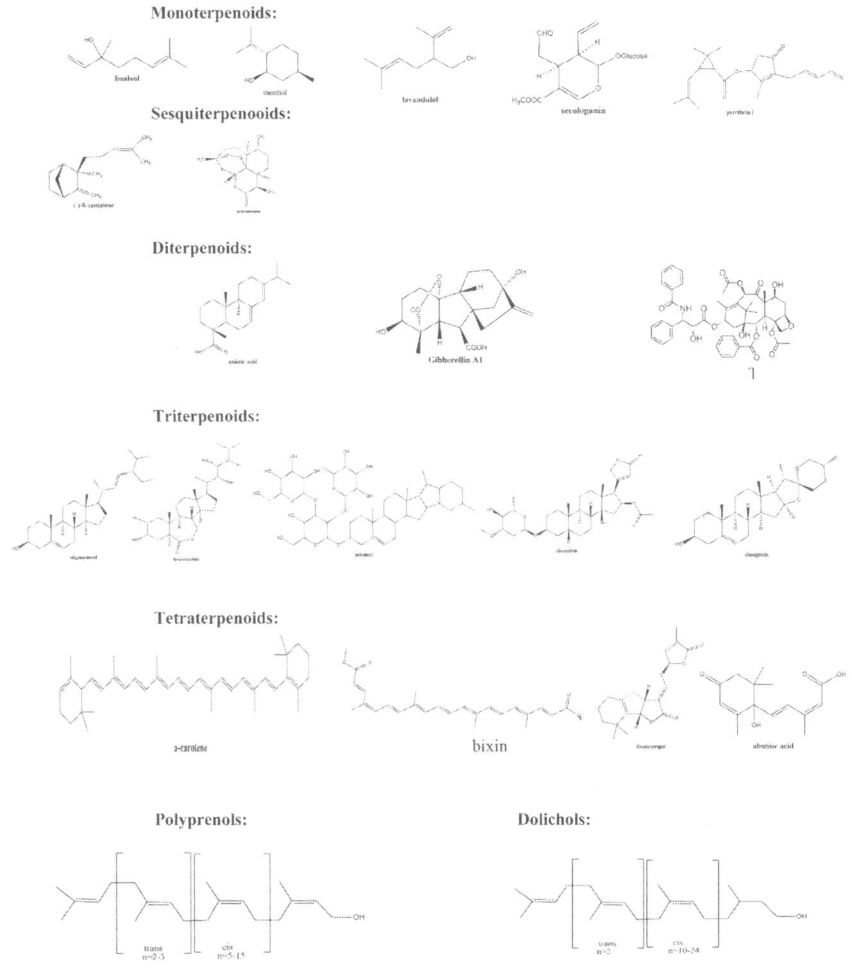

Figure 3.4 Examples of plant terpenoids. (From Pichersky and Raguso 2018.)

main sterol constituents are stigmasterol, sitosterol, and campesterol. Major components of photosynthetic tissues include carotenoids, such as a- and b-carotene. Degradation of carotenoids yields the hormones strigolactones and abscisic acid (Al-Babili and Bouwmeester 2015) (Figure 3.4).

While the head-to-tail and head-to-head condensations are regarded as "regular" terpene biosynthetic reactions, a "head to middle" linkage of isoprene units that is a component of the biosynthetic pathway of compounds like chrysanthemic acid and lavandulol is referred to as "irregular." Prenyl-diphosphates, the diphosphate-activated forms of isoprene units, are really where the condensation of isoprene units in plants takes place. There are two possible outcomes for head-to-tail condensations: trans-prenyl diphosphates and cis-prenyl diphosphates (Figure 3.2). The head-to-head condensation of two transsesquiterpenyl diphosphates yields the majority of plant triterpenes (sterols, C30), while the head-to-head condensation of two trans-diterpenyl diphosphates yields the majority of plant tetraterpenes (carotenoids, C40) (Li et al. 2021) (Figure 3.5).

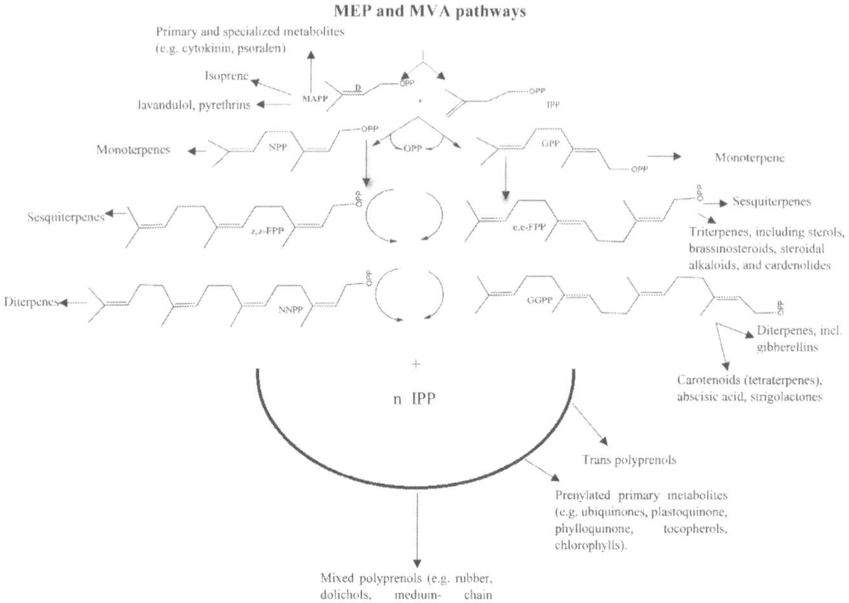

The classes of terpenes and their source from the isoprene building blocks dimethylallyl diphosphate (DMAPP) and isopentenyl diphosphate (IPP), the cis-prenyl diphosphate NPP (neryl diphosphate), z,z-FPP and NNPP (nerylneryl diphosphate), and the trans-prenyl diphosphate GPP (geranyl diphosphate) and e,e-FPP (farnesyl diphosphate) and GGPP (geranylgeranyl diphosphate Plants produce IPP and/or DMAPP through two different routes, and they also have isomerases that interconvert the two compounds. One process, known by its acronym MEP, begins with pyruvate and glyceraldehyde-4-phosphate and functions within plastids. The second pathway, which begins with acetyl-CoA, is known by its acronym, MVA. It functions in the cytosol and may perhaps function partially in peroxisomes.

Figure 3.5 Classes of terpenes and their biosynthesis.

Terpenoids are essential for a plant's physiology, reactivity to the environment, and growth and development. One crucial characteristic of main metabolite terpenes—which shield plants from abiotic stress—is the presence of double bonds. It is inevitable that plants, the environment, and insects will interact. Stress tolerance, both biotic and abiotic, is linked to secondary metabolites, especially the volatile terpenes and organic chemicals released into the atmosphere. Particularly important functions of terpenoids include reducing abiotic stressors, reducing herbivore infestation, and influencing relationships with other beneficial species including those engaged in tritrophic and multitrophic interactions. Certain terpenoids, such as gibberellins, work as plant hormones, controlling various physiological processes, while other terpenoids are secondary metabolites that help in host defense and shield plants and animals from possible infections.

It has been discovered that traditional Chinese herbal remedies with anti-inflammatory properties, such as *Paeonia lactiflora*, Inula flower, *Tripterygium wilfordii*, contain triptolide, tripterine, and triptonide. A monoterpene glycoside molecule called pedoniflorin prevents lipopolysaccharides (LPS) from inducing the formation of inflammatory factors (Yang et al. 2020).

Terpenoids have potent antibacterial properties and are primarily found in the *Mentha* genus. The antibacterial properties of menthol, patchouli alcohol, and artemisinin medications vary depending on the type of bacteria they target. It has been discovered that the diterpene lactone molecule andrographolide significantly inhibits the growth of *Pseudomonas aeruginosa* and *Staphylococcus aureus*. It has been discovered that oleanolic acid, a pentacyclic triterpenoid molecule derived from plants, inhibits *S. aureus*, *S. aureus* that is resistant to methicillin, and *Streptococcus mutans*.

Herpes simplex virus-1 is effectively inhibited by monoterpenoids such as borneol and isoborneol (HSV-1). Stronger even than acyclovir is the HSV-1 inhibitor tereticornate A. Putranjivain A, betulonic acid, moronic acid, and notoginsenoside ST-4i are among the other substances that exhibit antiviral properties. The hepatitis B and C viruses are significantly inhibited by artemisinin and its derivatives. Andrographolide can influence virus fusion and cell assembly and exhibits strong anti-DENV action.

Following pyrimethamine, chloroquine, and primaquine as the most effective antimalarial medication, artemisinin is a sesquiterpene lactone molecule. It has been synthesized into medications such as artesunate, arteether, and artemether due to its great efficiency and low toxicity. *Salvia miltiorrhiza*, a Chinese herbal medicine, contains tanshinone IIA (TS). Bunge inhibits the growth of cardiomyocytes and prevents atherosclerosis and heart damage, whereas ginsenosides control vascular function. It has been discovered that artemisinin ameliorates type 1 diabetes and advances both hypoglycemic activity and diabetes prevention. Other biological functions of terpenoids include neuroprotection, antioxidation, immunoregulation, insect resistance, and antiaging. It has been discovered that they possess antiviral, antibacterial, hypoglycemic, anticancer, and anti-inflammatory properties. In addition, it has been discovered that terpenoids have neuroprotective, immunomodulatory, and insecticidal properties (Yang et al. 2020).

3.1.1.6 Flavonoids

Plants, fruits, and seeds contain large amounts of flavonoids, which are secondary metabolites that give them their distinctive color, flavor, and scent. Flavonoids have a wide range of roles in plants, including controlling cell division, drawing pollinating insects, and providing defense against biotic and abiotic stressors (Rodríguez De Luna et al. 2020). Plants are omnipresent sources of flavonoids, a broad class of polyphenolic chemicals with a benzo-μ-pyrone structure (Kumar and Pandey 2013).

While flavonoids can be found in many foods and drinks, such as wine, beer, and tea, the foods and drinks that contain the highest natural flavonoid content include fruits, vegetables, flowers, and seeds. Nevertheless, a number of variables, including plant cultivar/genotype, growth environment, soil properties, harvest, and storage, affect the quantity of these chemicals (Rodríguez-García et al. 2019).

Aglycones, glycosides, and methylated derivatives are forms of flavonoids. Flavonoids aglycones, or flavonoids without sugar attached, exist in a range of structural configurations in plants. The fundamental molecular structure of flavonoids is a diphenyl propane skeleton with 15 carbon atoms in the main nucleus: 2 rings of 6 carbon atoms each connected by a unit of 3 carbons that could be a component of a third ring of 16 carbon atoms. Essentially, ring A and ring B of benzene are joined by a third heterocyclic oxygen-containing pyrene ring (Kuhnau 1976). Thus, this structure with the labels A, B, and C is also known as C6–C3–C6 (Figure 3.6).

Depending on the carbon of the C ring, which the B ring is linked to, as well as the degree of unsaturation and oxidation of the C ring, flavonoids can be further classified into various subgroups. Isoflavones are flavonoids in which the B ring is bonded to the C ring at position 3. Neo-flavonoids are defined as those in which the B ring is joined in position 4, whereas those in which the B ring is linked in position 2 can be further separated into multiple subgroups according to the structural characteristics of the C ring. These subgroups include anthocyanins, chalcones, flavones, flavanols, flavanones, and flavanonols, as well as flavanols or catechins (Panche et al. 2016) (Figure 3.7).

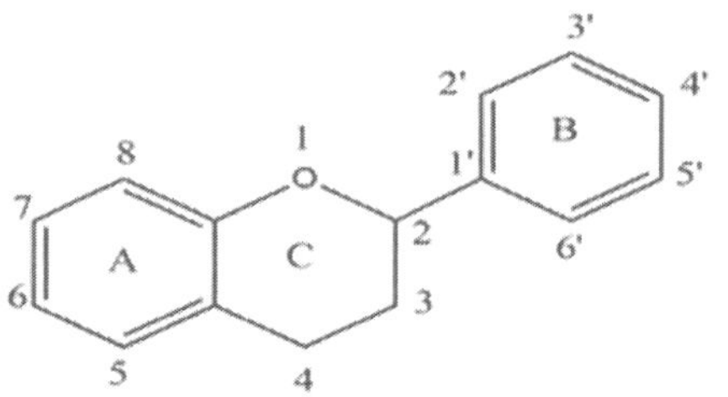

Figure 3.6 Basic flavonoid structure.

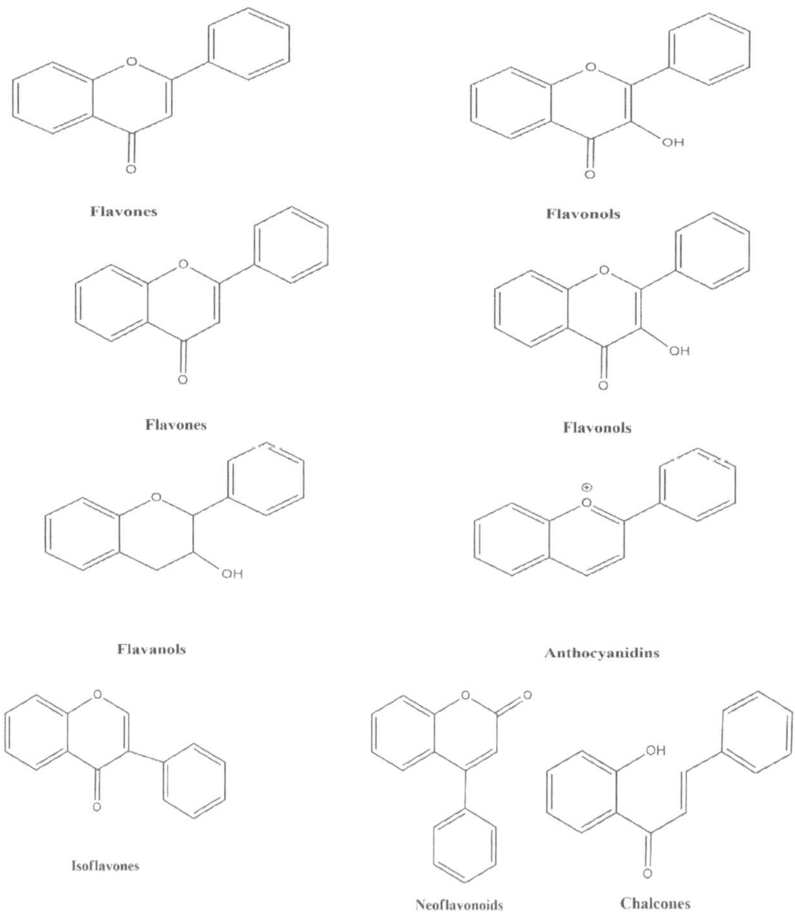

Figure 3.7 Different types of flavonoids and their chemical structure. (From He and Giusti 2010).

Plants create flavonoids using two distinct biosynthetic pathways: the polyketide process, which yields blocks for polymeric C2 units, and the phenylpropanoid pathway, which results in the phenylpropanoid skeleton (C6–C3). The 20-hydroxychalcone scaffold, scientifically known as (E)-1-(2-hydroxyphenyl)-3-phenylprop-2-en-1-one, is created by the enzyme chalcone synthase from p-coumaroyl CoA and malonyl CoA. These scaffolds are subsequently utilized in a series of enzymatic processes to generate more flavonoids. The expression of the genes involved in flavonoid biosynthesis is influenced by a number of factors, including physical injuries, hormones (such jasmonic acid), and environmental conditions (e.g., light, water availability, and temperature), which can modify the availability of flavonoids (Dias et al. 2021) (Figure 3.8).

Human studies have demonstrated the health benefits of flavonoids, which include their capacity to scavenge free radicals and function as antioxidants. Their antioxidant activity mechanisms include the direct scavenging of reactive oxygen species (ROS), the prevention of ROS synthesis by trace element chelation, and the inhibition of enzymes involved in the production of free radicals. Moreover, flavonoids contain antiviral, antibacterial, cardioprotective, anti-inflammatory, and anticancer properties. Numerous factors, including physical damage to tissue, chemical exposure, and microbial infection, can produce inflammation. Flavonoids

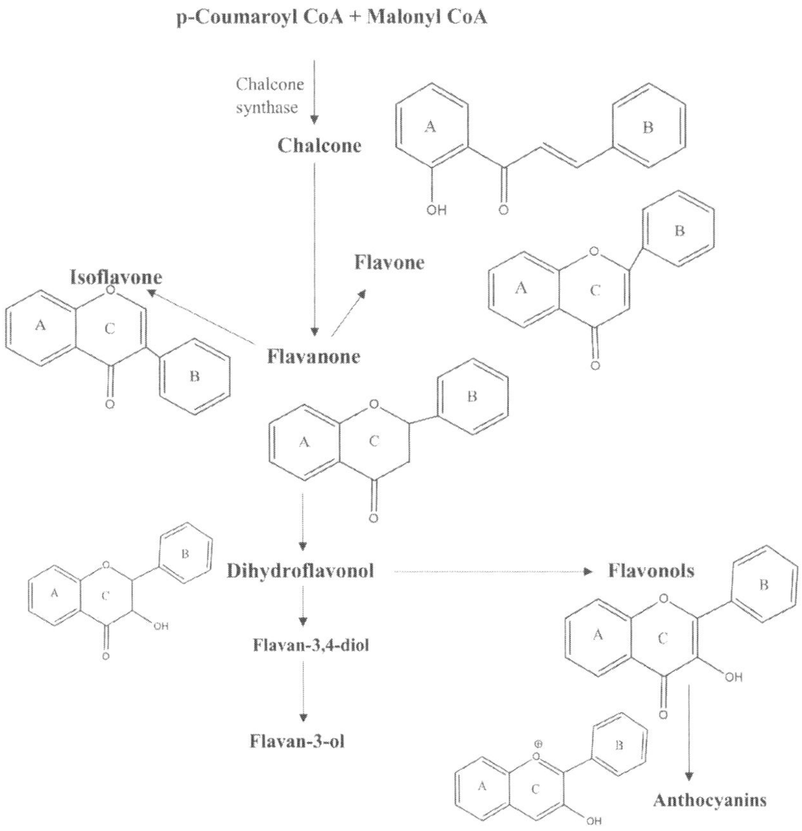

Figure 3.8 Flavonoid biosynthesis.

have the ability to modulate immune cell function, inhibit regulatory enzymes, and act as antioxidants. Inflammatory illnesses can be avoided by leading an active lifestyle with a diet rich in fruits, vegetables, whole foods, and low in sugar. Flavonoids also have anticancer properties through neutralizing carcinogens, inducing apoptosis, starting cell cycle arrest, and stopping angiogenesis. They can also act as cardio-protective medications by lowering oxidative stress and inflammation, encouraging vasodilation, and controlling apoptotic processes in the endothelium. By disrupting lipid bilayers and preventing the formation of biofilms, cell envelope creation, nucleic acid synthesis, electron transport chain, and ATP generation, flavonoids can also have antibacterial effects (Dias et al. 2021).

3.1.1.7 Tannins

Proteins precipitate due to phenolic molecules called tannins. They consist of an extremely varied collection of polymers and oligomers. They can combine with minerals, carbohydrates, cellulose, and proteins to form complexes. Except for a few higher molecular weight structures, tannins are water-soluble substances (Chung et al. 1998). Tree bark, wood, leaves, buds, stems, fruits, seeds, roots, and plant galls are common places to find tannins. Pomegranate ellagitannin, gallic tannins (found in gallnuts, lacquer leaves, and Cotinus leaves), and ellagic tannins (found in oaks, blackberries, and pomegranates) are a few examples of tannins. Condensed and hydrolysable tannins are both present in strawberries (Khanbabaee and Van Ree 2001).

For tannin molecules to bind proteins, they usually need a minimum of 12 hydroxyl groups and five phenyl groups (Figure 3.9).

Tannic acid is a particular tannin with a glucose center surrounded by 10 galloyl (3,4,5-trihydroxyphenyl) units. Tannic acid has a lot of special qualities. It possesses anticancer and antimutagenic qualities. Tannic acid exhibits antimicrobial properties (fights off germs and viruses).

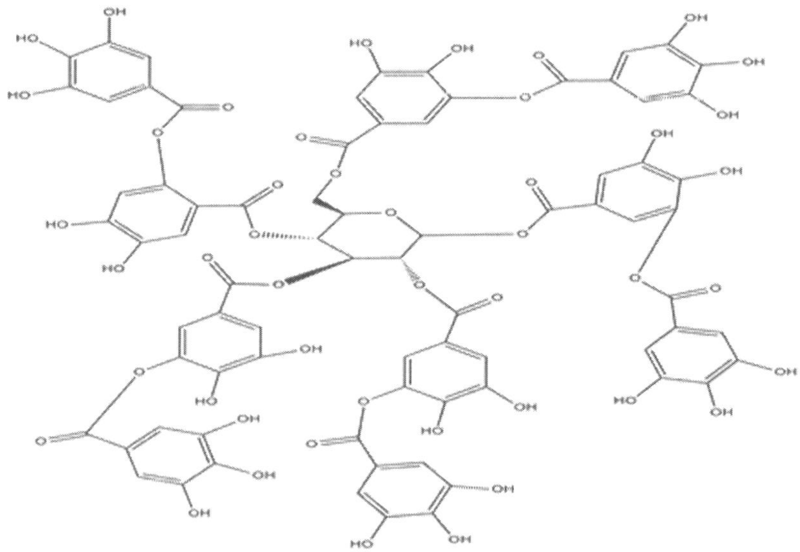

Figure 3.9 Chemical structure of tannic acid.

Tannins are classified into phlorotannins, pseudotannins, complex tannins, hydrolysable tannins, and condensed tannins. **Hydrolysable tannins** are produced when acids or enzymes hydrolyse them, yielding gallic and ellagic acid. These are, chemically speaking, phenolic acid esters, similar to gallic and ellagic acids. The tannins obtained from ellagic acid are referred to as ellagitannins, and those from gallic acid as gallotannins (Kaur et al. 2022). **Condensed tannins,** Flavanols, catechins, and flavan-3, 4-diols are the sources of tannins, which are resistant to hydrolysis. Phlobaphenes are produced when they are treated with acids or enzymes. Catechol is produced by condensed tannin during dry distillation. We refer to these tannins as catechol tannins. **Complex tannins** are a group of substances made up of partially carbon–carbon bonded hydrolysed tannins and flavane-3-ol, the unit of condensed tannins. The oligomers of phloroglucinol that make up **phlorotannins** are different from hydrolysable or condensed tannins (Chang et al. 2019). **Pseudo-**low-molecular-weight substances linked to other substances are called tannins. Unlike hydrolysable and condensed tannins, they do not change colour during the Goldbeater's skin test and are not suitable for use as tanning agents (Figures 3.10 and 3.11).

Tannins have been shown to have antibacterial, antioxidant, virucide, and anticancer properties. Applications of tannins as anti-diabetic, anti-wrinkle, anti-cardiac, and anti-diarrheics are described. Neurodegenerative disorders and other problems are preventable by tannins. The gut microbiota is influenced by tannins and polyphenols. Tannins are used in biotechnology for a variety of purposes, including pollution removal. Tannins are directly utilised in wine. Tannins cause liver necrosis, lower blood pressure, lower serum cholesterol levels, quicken blood clotting, and alter immune responses (Lall et al. 2015, Squillaro et al. 2018).

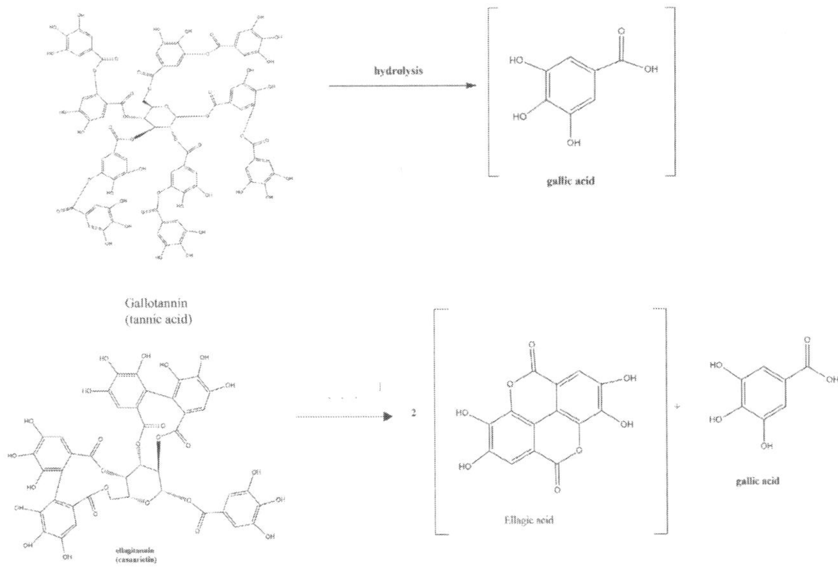

Figure 3.10 Hydrolysis of tannins.

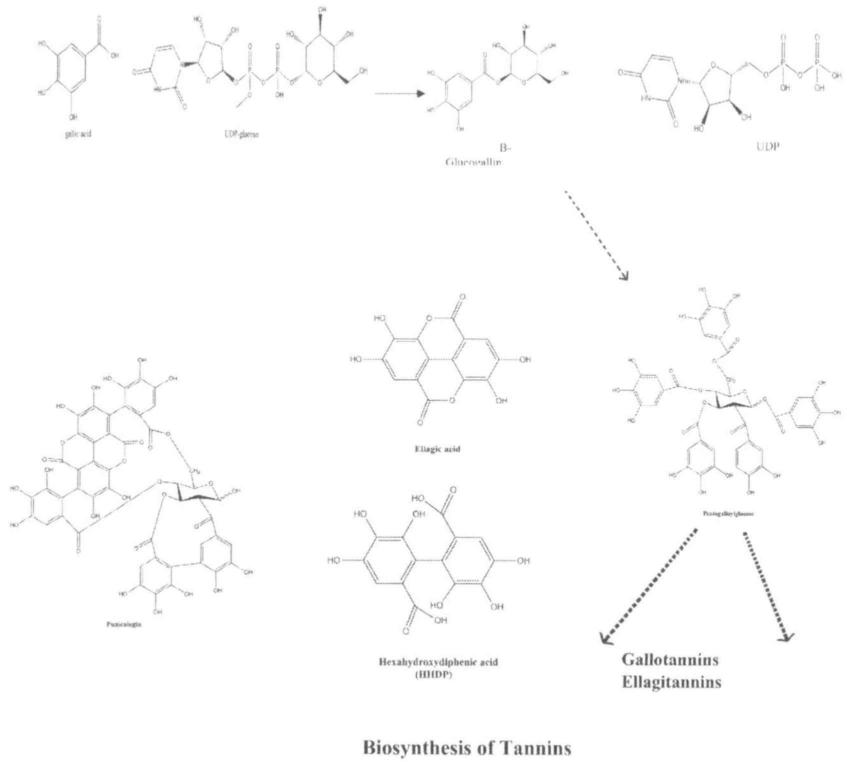

Figure 3.11 Biosynthesis of tannins.

3.1.2 Plant Hormones and Growth Regulators

Small chemical molecules known as phytohormones, or plant hormones, have a major impact on a plant's ability to grow, develop, defend itself, produce, and use its physiological systems. Both vascular and nonvascular plants contain them (Bajguz 2023). Different definitions exist for plant hormones; the word was first applied to chemicals that mammals secrete (Yamaguchi 2010).

Major types of plant hormones include auxins, gibberellins, cytokinins, ethylene, and abscisic acid (ABA).

Biosynthesis and regulation: Primary metabolism is crucial for plant growth regulators, regulating specialized metabolites and pathways for hormone biosynthesis. It contributes to the synthesis of ethylene, auxin, and salicylic acid, which stimulates fruit ripening and senescence. The regulation of ACC synthase controls ethylene production, as shown in Figure 3.12.

Important mechanisms that control apical dominance, shoot elongation, and root initiation are YUCCA flavin monooxygenase and tryptophan aminotransferase. The primary hormone IAA, auxin and SA, are produced from precursors, which are aromatic amino acids (Bajguz 2023) (Figure 3.13).

Gibberellins, ABA, and cytokinins are synthesised with the help of metabolites from lipid and isoprenoid pathways, leading to a variety of chemical compounds and biological effects. Diterpenoid carboxylic acids, or gibberellins, are made by bacteria, fungus, and plants (Bajguz 2023) (Figure 3.14).

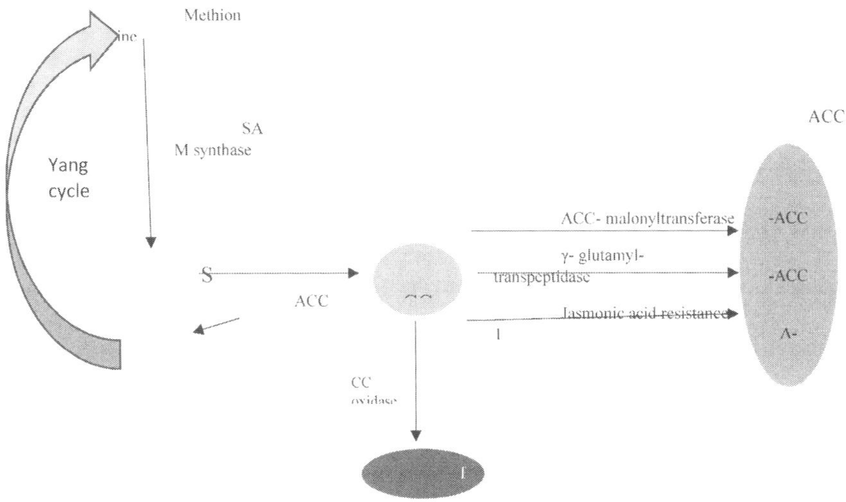

Figure 3.12 Production of ethylene.

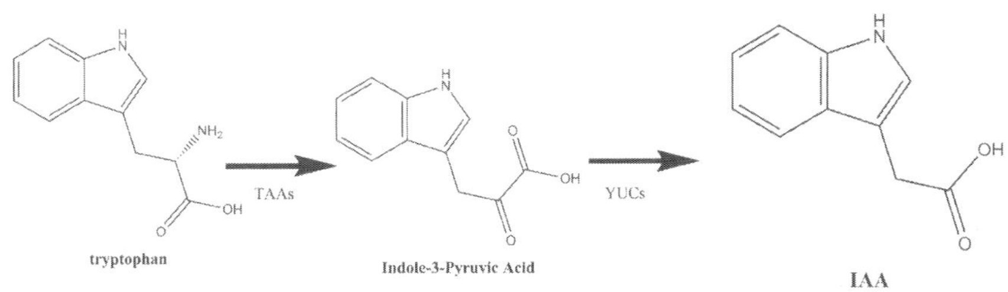

Figure 3.13 Production of auxins.

ABA, mostly produced in vascular tissues and transported by xylem and phloem, enables bidirectional transmission between roots and shoots. There are two distinct paths in plants: direct and indirect (Yamaguchi 2010) (Figure 3.15).

Cytokinins, also known as N6-substituted adenine derivatives, are mostly produced close to the tip of the root but are also found in the cambium, the apex of the shoot, and immature seeds. They can be made as mono-, di-, or triphosphates of nucleotides and have minimal biological effect. Enzymes catalyse the conversion of nucleotides, nucleosides, and free bases (Bajguz 2023) (Figure 3.16).

3.1.2.1 Functions in Plant Growth and Development

Role in cell division, elongation, and differentiation: Plants go through periods of cell division, elongation, and differentiation in meristems, where undifferentiated cells generate new ones. These cells undergo both physiological and morphological changes after differentiating. Dedifferentiation is the process by which differentiated cells revert to their pluripotent state. Plant hormones regulate growth, maturation, reproduction, life expectancy, and demise. The

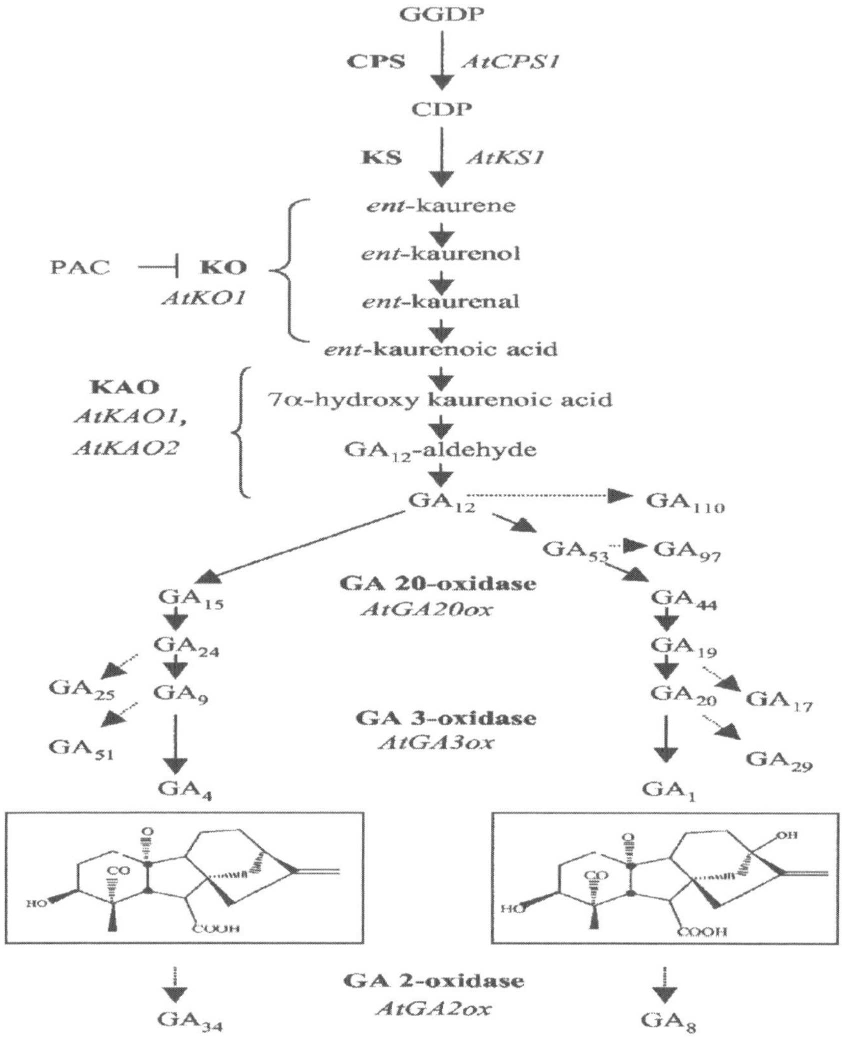

Figure 3.14 Production of gibberellins.

process of secondary metabolism produces them. These hormones enable plants to react to their environment and regulate a wide range of functions. However, isolation and laboratory testing are challenging due to their minute numbers (Ali 2020 #70).

Response to environmental stimuli: ABA, a phytohormone essential to plant growth, regulates how plants respond to stress. Heat shock increases ABA levels, which increases antioxidant capacity. Plants go through a process called thermomorphogenesis, which involves auxin. Plants that are transgenic or mutants that produce auxin signalling show limited growth driven by heat. Applying cytokinins increases heat stress resilience, which could be good for crop breeding and productivity.

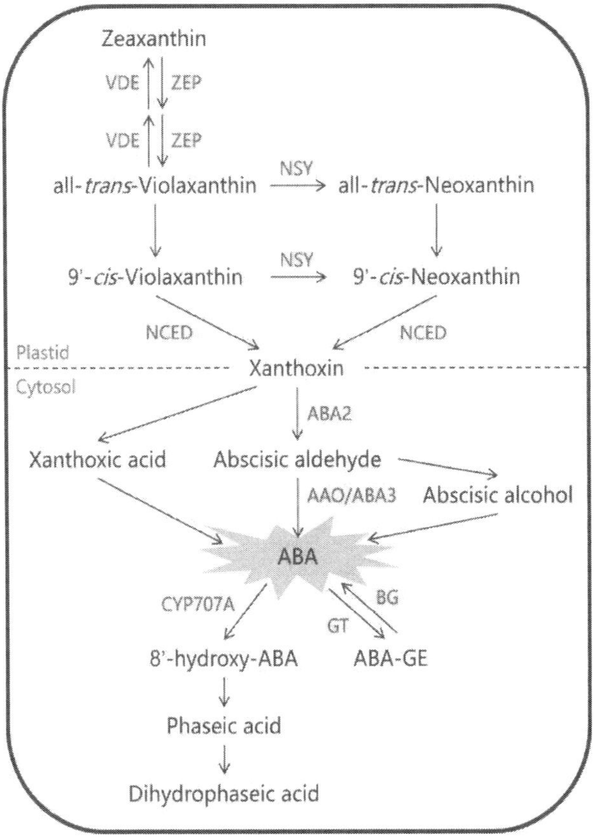

Figure 3.15 Production of abscisic acid.

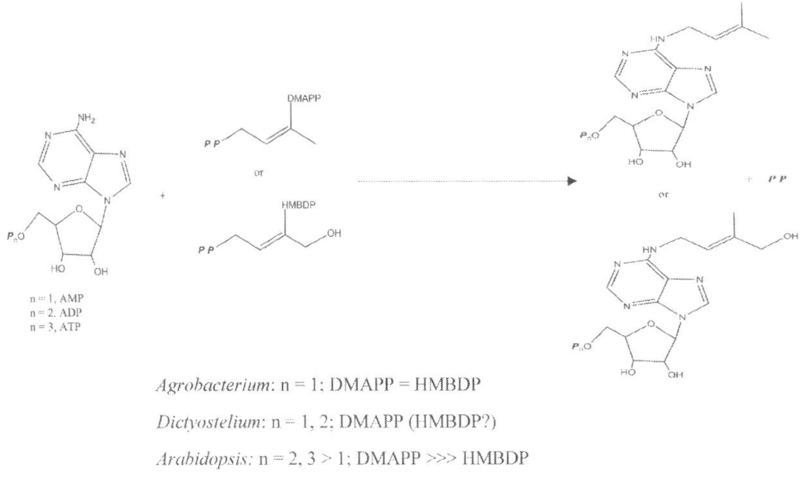

Figure 3.16 Production of cytokinins.

REFERENCES

Aiyegoro, O. and A. Okoh (2009). "Use of bioactive plant products in combination with standard antibiotics: Implications in antimicrobial chemotherapy." *Journal of Medicinal Plants Research* **3**(13): 1147–1152.

Al-Babili, S. and H. J. Bouwmeester (2015). "Strigolactones, a novel carotenoid-derived plant hormone." *Annual Review of Plant Biology* **66**(1): 161–186.

Bajguz, A. and A. Piotrowska-Niczyporuk (2023). "Biosynthetic pathways of hormones in plants." *Metabolites* **13**(8): 884.

Bartnik, M. (2023). "GC-MS analysis of essential oil and volatiles from aerial parts of *Peucedanum tauricum* MB during the phenological period." *Separations* **10**(9): 484.

Boeckler, G. A., et al. (2011). "Phenolic glycosides of the *Salicaceae* and their role as anti-herbivore defenses." *Phytochemistry* **72**(13): 1497–1509.

Chang, Z., et al. (2019). "A comprehensive review of the structure elucidation of tannins from terminalia Linn." *Evidence-Based Complementary and Alternative Medicine* **2019**(1): 8623909.

Chen, I.-C., et al. (2011). "Rapid range shifts of species associated with high levels of climate warming." *Science* **333**(6045): 1024–1026.

Chung, K.-T., et al. (1998). "Tannins and human health: A review." *Critical Reviews in Food Science and Nutrition* **38**(6): 421–464.

Cribb, A. B. and J. W. Cribb (1981). Useful wild plants in Australia.

De Pooter, H. L., et al. (1985). "The essential oil of greater galanga (*Alpinia galanga*) from Malaysia." *Phytochemistry* **24**(1): 93–96.

Dias, M. C., et al. (2021). "Plant flavonoids: Chemical characteristics and biological activity." *Molecules* **26**(17): 5377.

Eagleson, M., et al. (1994). Concise encyclopedia chemistry.

Erb, M. and D. J. Kliebenstein (2020). "Plant secondary metabolites as defenses, regulators, and primary metabolites: The blurred functional trichotomy." *Plant Physiology* **184**(1): 39–52.

Field, B., et al. (2006). "First encounters: deployment of defence-related natural products by plants." *New Phytologist* **172**(2): 193–207.

Finar, I. L. (1956). Stereochemistry and the chemistry of natural products.

Fraenkel, G. S. (1959). "The raison d'etre of secondary plant substances: These odd chemicals arose as a means of protecting plants from insects and now guide insects to food." *Science* **129**(3361): 1466–1470.

Hartmann, T. (1999). "Chemical ecology of pyrrolizidine alkaloids." *Planta* **207**(4): 483–495.

He, J. and M. M. Giusti (2010). "Anthocyanins: Natural colorants with health-promoting properties." *Annual Review of Food Science and Technology* **1**(1): 163–187.

Hill, A. F. (1937). Economic botany. A textbook of useful plants and plant products.

Jaeger, R. and E. Cuny (2016). "Terpenoids with special pharmacological significance: A review." *Natural Product Communications* **11**(9): 1373–1390.

Jain, C., et al. (2019). "Bioactivity of secondary metabolites of various plants: a review." *International Journal of Pharmaceutical Sciences and Research* **10**(2): 494–504.

Kaur, G., et al. (2022). "Chemistry, processing, and utilization." Handbook of fruit wastes and by-products: Chemistry, processing technology, and utilization: 223.

Khanbabaee, K. and T. Van Ree (2001). "Tannins: Classification and definition." *Natural Product Reports* **18**(6): 641–649.

Kuhnau, J. (1976). "The flavonoids. A class of semi-essential food components: their role in human nutrition." *World Review of Nutrition and Dietetics* **24**: 117–191.

Kumar, S. and A. K. Pandey (2013). "Chemistry and biological activities of flavonoids: An overview." *The Scientific World Journal* **2013**(1): 162750.

Kytidou, K., et al. (2020). "Plant glycosides and glycosidases: A treasure-trove for therapeutics." *Frontiers in Plant Science* **11**: 357.

Lall, R. K., et al. (2015). "Dietary polyphenols in prevention and treatment of prostate cancer." *International Journal of Molecular Sciences* **16**(2): 3350–3376.

Langenheim, J. H. (2003). Plant resins: Chemistry, evolution, ecology, and ethnobotany.

Lattanzio, V. (2013). "Phenolic compounds: Introduction 50." *Natural Products*: 1543–1580.

Li, Z.-J., et al. (2021). "Advanced strategies for the synthesis of terpenoids in Yarrowia lipolytica." *Journal of Agricultural and Food Chemistry* **69**(8): 2367–2381.

Maciel, M., et al. (2010). "Chemical composition of *Eucalyptus spp.* essential oils and their insecticidal effects on *Lutzomyia longipalpis*." *Veterinary Parasitology* **167**(1): 1–7.

Meland, M., et al. (2024). "Profiling of metabolites in organically grown plums from Norway: Does location or cultivar matter?" *Antioxidants* **13**(5): 526.

Murthy, H. N. (2021). "Chemical constituents and applications of gums, resins, and latexes of plant origin." Gums, resins and latexes of plant origin: Chemistry, biological activities and uses: 1–21.

Panche, A. N., et al. (2016). "Flavonoids: An overview." *Journal of Nutritional Science* **5**: e47.

Pichersky, E. and R. A. Raguso (2018). "Why do plants produce so many terpenoid compounds?" *New Phytologist* **220**(3): 692–702.

Roberts, M. F., et al. (2010). "Biosynthesis of alkaloids and betalains." Annual plant reviews volume 40: *Biochemistry of plant secondary metabolism*: 20–91.

Rodríguez De Luna, S. L., et al. (2020). "Environmentally friendly methods for flavonoid extraction from plant material: Impact of their operating conditions on yield and antioxidant properties." *The Scientific World Journal* **2020**(1): 6792069.

Rodríguez-García, C., et al. (2019). "Dietary flavonoids as cancer chemopreventive agents: An updated review of human studies." *Antioxidants* **8**(5): 137.

Rojas, J. and A. Buitrago (2015). "Essential oils and their products as antimicrobial agents: Progress and prospects." *Therapeutic Medicinal Plants from Lab to Mark*, p. 253.

Rowell-Rahier, M. (1984). "The presence or absence of phenolglycosides in *Salix* (*Salicaceae*) leaves and the level of dietary specialisation of some of their herbivorous insects." *Oecologia* **62**: 26–30.

Roy, A. (2017). "A review on the alkaloids an important therapeutic compound from plants." *IJPB* **3**(2): 1–9.

Sebei, K., et al. (2015). "Chemical composition and antibacterial activities of seven *Eucalyptus* species essential oils leaves." *Biological Research* **48**: 1–5.

Šernaitė, L. (2017). "Plant extracts: Antimicrobial and antifungal activity and appliance in plant protection." *Sodinink Daržinink* **36**: 58–68.

Squillaro, T., et al. (2018). "Nano-delivery systems for encapsulation of dietary polyphenols: An experimental approach for neurodegenerative diseases and brain tumors." *Biochemical Pharmacology* **154**: 303–317.

Tiwari, R. and C. Rana (2015). "Plant secondary metabolites: A review." *International Journal of Engineering Research and General Science* **3**(5): 661–670.

Van Wyk, B.-E. and M. Wink (2018). *Medicinal plants of the world.*

Wink, M. (1988). "Plant breeding: Importance of plant secondary metabolites for protection against pathogens and herbivores." *Theoretical and Applied Genetics* **75**: 225–233.

Wink, M. (1993). Allelochemical properties or the raison d'etre of alkaloids. *The alkaloids: Chemistry and pharmacology.* **43**: 1–118.

Wink, M. (2000). "Interference of alkaloids with neuroreceptors and ion channels." *Studies in Natural Products Chemistry* **21**: 3–122.

Wink, M. (2008). "Ecological roles of alkaloids." Modern alkaloids: 3–52.

Yang, W., et al. (2020). "Advances in pharmacological activities of terpenoids." *Natural Product Communications* **15**(3): 1934578X20903555.

Zhu, D. and B. Yu (2018). "Synthesis of the diverse glycosides in traditional Chinese medicine." *Chinese Journal of Chemistry* **36**(8): 681–691.

4 Secondary Metabolites and Drug Discovery

Malik Saadullah, Nasir Ali, Muhammad Fakhar-e-Alam, Abid Mahmood, Rizwan Rasheed Bazmi, and Azhar Rasool*

4.1 INTRODUCTION

The intermediaries and end products of metabolism are known as metabolites. Metabolites are mostly small molecules. Stimulating and inhibiting effects on enzymes, catalytic activity on their own (often as a cofactor to an enzyme), defense, and interactions with other organisms are just a few of the many roles that metabolites play. Plants generate a wide range of chemical molecules, the majority of which are not directly involved in the development or growth of plants producing them. These compounds are referred to as secondary metabolites (SMs) (Tiwari & Rana, 2015; Vigushin et al., 1998).

SMs determine how an organism interacts with its surroundings but are not necessary for an organism to live. These substances are frequently playing a role in plants' defense against biological as well as abiotic stressors. These belong to many groups of metabolites and can be elevated in reaction to stressful situations. Primary metabolites are substances that take part in nutrition and reproduction to fulfill crucial metabolic functions. Some SMs have a high economic value because they are employed as chemicals, such as medicines, flavors, perfumes, pesticides, and dyes. New technologies will extend and enhance the utilization of higher plants as a source of new chemicals, particularly therapeutic compounds (Pagare, Bhatia, Tripathi, Pagare, & Bansal, 2015).

Shikimic acid or aromatic amino acids are the source of a huge range of secondary SMs, many of which play crucial roles in defense mechanisms of plants. Though their precise roles and interconnections in signaling and development are not fully known, salicylates and jasmonates have a role in producing intracellular signals as well as numerous other physiological processes (Bennett & Wallsgrove, 1994).

SMs play a vital role in drug development. The natural plant alkaloid morphine was the first SM isolated and used therapeutically by Merck in 1827. In 1899, Bayer released aspirin, the first semi-synthetic pure medicine, derived from the salicin, a natural substance obtained from *Salix alba* (Figure 4.1). Because of this, early drugs like cocaine, codeine, digoxin, quinine, and pilocarpine were isolated, some of which continue to be used today (Newman & Cragg, 2012).

The use of medicinal plants to cure various diseases is a practice that dates back to the start of civilization. The majority of the healing properties of plants come from metabolites that are secondary in nature. SMs are natural substances that are found in plants and have a significant pharmacological and toxicological impact on humans (Velu, Palanichamy, & Rajan, 2018).

*Corresponding author

 DOI: 10.1201/9781032707273-4

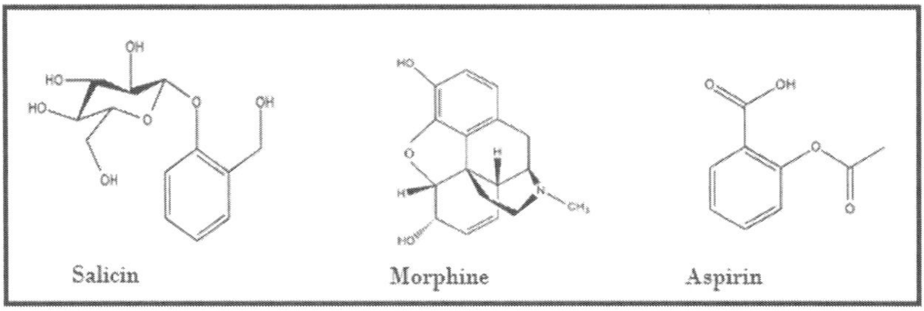

Figure 4.1 Structures of few earlier drugs from secondary metabolites (salicin, morphine, and aspirin).

4.2 CLASSIFICATION OF SECONDARY METABOLITES

SMs may be classified into following classes (Kabera, Semana, Mussa, & He, 2014):

1. Terpenoids
2. Alkaloids
3. Glycosides
4. Tannins
5. Phenols
6. Saponins
7. Terpenoids

4.2.1 Terpenoids

These are most variable class of SMs found in plants. The name terpene is derived from turpentine. All terpenoids are obtained by the polymerization of branched five carbon units called isoprenoid or isoprene unit (Figure 4.2). The classification of terpenes is shown in Table 4.1.

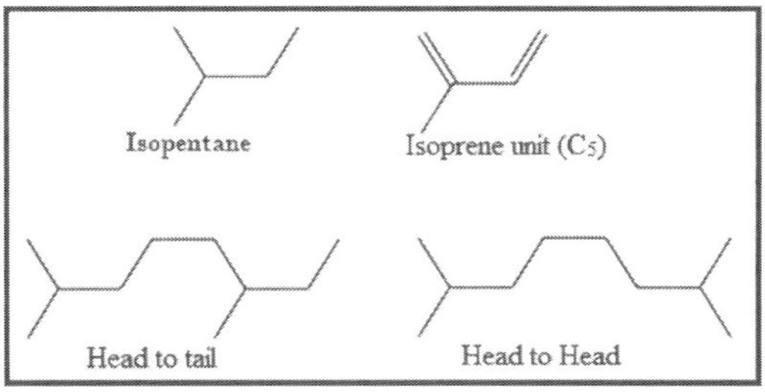

Figure 4.2 Structures of isoprene unit and their combinations.

Table 4.1: Classification of Terpenes

Carbon	Name	Example
C_5	Hemiterpene	Isoprene, prenol, isovaleric acid
C_{10}	Monoterpene	Limonene, eucalyptol, pinene
C_{15}	Sesquiterpene	Abscisic acid
C_{20}	Diterpene	Gibberellin
C_{25}	Sesterterpene	
C_{30}	Triterpene	Bassinosteroids, squalen, lanosterol
C_{40}	Tetraterpene	Carotenoids, lycopene
$C_{>40}$	Polyterpene	Rubber, vitamin E, cytokonines

Source: Kogan, Kaliya, & Froumin, 2006.

4.2.1.1 Biosynthesis of Isopentenyl Pyrophosphate and Lanosterol

Isopentenyl pyrophosphate (IPP) is synthesized through acetate mevalonate pathway (Figures 4.3 and 4.4).

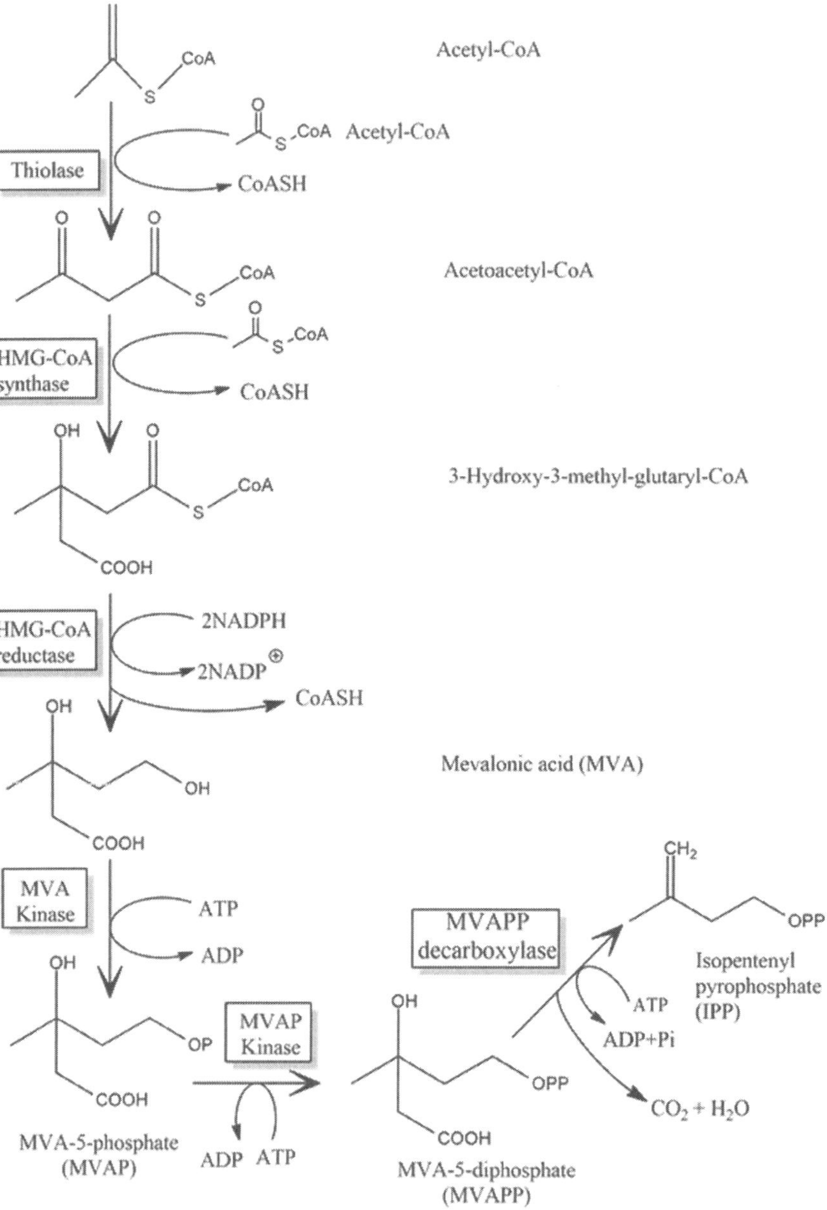

Figure 4.3 The mevalonate pathway.

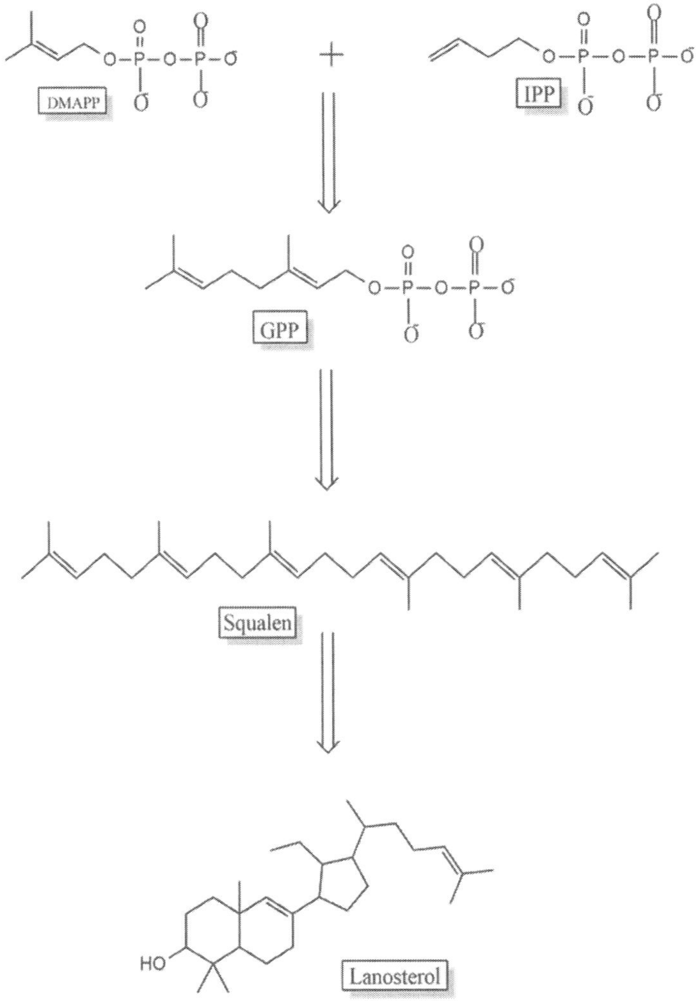

Figure 4.4 Synthesis of lanosterol.

4.2.1.2 Pharmacological Actions of Terpenes

There are several physiological and pharmacological features associated with terpenes (Table 4.2). They have become increasingly important in medicine (Maffei, 2010). The latest research has shown that terpenes have antihypertensive properties. They are employed in both agriculture and horticulture as insecticides and fungicides (Böhme, Barros-Velázquez, Calo-Mata, & Aubourg, 2014; Kataev et al., 2011) (Table 4.2).

4.2.1.2.1 Anticancer Property of Terpenes

D-limonene and perillyl alcohol, two dietary monoterpenes, have been demonstrated to have chemo-preventive and therapeutic effects on a variety of

Table 4.2: Pharmacological Actions of Terpenes

Terpene	Location	Pharmacological Activity	References
Isoprenol	Synthetic	Used as alcohol	(Eadie, 2004)
Isovaleric acid	Essential oils	Anticonvulsant	(Ara et al., 2006; Elson, Maltzman, Boston, Tanner, & Gould, 1988)
Limonene	Essential oils	Anticancer, antibacterial	(Coelho et al., 2013; Espina, Gelaw, de Lamo-Castellvi, Pagán, & Garcia-Gonzalo, 2013)
Linalool	Essential oils	CNS effects, antibacterial	(Taniguchi et al., 2014; Zhang, Schurr, & Davies, 1987)
ABA		Antitranspirant	(Hakoshima, Murase, Hirano, & Sun, 2011)
Phytoalexins	At the site of infection	Anti-oxidant, antimicrobial, anti-tumor, neurotrophic effects	(Reid & Ross, 2003; Sun, 2010)
Gibberelin	*Gibberella fujikuroi*	Role in plant growth	
Brassinosteroid	*Brassica napus*	Protects plants	(Hacham et al., 2011; Iwu, 2014)
Carotenoids	Carrots	Anti-oxidant	

human malignancies (Kris-Etherton et al., 2002). Currently, it is said that they limit the growth of breast, liver, skin, lung, colon, forestomach, prostate, and pancreatic malignancies in a manner that is dose-dependent (Bardon, Foussard, Fournel, & Loubat, 2002; Crowell, 1999; Gupta & Myrdal, 2004). Preclinical research has found that D-limonene is a naturally existing monoterpene having single ring with minimal toxicity, chemo-preventive and cytotoxic properties (Vigushin et al., 1998). Monoterpenes prevent cancer through more than one way. They may interfere with initial phase of carcinogen action by inhibiting its binding with the DNA or they may inhibit the cancer cell development and movement within the body. They cause the cancer cells to be died through apoptosis rather than necrosis. Cancer cells may be re-differentiated. Mechanisms involving regulation of cancer cells may be inhibited. The most accepted mechanism is that monoterpenes add isoprenyl unit to growth regulators of cancer cells immediately after translation (Crowell, 1999).

One of such growth regulatory protein of cancer cells is p21ras oncoprotein that regulates signals transduction and cell growth in cancer cells. Isoprenylation of p21ras leads to apoptosis and tumor suppression (Crowell, 1997) (Figure 4.5).

Terpenes are readily obtainable and belong to a potential new family of substances that might be authorized as additional medications in modern oncology (Paduch, Kandefer-Szerszeń, Trytek, & Fiedurek, 2007).

4.2.1.2.2 Antimicrobial and Antifungal Activities

It has been discovered that several terpenes show effectiveness against a vast variety of bacteria, both gram-positive and gram-negative as well as fungi. It has been demonstrated that bacteria exhibit varied levels of sensitivity to substances

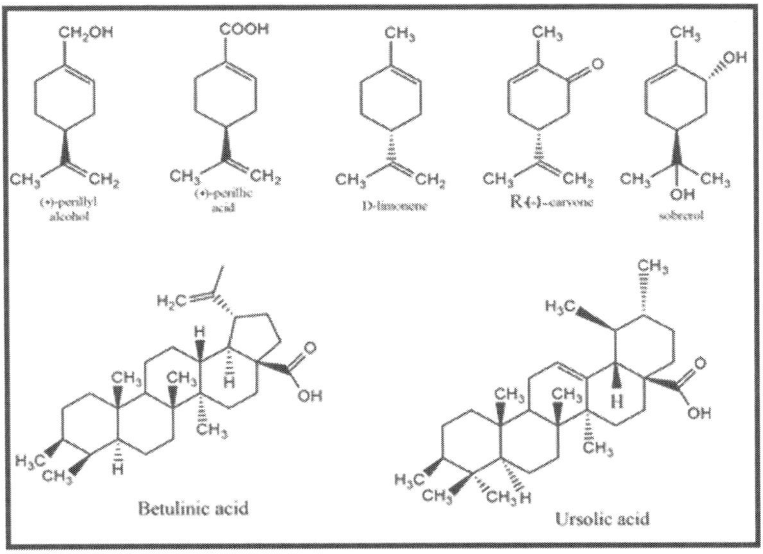

Figure 4.5 Structures of few anticancer terpenes ((+)-perillyl, (+)-perillic acid, D-limonene, R-(-)-carvone, sobrerol, betulinic acid, ursolic acid).

originating from plants. Gram-positive bacteria are often more vulnerable. Terpenes' lipophilic nature and their ability to kill bacteria are closely related. The topology of membrane proteins is altered by monoterpenes, increasing the fluidity and permeability of membranes and disrupting the respiration chain (Trombetta et al., 2005) (Figure 4.6).

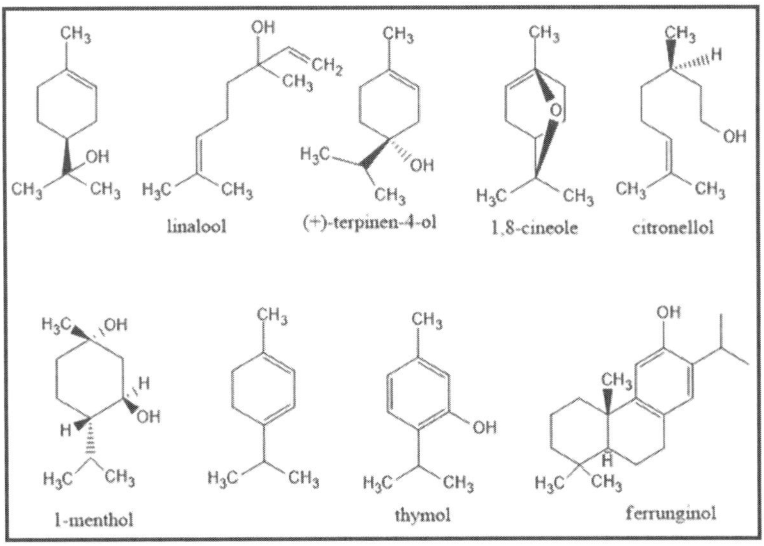

Figure 4.6 Some important antibacterial and antifungal terpenes (linalool, (+)-terpinen-4-ol, 1,8-cineole, citronellol, 1-menthol, thymol, ferruginol).

Derivatives of terpenoids have also been discovered to have antimycobacterial properties. The most effective metabolite, ferruginol (diterpene), was investigated. It showed inhibitory efficacy against many types of mycobacteria. On the other hand, it has been demonstrated that *M. tuberculosis* development is inhibited by benzoxazole-containing diterpenes and triterpenoids. Sesquiterpenes and related lactones in a wide variety also demonstrated efficacy against mycobacteria. In addition, sesquiterpene lactones with antimycobacterial action included costunolide, parthenolide, and others (Copp, 2003). Oleanolic acid and ursolic acid, two triterpenes, were similarly effective against gram-positive and gram-negative bacteria (Mallavadhani, Mahapatra, Jamil, & Reddy, 2004), as well as *M. intracellulare*.

Terpenes have antifungal properties as well. *Saccharomycetes* and mildews, which may be dangerous to humans, have been the subjects of experiments. It was discovered that carvone's two optical isomers are both active against a variety of human pathogenic fungi. The change of *Candida albicans* from the coccal to the filamentous form, which is what gives the fungus its virulence, was prevented by carvone and perillaldehyde (De Carvalho & Da Fonseca, 2006).

Monoterpenes containing 4-hydroxyterpinene, pinene, α-terpineol, linalool, cineole in a similar way inhibited the growth of *Candida* species (Hammer, Carson, & Riley, 2003). Additionally, they prevented the growth of dermatophytes as *Trichophyton rubrum*, *T. mentagrophytes* and *Microsporum gypsum*. Activity against fungi has also been associated with α-terpinene (Hammer et al., 2003).

The use of terpenes and their derivatives in antifungal therapy is generally extremely promising but may be challenging to implement since substantial concentrations of terpenes or terpenoids must be utilized. Because of this, such therapy may have negative side effects. Nevertheless, these chemicals could act as supplemental agents that advance regular, traditional antifungal treatments.

All this discussion shows that terpenes may be used as an important source of new antimicrobial and antifungal agents thus helpful in drug discovery process.

4.2.1.2.3 Antiviral Activity

Terpenoids' antiviral action remains poorly known. As a result, there is currently a lot of research being conducted to find substances, including those from natural sources, that may have strong antiviral action. The novel antiviral substances shouldn't affect the typical eukaryotic cell metabolism and should instead selectively hinder viral operations. Isoborneol was discovered after the discovery of many natural antiviral drugs.

It has been demonstrated that this monoterpene has a rather high anti-herpes simplex virus-1 (anti-HSV-1) effect and minimal cytotoxicity. The interaction of the hydroxyl groups in isoborneol with the lipids in the viral envelope is the basis for its mode of action. Isoborneol may also prevent viral proteins from being glycosylated and from replicating. As a consequence, HSV-1 stops being contagious (Armaka, Papanikolaou, Sivropoulou, & Arsenakis, 1999). Monoterpenes like cineol as well as borneol, a stereoisomer of isoborneol, have also been found to have strong anti-HSV-1 properties. In addition, putranjivain A, isolated from *Euphorbia jolkini* (*Euphorbiaceae*), has been shown by diterpene analysis to potentially block viral attachment and cell penetration as well as the late phases of HSV-2 reproduction (Khan, Ather, Thompson, & Gambari, 2005).

Moronic acid has anti HSV-1 activity which is obtained from *Rhus javanica* and *Melia azeolarch*. They blocked polypeptides generated by infected cells and lowered the replication of viral DNA. Lupenone was shown to have activity against virus. They had an effect on HSV-1 replication and inhibited viral plaque development. Anti-HSV-1 action was found in saponins made from triterpenoid saponins like glycyrrhizin and oleanane-type triterpenoid saponins like glycyrrhizic acid. Surprisingly, glycyrrhizin also efficiently suppressed the replication of the SARS-associated virus (Ikeda et al., 2005).

Additionally, it altered the fluidity of the HIV (human immunodeficiency virus) envelope (Harada, 2005). Numerous triterpene compounds' antiviral properties have also been reported in recent years. It has been discovered that moronic and betulinic acids have anti-HSV-1 actions. Moreover, betulin, and a number of its derivatives have been identified as potent inhibitors of reverse transcriptase, the process of budding and viral structures. Vesicular stomatitis and encephalomyocarditis viruses have also been demonstrated to be inhibited by betulin, betulinic acid, and oleanolic acid (Kaminska et al., 2004). HSV was also effectively inhibited by the terpene components of *Ganoderma pfeifferi* oil, lucialdehyde D, and ganoderon A and C (Niedermeyer et al., 2005). To sum up, terpenoids are an intriguing class of naturally occurring substances with a variety of antiviral actions that might be exploited to boost the therapeutic potency of conventional antiviral medication (Figure 4.7).

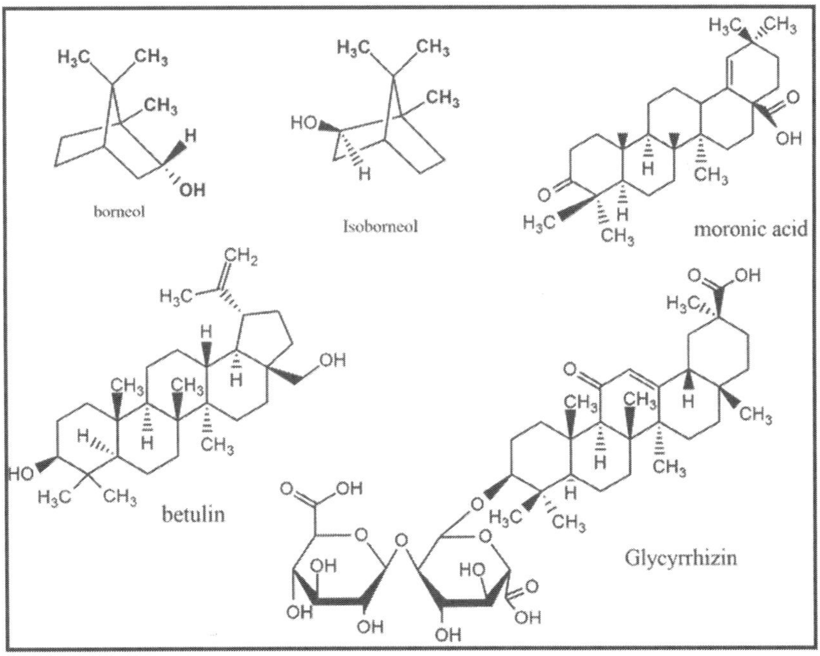

Figure 4.7 Antiviral terpenes (borneol, isoborneol, moronic A, betulin, and glycyrrhizin).

4.2.1.2.4 Hypoglycemic Function

Diabetes mellitus type 2 is caused by persistent metabolic abnormalities including reduced first-phase insulin biosynthesis, reduced function of β-cells of islets of Langerhans's or extra production of glucagon (Jeppesen et al., 2003; Jeppesen, Gregersen, Alstrup, & Hermansen, 2002).

The most potent physiological stimulus for the cell is glucose. Cell function is impaired as persistent blood glucose levels increase. Nutrition and physical activity are adequate to maintain glucose homeostasis in diabetes type 2 in its early stages (Chen, Jeppesen, Abudula, et al., 2006). But with time, oral diabetes medications and insulin replacement are frequently required. The hypoglycemic effect of the medicines is gradually reduced due to desensitization caused by them (J. Chen, Jeppesen, Nordentoft, & Hermansen, 2006). As a result, the search for and usage of natural antidiabetic medicines is receiving more and more scientific and clinical support. The plant *Stevia rebaudiana* produces stevioside, a diterpene steviol glycoside with insulinotropic, glucagonostatic, and antihyperglycemic properties. Stevioside and aglucon steviol have both been found to increase insulin production based on both; the dose and an increase in blood glucose levels (Jeppesen, Gregersen, Poulsen, & Hermansen, 2000).

Stevioside and steviol's antihyperglycemic effects may be partially attributed to their induction of glycolysis-related genes or their inhibitory effects on the addition of phosphorus and hepatic oxidation, which increase glucose breakdown and suppress its synthesis (Chen et al., 2005).

Similar to stevioside, rebaudioside A has insulinotropic actions but does not promote insulin secretion at close to normal blood glucose levels (Abudula, Jeppesen, Rolfsen, Xiao, & Hermansen, 2004).

Stevioside has two beneficial effects since it lowers blood pressure in addition to being an antihyperglycemic drug. Stevioside has been demonstrated to cause diuresis and lower blood pressure in rats. As a result, the plant glycoside's antihypertensive impact is directly linked to the lowering of cardiovascular risk factors, which manifests through a calcium-antagonist mechanism (Chan et al., 1998; Hsieh et al., 2003). Additionally, it has been demonstrated that isosteviol blocks endothelin-1 release and angiotensin-II-induced cell growth. These findings could imply that isosteviol has significant advantages for the circulatory system (Wong et al., 2006).

4.2.1.2.5 Anti-inflammatory Function

The natural enantiomer (-)-linalool has anti-inflammatory properties. In addition, (-)-linalool and linalyl acetate, an ester of the compound, showed analgesic and edema-reducing properties. Eucalyptus oil contains a monoterpene called euclyptol which reduces the arachidonic acid metabolic activity and pro-inflammatory production of cytokines in stimulated human monocytes. While 1,8-cineole was quickly absorbed by tissues, it was released into plasma more gradually. In addition, because of its prolonged terminal half-life, large amounts of it persisted in the tissues even after the treatment period had ended. As a result, it was shown that this monoterpene was particularly helpful in treating chronic conditions related to asthma, as well as acting as a preventative agent for recurring respiratory infections (Peana et al., 2003). Nuclear factor (NF)-β inhibition is the activity of the sesquiterpene-monoterpene artemisolide (lactone), derived from *Artemisia asiatica* (Compositae). Within cultured macrophages, artemisolide reduced the prostaglandin E_2 (PGE_2) and NO generation that was triggered by lipopolysaccharide (LPS) (Reddy et al., 2006). Many terpenes found in tea tree oil similarly inhibited the synthesis of a number of proinflammatory molecules, including PGE_2, TNF-α, and interleukin 1 (Habtemariam, 2000; Hart et al., 2000; Santos & Rao, 2000). In paw homogenates of rats with adjuvant-induced arthritis, Escandell et al. (Escandell et al., 2007) showed that synthesis of nitric oxide, PGE_2. Furthermore, a natural diterpene pepluanone, separated from *Euphorbia peplus*, inhibits iNOS, COX-2 and NF-kB, hence lowering the compounds or regulators produced by these (Corea et al., 2005). A well-known protein called NF-kB is in charge of managing a variety of intricate processes, including the gene expression leading to inflammation causing cytokines and other types of mediators as well as immunological receptors. These all have important roles to play in inflammatory processes. Excisanin, Kamebakaurin, sesquiterpenes and sesquiterpene lactones, celaphanol A, celastrol and aglycones of saponins have all been found to naturally occur as NF-B inhibitors (Nam, 2006; Ralstin et al., 2006; Safayhi & Sailer, 1997). A pentacyclic terpene acetyl-11-keto-boswellic acid (AKBA), displayed substantial effect against inflammation. It binds to the receptors other than normal active site for its substrate. t selectively inhibits the 5-lipoxygenase. As a result, AKBA is regarded as a treatment for inflammation (Máñez, Recio, Giner, & Ríos, 1997) (Figure 4.8).

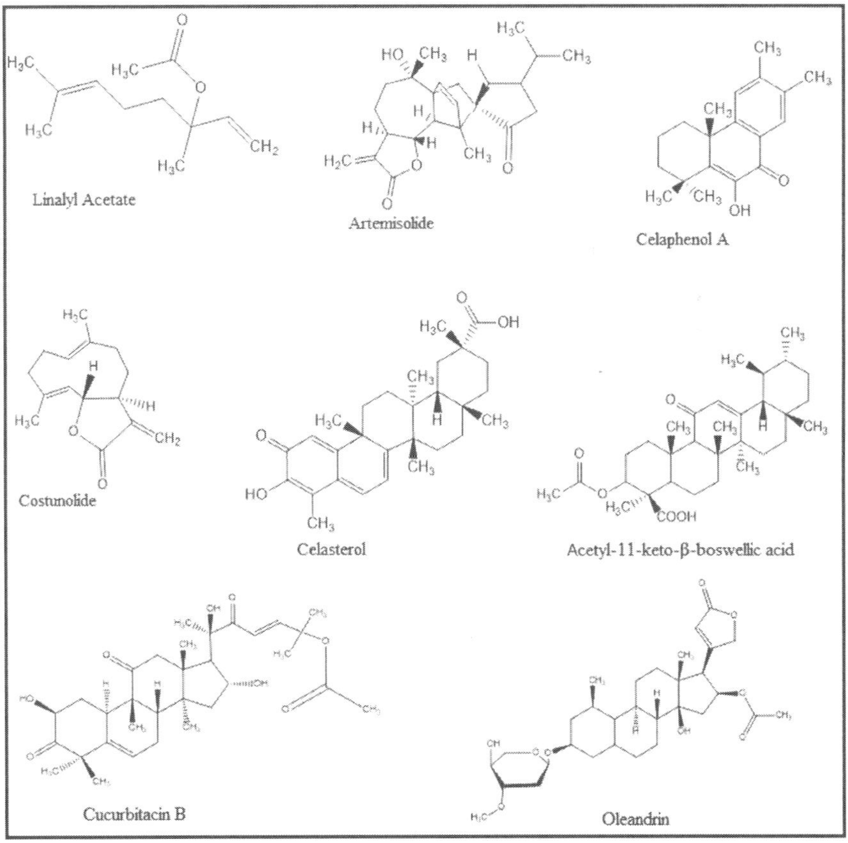

Figure 4.8 Chemical structure of important anti-inflammatory terpenes (linalyl acetate, artemisolode, celaphanol A, costunolide, celestarol, acetyl-11-keto-β-boswellic acid, cucurbitacin B & oleandrin).

4.3 ALKALOIDS

A class of chemical compounds that occur naturally, known as alkaloids, includes the majority of the fundamental nitrogen atoms. Some of the related chemicals in this category also have neutral or even mildly acidic characteristics. Some synthesized substances with a similar structure are likewise referred to as alkaloids. Alkaloid molecules may also include sulfur, and they seldom ever include chlorine, bromine, or phosphorus. Alkaloids are a class of natural compounds that are produced by a variety of organisms, including bacteria, fungi, plants, and mammals (Badri, Basu, Chandra, & Anasuya, 2019).

Carl F. W. Meissner, a German chemist, first used the term "alkaloid" in 1819. It was taken from the Arabic name *al-qali*, which is linked to the plant from which soda was initially isolated (Croteau, Kutchan, & Lewis, 2000). Alkaloids are low-molecular-weight compounds that make about 20% of secondary metabolites found in plants. Approximately 12,000 alkaloids have been isolated so far from different plant genera (Kaur & Arora, 2015).

4.3.1 Classification of Alkaloids

4.3.1.1 Tropane Alkaloids

Tropane (C_4N skeleton) is the nucleus of this class of alkaloids. They are widely distributed across the Solanaceae family. They come from acetoacetate and ornithine. Pyrrolines are these types of alkaloids' structural forerunners. The majority of them are mono, di, and trihydroxytropane esters with various hydroxylation patterns. Due to the significant pharmacological effects of cocaine, atropine, scopolamine, and their derivatives, much research has been done on these substances since the 19th century (Biastoff & Dräger, 2007) (Figure 4.9).

4.3.1.2 Pyrrolizidine Alkaloids

This class of alkaloids is characterized by their pyrrolizidine nucleus. They are found in members of the Fabaceae and Asteraceae families of plants. The

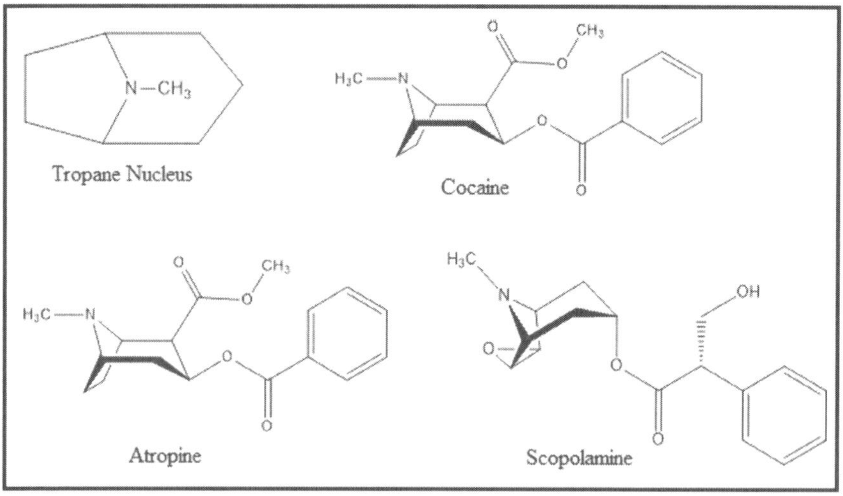

Figure 4.9 Chemical structure of tropane alkaloids (tropane nucleus, cocaine, atropine, and scopolamine).

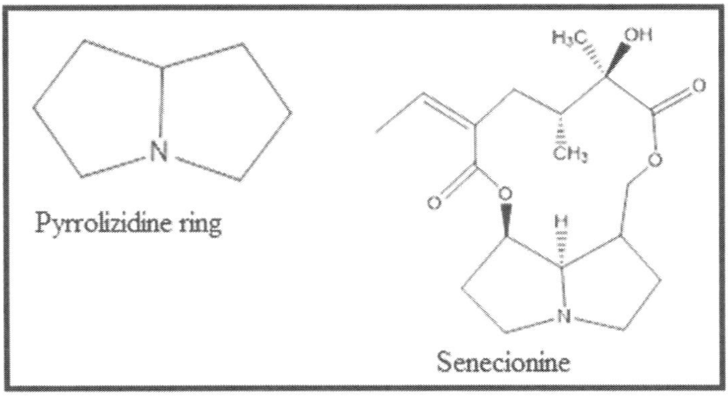

Figure 4.10 Chemical structure of senecionine and pyrrolizidine ring.

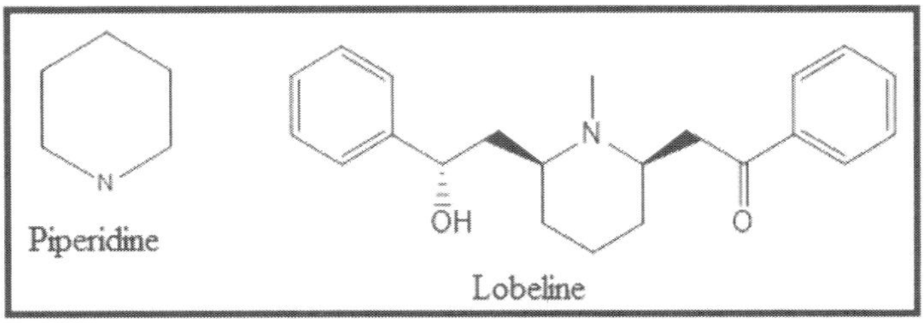

Figure 4.11 Chemical structures of piperidine alkaloids (piperidine, lobeline).

most of alkaloids belonging to this class are found in plants in the form of N-oxides; however, their function is lost upon separation. Alkaloids have undergone a thorough study due to their hazardous effects, particularly hepatic toxicity. These alkaloids may change the animals into antifeedants. The widely used alkaloid of this kind is senecionine (Hartmann & Ober, 2000) (Figure 4.10).

4.3.1.3 Piperidine Alkaloids

These contain piperidine as an important structural part. True piperidine alkaloids have monocycle molecules with the C5N nucleus as a defining characteristic. Similar characteristic of these alkaloids is their odor. These cause long-term neurotoxicity. These alkaloids may be obtained from lysine or acetoacetate, much like the straightforward pyrrolidine alkaloids. One of the significant alkaloids in this category is lobeline (Strunz & Findlay, 1985) (Figure 4.11).

4.3.1.4 Quinoline Alkaloids

These were obtained from the bark of Cinchona plant. The 4,8-quinolinediol from squid ink and the 2-heptyl-4 hydroxyquinoline from a marine pseudomonas are two examples of the simple heteroaromatic quinolines that have been obtained from diverse marine sources. In this particular category, cinchonine, cinchonidine, quinine, and quinidine are the main alkaloids (Michael, 2008) (Figure 4.12).

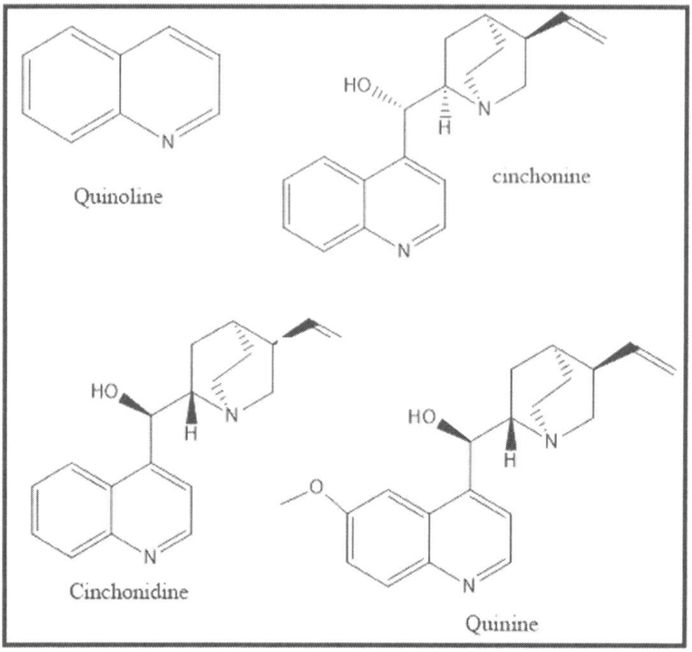

Figure 4.12 Chemical structures of quinoline alkaloids (quinoline, cinchonine, cinchonidine, and quinine).

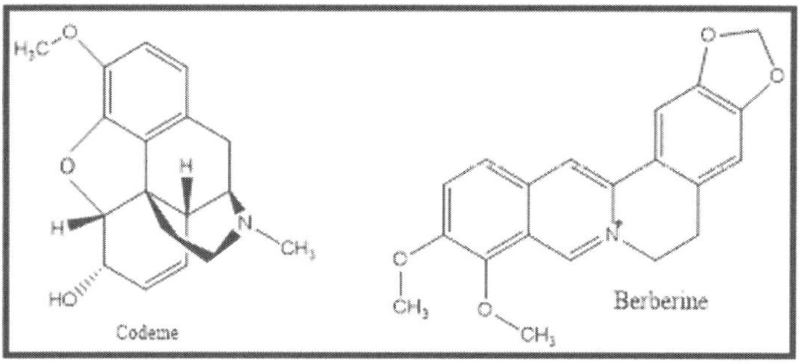

Figure 4.13 Chemical structure of codeine and berberine.

4.3.1.5 Isoquinoline Alkaloids

They exhibit a broad spectrum of structural variation and are probably formed from the tyrosine or phenylalanine. Since the poppy plant's morphine, the first bioactive isoquinoline alkaloid, was discovered in the early 19th century, this family of compounds has garnered significant scientific interest. Numerous studies have documented the anticancer, antimalarial, antibacterial, antifungal, antiparasitic and insecticidal, antiviral, anti-inflammatory, antiplatelet, and other actions of isoquinoline alkaloids that have been isolated and discovered from natural sources. Isoquinoline alkaloids have a high likelihood of success as lead compounds in the drug discovery and development process, as demonstrated by a number of ground-breaking medications, including the analgesic morphine, the antibacterial berberine, the antitussive codeine, the antirheumatic sinomenine, and the acetylcholinesterase inhibitor galanthamine 16. As a result, natural product chemistry research continues to be active in the hunt for new isoquinolines that may be effective therapeutic leads (Koleva, van Beek, Soffers, Dusemund, & Rietjens, 2012) (Figure 4.13).

4.3.1.6 Imidazole Alkaloids

This kind of alkaloid is characterized by its imidazole ring structure. These alkaloids are an exception to the structure-transformation process since their imidazole ring was already created at the precursor stage. Many structurally diverse alkaloids of this kind can be found. They exhibit a wide range of biological actions and have a sizable potential for medicinal use. The most important imidazole alkaloid from a pharmacological perspective is pilocarpine (Jin, 2006; Maat & Beyerman, 1984) (Figure 4.14).

4.3.1.7 Purine Alkaloids

The nitrogenous base of nucleotides, which make up the building blocks of DNA and RNA, is called purine. It is made up of a purine ring, pentose sugar, and pyrimidine. These include theobromine, theophylline, and caffeine. These may also be obtained from marine source e.g Phidolopin and terpenoid based purine alkaloids (Ashihara, Yokota, & Crozier, 2013; Rosemeyer, 2004) (Figure 4.15).

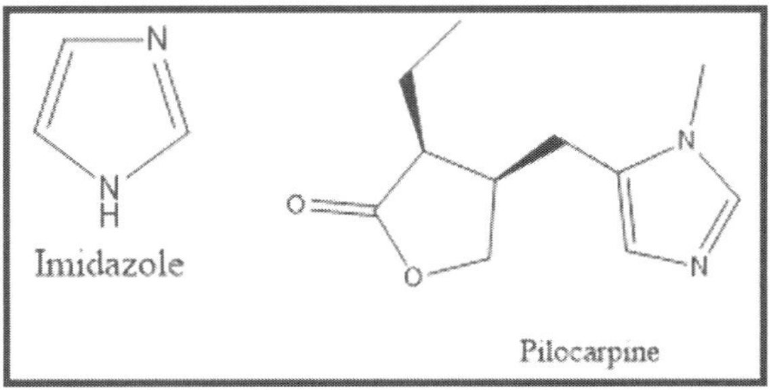

Figure 4.14 Chemical structure of imidazole and pilocarpine.

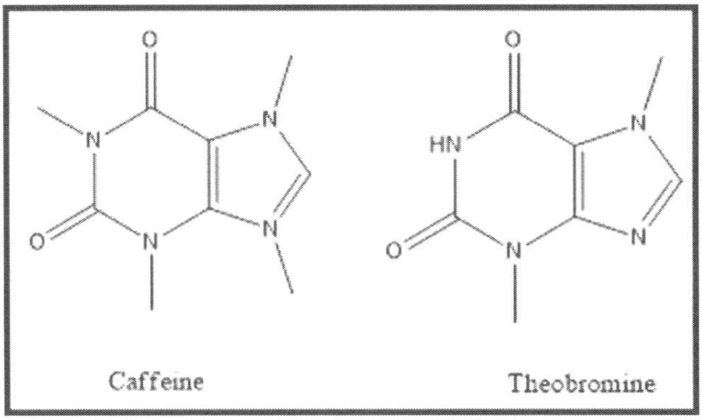

Figure 4.15 Chemical structure of caffeine and theobromine.

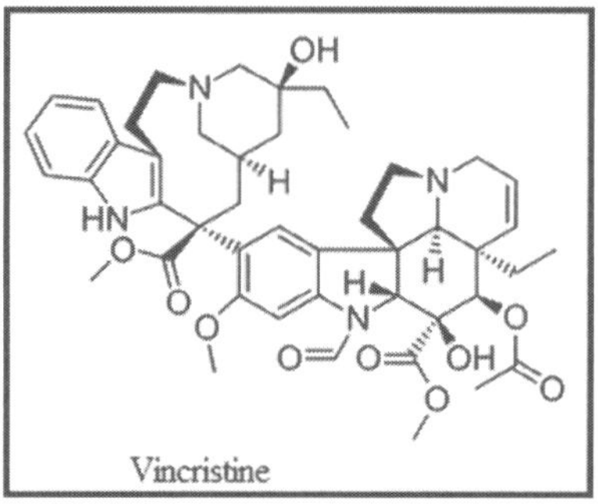

Figure 4.16 Chemical structure of vincristine.

4.3.2 Pharmacological Actions of Alkaloids

4.3.2.1 Anticancer Activity

Vincristine and vinblastine, two of the vinca alkaloids, are primarily used as chemotherapy drugs to treat cancer. They may suppress the proliferation of a variety of cell lines of cancer cells, including mouse neuroblastoma cells and L1210. Vincristine and vinblastine cause the disruption of mitotic spindle through interactions with tubulin that makes up the mitotic spindles and thus lead to blockade of metaphase (Ferguson, Phillips, Selner, & Cass, 1984).

The compound vallesiachotamine, which was isolated from *Palicourea rigida* leaves, exhibits potent anticancer action and may do so by inducing apoptosis. According to flow cytometry analysis, vallesiachotamine inhibits the G0/G1 phase, thus raising the proportion of cells that remain in a phase earlier than G1, and the results were independent of the amount of time that the drug was incubatedAutho. A 24-hour exposure to vallesiachotamine at a dosage of 50 M resulted in severe cytotoxicity and necrosis (Soares, de Oliveira, de Oliveira, Kato, & Guillo, 2012) (Figure 4.16).

The deep-water New Caledonian sponge compound gelliusine A is a brominated tris-indole alkaloid. Different cell lines are moderately sensitive to the sponge (IC50 values range from 10 to 20 g/mL for each test method). Concentration of 50 M for 24 hours resulted in significant necrosis and cytotoxicity (Soares et al., 2012).

4.3.2.2 Antimicrobial Activity

A large number of antibiotics have been developed from alkaloids. Traditional miswak contains antimicrobial agents against a large number of bacteria (Othman, Sleiman, & Abdel-Massih, 2019). The glands present in the skin of *Salamandra maculosa* produce some alkaloids which have shown bacteria and fungi killing feature (Preusser, Habermehl, Sablofski, & Schmall-Haury, 1975). A large number of isoquinoline alkaloids have antimicrobial activity (Abbasoglu, Sener, Günay, & Temizer, 1991). Two important alkaloids present in *Sida acuta* include quindoline and cryptolepine which have significant antimicrobial activities (Karou et al., 2006) (Figure 4.17).

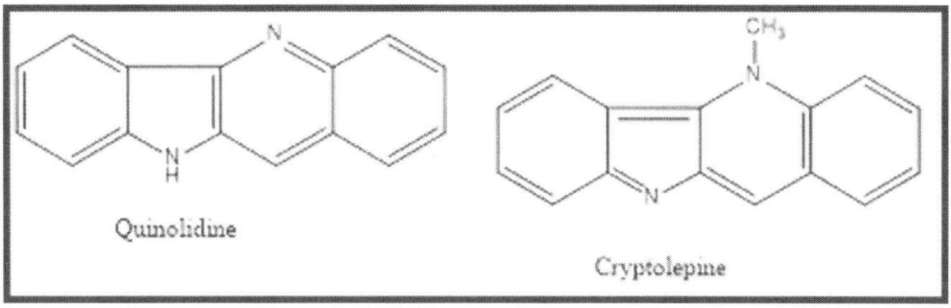

Figure 4.17 Chemical structures of quinolidine and cryptolepine.

4.3.2.3 Anti-inflammatory Activity

Whenever an infection-causing agent enters the human body, it either produces antigens or even causes death of cells. It results into immunological response in the form of inflammation. It may indicate a disease state as well. Anti-inflammatory action is attributed to indole, isoquinoline, and quinoline alkaloids. Berberine has profound action against inflammation. It also reduces edema in experimental models (Souto et al., 2011).

REFERENCES

Abbasoglu, U., Sener, B., Günay, Y., & Temizer, H. (1991). Antimicrobial activity of some isoquinoline alkaloids. *Archiv der Pharmazie*, *324*(6), 379–380.

Abudula, R., Jeppesen, P. B., Rolfsen, S. E. D., Xiao, J., & Hermansen, K. (2004). Rebaudioside A potently stimulates insulin secretion from isolated mouse islets: Studies on the dose-, glucose-, and calcium-dependency. *Metabolism*, *53*(10), 1378–1381.

Ara, K., Hama, M., Akiba, S., Koike, K., Okisaka, K., Hagura, T., … Tomita, F. (2006). Foot odor due to microbial metabolism and its control. *Canadian Journal of Microbiology*, *52*(4), 357–364.

Armaka, M., Papanikolaou, E., Sivropoulou, A., & Arsenakis, M. (1999). Antiviral properties of isoborneol, a potent inhibitor of herpes simplex virus type 1. *Antiviral Research*, *43*(2), 79–92.

Ashihara, H., Yokota, T., & Crozier, A. (2013). Purine alkaloids, cytokinins and purine-like neurotoxin alkaloids.

Badri, S., Basu, V. R., Chandra, K., & Anasuya, D. (2019). A review on pharmacological activities of alkaloids. *World Journal of Current Medical and Pharmaceutical Research*, 1(6), 230–234.

Bardon, S., Foussard, V., Fournel, S., & Loubat, A. (2002). Monoterpenes inhibit proliferation of human colon cancer cells by modulating cell cycle-related protein expression. *Cancer Letters*, *181*(2), 187–194.

Bennett, R. N., & Wallsgrove, R. M. (1994). Secondary metabolites in plant defence mechanisms. *New Phytologist, 127*(4), 617–633.

Biastoff, S., & Dräger, B. (2007). Calystegines. *The Alkaloids: Chemistry and Biology, 64*, 49–102.

Böhme, K., Barros-Velázquez, J., Calo-Mata, P., & Aubourg, S. P. (2014). Antibacterial, antiviral and antifungal activity of essential oils: Mechanisms and applications. In *Antimicrobial Compounds: Current Strategies and New Alternatives* (pp. 51–81).

Chan, P., Xu, D.-Y., Liu, J.-C., Chen, Y.-J., Tomlinson, B., Huang, W.-P., & Cheng, J.-T. (1998). The effect of stevioside on blood pressure and plasma catecholamines in spontaneously hypertensive rats. *Life Sciences, 63*(19), 1679–1684.

Chen, J., Jeppesen, P. B., Abudula, R., Dyrskog, S. E., Colombo, M., & Hermansen, K. (2006). Stevioside does not cause increased basal insulin secretion or β-cell desensitization as does the sulphonylurea, glibenclamide: Studies in vitro. *Life Sciences, 78*(15), 1748–1753.

Chen, J., Jeppesen, P. B., Nordentoft, I., & Hermansen, K. (2006). Stevioside counteracts the glyburide-induced desensitization of the pancreatic beta-cell function in mice: Studies in vitro. *Metabolism, 55*(12), 1674–1680.

Chen, T.-H., Chen, S.-C., Chan, P., Chu, Y.-L., Yang, H.-Y., & Cheng, J.-T. (2005). Mechanism of the hypoglycemic effect of stevioside, a glycoside of *Stevia rebaudiana*. *Planta Medica, 71*(02), 108–113.

Coelho, V., Mazzardo-Martins, L., Martins, D. F., Santos, A. R. S., da Silva Brum, L. F., Picada, J. N., & Pereira, P. (2013). Neurobehavioral and genotoxic evaluation of (–)-linalool in mice. *Journal of Natural Medicines, 67*, 876–880.

Copp, B. R. (2003). Antimycobacterial natural products. *Natural Product Reports, 20*(6), 535–557.

Corea, G., Fattorusso, E., Lanzotti, V., Di Meglio, P., Maffia, P., Grassia, G., … Ianaro, A. (2005). Discovery and biological evaluation of the novel naturally occurring diterpene pepluanone as anti-inflammatory agent. *Journal of Medicinal Chemistry, 48*(22), 7055–7062.

Croteau, R., Kutchan, T. M., & Lewis, N. G. (2000). Natural products (secondary metabolites). *Biochemistry and Molecular Biology of Plants, 24*, 1250–1319.

Crowell, P. L. (1997). Monoterpenes in breast cancer chemoprevention. *Breast Cancer Research and Treatment, 46*, 191–197.

Crowell, P. L. (1999). Prevention and therapy of cancer by dietary monoterpenes. *The Journal of Nutrition, 129*(3), 775S–778S.

De Carvalho, C. C., & Da Fonseca, M. M. R. (2006). Carvone: Why and how should one bother to produce this terpene. *Food Chemistry, 95*(3), 413–422.

Eadie, M. J. (2004). Could valerian have been the first anticonvulsant? *Epilepsia, 45*(11), 1338–1343.

Elson, C. E., Maltzman, T. H., Boston, J. L., Tanner, M. A., & Gould, M. N. (1988). Anti-carcinogenic activity of d-limonene during the initiation and promotion/progression stages of DMBA-induced rat mammary carcinogenesis. *Carcinogenesis, 9*(2), 331–332.

Escandell, J. M., Recio, M. C., Máñez, S., Giner, R. M., Cerdá-Nicolás, M., & Ríos, J. L. (2007). Cucurbitacin R reduces the inflammation and bone damage associated with adjuvant arthritis in Lewis rats by suppression of tumor necrosis factor-α in T lymphocytes and macrophages. *Journal of Pharmacology and Experimental Therapeutics, 320*(2), 581–590.

Espina, L., Gelaw, T. K., de Lamo-Castellvi, S., Pagán, R., & Garcia-Gonzalo, D. (2013). Mechanism of bacterial inactivation by (+)-limonene and its potential use in food preservation combined processes. *PLoS One, 8*(2), e56769.

Ferguson, P. J., Phillips, J. R., Selner, M., & Cass, C. E. (1984). Differential activity of vincristine and vinblastine against cultured cells. *Cancer Research, 44*(8), 3307–3312.

Gupta, A., & Myrdal, P. B. (2004). Development of a perillyl alcohol topical cream formulation. *International Journal of Pharmaceutics, 269*(2), 373–383.

Habtemariam, S. (2000). Natural inhibitors of tumour necrosis factor-α production, secretion and function. *Planta Medica, 66*(04), 303–313.

Hacham, Y., Holland, N., Butterfield, C., Ubeda-Tomas, S., Bennett, M. J., Chory, J., & Savaldi-Goldstein, S. (2011). Brassinosteroid perception in the epidermis controls root meristem size. *Development, 138*(5), 839–848.

Hakoshima, T., Murase, K., Hirano, Y., & Sun, T.-P. (2011). Gibberellin perception by the gibberellin receptor and its effector recognition. *Nihon Kessho Gakkaishi, 52*(1), 37–41.

Hammer, K., Carson, C., & Riley, T. (2003). Antifungal activity of the components of *Melaleuca alternifolia* (tea tree) oil. *Journal of Applied Microbiology, 95*(4), 853–860.

Harada, S. (2005). The broad anti-viral agent glycyrrhizin directly modulates the fluidity of plasma membrane and HIV-1 envelope. *Biochemical Journal, 392*(1), 191–199.

Hart, P., Brand, C., Carson, C., Riley, T., Prager, R., & Finlay-Jones, J. (2000). Terpinen-4-ol, the main component of the essential oil of *Melaleuca alternifolia* (tea tree oil), suppresses inflammatory mediator production by activated human monocytes. *Inflammation Research, 49*, 619–626.

Hartmann, T., & Ober, D. (2000). Biosynthesis and metabolism of pyrrolizidine alkaloids in plants and specialized insect herbivores. In *Biosynthesis: Aromatic Polyketides, Isoprenoids, Alkaloids* (pp. 207–243).

Hsieh, M.-H., Chan, P., Sue, Y.-M., Liu, J.-C., Liang, T. H., Huang, T.-Y., … Chen, Y.-J. (2003). Efficacy and tolerability of oral stevioside in patients with mild essential hypertension: A two-year, randomized, placebo-controlled study. *Clinical Therapeutics*, *25*(11), 2797–2808.

Ikeda, T., Yokomizo, K., Okawa, M., Tsuchihashi, R., Kinjo, J., Nohara, T., & Uyeda, M. (2005). Anti-herpes virus type 1 activity of oleanane-type triterpenoids. *Biological and Pharmaceutical Bulletin*, *28*(9), 1779–1781.

Iwu, M. M. (2014). Handbook of African Medicinal Plants.

Jeppesen, P. B., Gregersen, S., Alstrup, K., & Hermansen, K. (2002). Stevioside induces antihyperglycaemic, insulinotropic and glucagonostatic effects in vivo: Studies in the diabetic Goto-Kakizaki (GK) rats. *Phytomedicine*, *9*(1), 9–14.

Jeppesen, P. B., Gregersen, S., Poulsen, C., & Hermansen, K. (2000). Stevioside acts directly on pancreatic β cells to secrete insulin: Actions independent of cyclic adenosine monophosphate and adenosine triphosphate—sensitivie K+-channel activity. *Metabolism*, *49*(2), 208–214.

Jeppesen, P., Gregersen, S., Rolfsen, S., Jepsen, M., Colombo, M., Agger, A., … Hermansen, K. (2003). Antihyperglycemic and blood pressure-reducing effects of stevioside in the diabetic Goto-Kakizaki rat. *Metabolism*, *52*(3), 372–378.

Jin, Z. (2006). Imidazole, oxazole and thiazole alkaloids. *Natural Product Reports*, *23*(3), 464–496.

Kabera, J. N., Semana, E., Mussa, A. R., & He, X. (2014). Plant secondary metabolites: Biosynthesis, classification, function and pharmacological properties. *Journal of Pharmacy and Pharmacology*, 2(7), 377–392.

Kaminska, T., Kaczor, J., Rzeski, W., Wejksza, K., Kandefer-Szerszen, M., & Witek, M. (2004). *A comparison of the antiviral activity of three triterpenoids isolated from Betula alba bark*. Paper presented at the Annales Universitatis Mariae Curie-Sklodowska. Sectio C, Biologia.

Karou, D., Savadogo, A., Canini, A., Yameogo, S., Montesano, C., Simpore, J., … Traore, A. S. (2006). Antibacterial activity of alkaloids from *Sida acuta*. *African journal of Biotechnology*, *5*(2), 195–200.

Kataev, V., Strobykina, I. Y., Andreeva, O., Garifullin, B., Sharipova, R., Mironov, V., & Chestnova, R. (2011). Synthesis and antituberculosis activity of derivatives of *Stevia rebaudiana* glycoside steviolbioside and diterpenoid isosteviol containing hydrazone, hydrazide, and pyridinoyl moieties. *Russian Journal of Bioorganic Chemistry*, *37*, 483–491.

Kaur, R., & Arora, S. (2015). Alkaloids-important therapeutic secondary metabolites of plant origin. *Journal of Critical Review*, 2(3), 1–8.

Khan, M. T. H., Ather, A., Thompson, K. D., & Gambari, R. (2005). Extracts and molecules from medicinal plants against herpes simplex viruses. *Antiviral Research*, *67*(2), 107–119.

Kogan, S., Kaliya, M., & Froumin, N. (2006). Liquid phase isomerization of isoprenol into prenol in hydrogen environment. *Applied Catalysis A: General, 297*(2), 231–236.

Koleva, I. I., van Beek, T. A., Soffers, A. E., Dusemund, B., & Rietjens, I. M. (2012). Alkaloids in the human food chain–natural occurrence and possible adverse effects. *Molecular Nutrition & Food Research, 56*(1), 30–52.

Kris-Etherton, P. M., Hecker, K. D., Bonanome, A., Coval, S. M., Binkoski, A. E., Hilpert, K. F., … Etherton, T. D. (2002). Bioactive compounds in foods: Their role in the prevention of cardiovascular disease and cancer. *The American Journal of Medicine, 113*(9), 71–88.

Maat, L., & Beyerman, H. (1984). The imidazole alkaloids. In *The Alkaloids: Chemistry and Pharmacology* (Vol. 22, pp. 281–333).

Maffei, M. E. (2010). Sites of synthesis, biochemistry and functional role of plant volatiles. *South African Journal of Botany, 76*(4), 612–631.

Mallavadhani, U. V., Mahapatra, A., Jamil, K., & Reddy, P. S. (2004). Antimicrobial activity of some pentacyclic triterpenes and their synthesized 3-O-lipophilic chains. *Biological and Pharmaceutical Bulletin, 27*(10), 1576–1579.

Máñez, S., Recio, M. C., Giner, R. M., & Ríos, J.-L. (1997). Effect of selected triterpenoids on chronic dermal inflammation. *European Journal of Pharmacology, 334*(1), 103–105.

Michael, J. P. (2008). Quinoline, quinazoline and acridone alkaloids. *Natural Product Reports, 25*(1), 166–187.

Nam, N.-H. (2006). Naturally occurring NF-κB inhibitors. *Mini Reviews in Medicinal Chemistry, 6*(8), 945–951.

Newman, D. J., & Cragg, G. M. (2012). Natural products as sources of new drugs over the 30 years from 1981 to 2010. *Journal of Natural Products, 75*(3), 311–335.

Niedermeyer, T. H., Lindequist, U., Mentel, R., Gördes, D., Schmidt, E., Thurow, K., & Lalk, M. (2005). Antiviral terpenoid constituents of *Ganoderma pfeifferi*. *Journal of Natural Products, 68*(12), 1728–1731.

Othman, L., Sleiman, A., & Abdel-Massih, R. M. (2019). Antimicrobial activity of polyphenols and alkaloids in middle eastern plants. *Frontiers in Microbiology, 10*, 911.

Paduch, R., Kandefer-Szerszeń, M., Trytek, M., & Fiedurek, J. (2007). Terpenes: Substances useful in human healthcare. *Archivum Immunologiae et Therapiae Experimentalis, 55*, 315–327.

Pagare, S., Bhatia, M., Tripathi, N., Pagare, S., & Bansal, Y. (2015). Secondary metabolites of plants and their role: Overview. *Current Trends in Biotechnology and Pharmacy, 9*(3), 293–304.

Peana, A. T., Paolo, S. D., Chessa, M. L., Moretti, M. D., Serra, G., & Pippia, P. (2003). (–)-Linalool produces antinociception in two experimental models of pain. *European Journal of Pharmacology*, *460*(1), 37–41.

Preusser, H.-J., Habermehl, G., Sablofski, M., & Schmall-Haury, D. (1975). Antimicrobial activity of alkaloids from amphibian venoms and effects on the ultrastructure of yeast cells. *Toxicon*, *13*(4), 285–288.

Ralstin, M. C., Gage, E. A., Yip-Schneider, M. T., Klein, P. J., Wiebke, E. A., & Schmidt, C. M. (2006). Parthenolide cooperates with NS398 to inhibit growth of human hepatocellular carcinoma cells through effects on apoptosis and G0-G1 cell cycle arrest. *Molecular Cancer Research*, *4*(6), 387–399.

Reddy, A. M., Lee, J.-Y., Seo, J. H., Kim, B. H., Chung, E. Y., Ryu, S. Y., … Kim, Y. (2006). Artemisolide from *Artemisia asiatica*: Nuclear Factor-κB (NF-κB) inhibitor suppressing prostaglandin E 2 and nitric oxide production in macrophages. *Archives of Pharmacal Research*, *29*, 591–597.

Reid, J., & Ross, J. (2003). *Journal of Plant Growth Regulation*, *22*, 122.

Rosemeyer, H. (2004). The chemodiversity of purine as a constituent of natural products. *Chemistry & Biodiversity*, *1*(3), 361–401.

Safayhi, H., & Sailer, E.-R. (1997). Anti-inflammatory actions of pentacyclic triterpenes. *Planta Medica*, *63*(06), 487–493.

Santos, F., & Rao, V. (2000). Antiinflammatory and antinociceptive effects of 1, 8-cineole a terpenoid oxide present in many plant essential oils. *Phytotherapy Research: An International Journal Devoted to Pharmacological and Toxicological Evaluation of Natural Product Derivatives*, *14*(4), 240–244.

Soares, P. R., de Oliveira, P. L., de Oliveira, C. M., Kato, L., & Guillo, L. A. (2012). In vitro antiproliferative effects of the indole alkaloid vallesiachotamine on human melanoma cells. *Archives of Pharmacal Research*, *35*, 565–571.

Souto, A. L., Tavares, J. F., Da Silva, M. S., Diniz, M. d. F. F. M., de Athayde-Filho, P. F., & Filho, J. M. B. (2011). Anti-inflammatory activity of alkaloids: An update from 2000 to 2010. *Molecules*, *16*(10), 8515–8534.

Strunz, G. M., & Findlay, J. A. (1985). Pyridine and piperidine alkaloids. In *The Alkaloids: Chemistry and Pharmacology* (Vol. 26, pp. 89–183).

Sun, T.-P. (2010). Gibberellin-GID1-DELLA: A pivotal regulatory module for plant growth and development. *Plant Physiology*, *154*(2), 567–570.

Taniguchi, S., Hosokawa-Shinonaga, Y., Tamaoki, D., Yamada, S., Akimitsu, K., & Gomi, K. (2014). Jasmonate induction of the monoterpene linalool confers resistance to rice bacterial blight and its biosynthesis is regulated by JAZ protein in rice. *Plant, Cell & Environment*, *37*(2), 451–461.

Tiwari, R., & Rana, C. (2015). Plant secondary metabolites: A review. *International Journal of Engineering Research and General Science*, *3*(5), 661–670.

Trombetta, D., Castelli, F., Sarpietro, M. G., Venuti, V., Cristani, M., Daniele, C., … Bisignano, G. (2005). Mechanisms of antibacterial action of three monoterpenes. *Antimicrobial Agents and Chemotherapy, 49*(6), 2474–2478.

Velu, G., Palanichamy, V., & Rajan, A. P. (2018). Phytochemical and pharmacological importance of plant secondary metabolites in modern medicine. In *Bioorganic Phase in Natural Food: An Overview* (pp. 135–156).

Vigushin, D. M., Poon, G. K., Boddy, A., English, J., Halbert, G. W., Pagonis, C., … Coombes, R. C. (1998). Phase I and pharmacokinetic study of D-limonene in patients with advanced cancer. *Cancer Chemotherapy and Pharmacology, 42*, 111–117.

Wong, K.-L., Lin, J.-W., Liu, J.-C., Yang, H.-Y., Kao, P.-F., Chen, C.-H., … Lin, J.-G. (2006). Antiproliferative effect of isosteviol on angiotensin-II-treated rat aortic smooth muscle cells. *Pharmacology, 76*(4), 163–169.

Zhang, J., Schurr, U., & Davies, W. (1987). Control of stomatal behaviour by abscisic acid which apparently originates in the roots. *Journal of Experimental Botany, 38*(7), 1174–1181.

5 Biosynthesis of Secondary Metabolites

*Malik Saadullah, Noor ul Amin Mohsin**, *Duryab, M. Rabi Raza, Muhammad Iqbal, Syeda Abida Ejaz, Hassan Nawaz, Ayesha Shehzadi, and Abeesha*

5.1 INTRODUCTION

Plants synthesize a wide range of organic molecules, including basic metabolites that support key functions such as growth, respiration, and photosynthesis. On the other hand, secondary metabolites are not directly involved in primary metabolism; rather, they have specific functions. While both types of compounds share common biosynthetic pathways and intermediates, secondary metabolites often exhibit more restricted distributions among specific plant families, genera, or species. Secondary metabolites that hold importance in the pharmaceutical industry are labeled as phytopharmaceuticals (Velu et al. 2018). This group includes a wide range of compounds, including volatile oils, alkaloids, flavonoids, glycosides, tannins, and resins. Secondary metabolites found in pharmaceuticals, scents, tastes, pigments, dyes, food additives, and pesticides have a significant impact on the global economy (Alamgir and Alamgir 2017).

A significant portion of pharmaceuticals available today are either synthetic derivatives or replicas of naturally occurring compounds. Secondary metabolites have recently received increased attention as a result of their expanding commercial value (Figure 5.1). Subsequent metabolites from plants are frequently taken into consideration by authors as possible alternatives to manufactured chemicals. Bioactive botanical chemicals derived from plants, either through extracts, essential oils, or both, are promising alternatives. Based on their chemical structure, these substances are divided into four classes: terpenes, phenolics, nitrogen-containing, and sulfur-containing compounds. Terpenoids are a wide and structurally diverse family of naturally occurring compounds that are formed by the head-to-tail joining of C5 isoprene units, which are the building blocks of terpenoids. Typical structures contain carbon skeletons represented by the (C5)n formula, and they are named and classified as hemiterpenes (C5), monoterpenes (C10), sesquiterpenes (C15), diterpenes (C20), sesquiterpenes (C25), triterpenes (C30), and tetraterpenes (C40). The structure of a terpenoid changes when five carbon atoms are added to a specified class. Higher polymers are found in products like rubber (Ashour et al. 2010).

Isoprene was identified as a decomposition result of naturally occurring cyclic hydrocarbons and considered as one of the building blocks for these chemicals, known as "isoprenoids" (Vickers and Sabri 2015). Isoprene occurs naturally but does not participate in the production of these compounds. Isoprene derivatives have been discovered as the diphosphate (pyrophosphate) esters dimethylallyl diphosphate (DMADP) and isopentenyl diphosphate (IDP). Few natural terpenoids, such as geraniol (C10), farnesol (C15), and geranylgeraniol (C20), strictly follow the simple rule in which isoprene units are combined head-to-tail in a linear manner. Despite being entirely composed of isoprene units, squalene (C30) and phytoene (C40), have a tail-to-tail linkage in the center of their molecules. Cyclization processes can further modify terpenoids, but their head-to-tail arrangement is mostly

*Corresponding author

DOI: 10.1201/9781032707273-5

retained, as seen in menthol, taxadiene, and bisabolene (McBee et al. 2024). During rearrangement reactions, linear configuration of isoprene units can get interred in a variety of forms, like those found in steroids. Furthermore, these structures frequently undergo changes, resulting in the loss of many carbons. Steroids, for example, are complex compounds with fused rings formed of isoprene units. However, due to biosynthetic processes and chemical reactions, the initial linear arrangement may be difficult to discern. Nonetheless, these molecules are produced from ordinary precursors of terpenoids. Terpenoid components are found in the structures of many natural compounds, also containing carbon skeletons from, for instance, the shikimate and acetate routes. Examples include various alkaloids, vitamins, and phenolic compounds. A single C5 unit, usually with a dimethylallyl substituent, is a common terpenoid fragment in such instances, and meroterpenoids are compounds having these isoprene subunits. Examples include rotenoids, ergot alkaloids, and furocoumarins. The alternate term "prenyl" is commonly used to refer to the dimethylallyl substituent. The biological isoprene units can be produced in two ways: via mevalonic acid (MVA) or 1-deoxy-D-xylulose 5-phosphate (Li et al. 2018). MVA itself is a product of acetate metabolism. The highly functionalized carbon backbone of terpenoid natural products is formed by the condensation of two essential metabolic intermediates: IDP and DMADP. Nature produces these molecules by two separate pathways: the 2-C-methyl-d-erythritol-4-phosphate (MEP) pathway, also known as the mevalonate independent process, and the MVA pathway. It was once thought that the early phases of the mevalonate pathway applied to the entire variety of naturally occurring terpenoid derivatives. However, subsequent research has revealed the existence of an alternate pathway that bypasses IDP and DMAPP and is extensively employed in nature and this pathway is mediated by deoxyxylulose phosphate. Most eukaryotes and archaea rely on the cytosolic MVA pathway for IDP and DMADP, whereas eubacteria and the plastids of algae, higher plants, and protists use the MEP route (Bergman et al. 2019). Although photosynthetic organisms use both precursor processes, they separate them into distinct cellular compartments. While there is some evidence that plants' plastids and cytosol exchange IDP/DMADP pools, this is usually only seen in the context of secondary metabolism. In general, the MEP pathway generates C5 prenyl diphosphates, which are used to synthesize C10 monoterpenes, C20 diterpenes, and C40 tetraterpenes, whereas the MVA system produces universal precursors for C15 sesquiterpenes, C27-29 sterols, C30 triterpenes, and their saponin derivative (Tholl 2015).

Shikimic acid, the precursor of the shikimate pathway, is produced from a combination of phosphoenolpyruvate (PEP from the glycolytic pathway) and erythrose 4-phosphate or (from the pentose phosphate pathway). Phenylalanine, tyrosine, and tryptophan are produced in the shikimate pathway; these are the building blocks of protein synthesis and common precursors for plant secondary metabolites such as phenolics and nitrogen-containing compounds (Crozier et al. 2006). Phenylalanine is the common precursor of flavonoids, lignans, lignins, condensed tannins, and phenylpropanoid/benzenoid volatiles. Tyrosine further produces isoquinoline alkaloids, pigment betalains, and quinones (e.g., tocochromanols and plastoquinone), while tryptophan is the precursor of alkaloids, phytoalexins, and many extended products (Zeng et al. 2020).

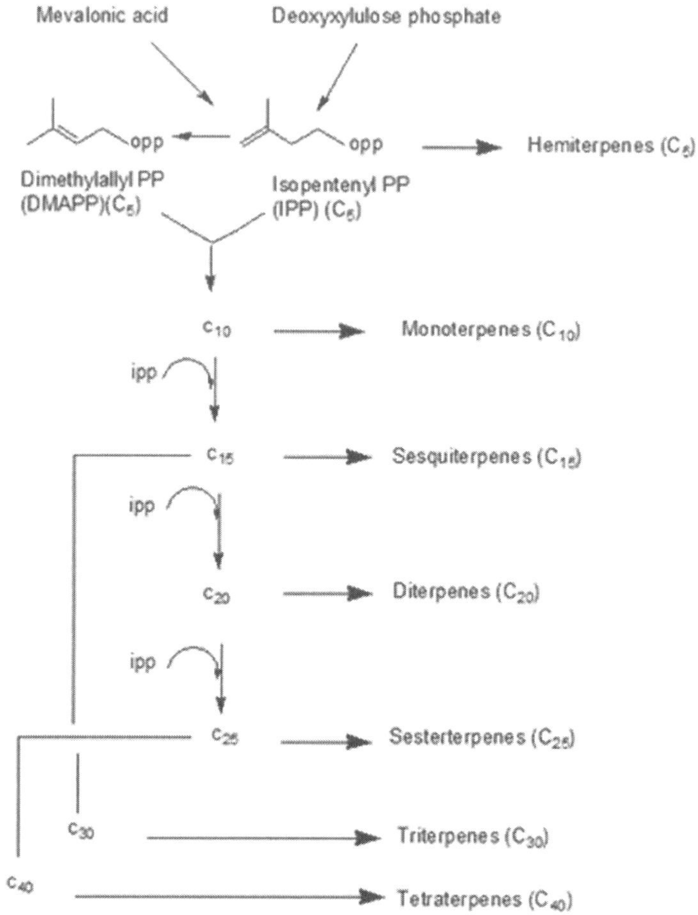

Figure 5.1 Biosynthesis of secondary metabolites.

5.1.1 The MVA Pathway

MVA, the precursor to the MVA pathway is formed by the interaction of three molecules of acetyl-coenzyme A (CoA) (Liao et al. 2014). Two molecules interact in a Claisen condensation to produce acetoacetyl-CoA, followed by a stereospecific aldol addition to produce branched-chain ester β-hydroxy-β-methylglutaryl-CoA (HMG-CoA). This third acetyl-CoA molecule appears to be linked to the enzyme via a thiol group, which is then hydrolyzed to produce the free acid group of HMG-CoA (Okamura et al. 2010). In the acetate pathway, the more nucleophilic malonic acid thioester would have been used to create an acetoacetic acid thioester (attached to the acyl carrier protein). However, the malonyl derivatives are not used in the mevalonate pathway and hence at the very first step gets diverged from the acetate pathway (Kang and Lee 2016). In the following step, it is noteworthy that the acidic substrate acetoacetyl-CoA acts as the nucleophile but not as a third acetyl-CoA. Thus, the enzyme obtains a less favorable reaction. The conversion of HMG-CoA to (3R)-MVA requires a two-step reduction of the thioester group to a primary alcohol, resulting in an essentially irreversible and rate-limiting transformation. Drug-mediated inhibition of this enzyme (HMG-CoA reductase) can be utilized to control the biosynthesis of mevalonate and, eventually, steroid cholesterol. Phosphorylation of the main alcohol group initiates a sequence of events that convert the six-carbon molecule MVA into the five-carbon phosphorylated isoprene units (Swiezewska and Danikiewicz 2005). Two distinct ATP-dependent enzymes are involved, producing MVA diphosphate, which is then decarboxylated/dehydrated to yield IPP. Although the final transition necessitates another molecule of ATP, there is no proof that the tertiary hydroxyl has been phosphorylated, even though this would have improved the leaving group. Perhaps ATP helps in hydroxyl loss. An isomerase enzyme converts IPP to the other isoprene unit, DMAPP, by stereo-specifically removing the pro-R proton (HR) from C-2 and incorporating a proton from water onto C-4. Although the isomerization is reversible, the equilibrium favors DMAPP. This reaction produces a reactive electrophile, which makes it an effective alkylating agent. DMAPP has an excellent leaving group, the diphosphate, and can produce allylic carbocation stabilized by charge delocalization via the SN1 method. In contrast, IPP, with its terminal double bond, is more likely to behave as a nucleophile, particularly against the electrophilic DMAPP. These varying reactivities serve as the foundation for terpenoid biosynthesis, and carbocations play an important role in mechanistic pathways (Zografos and Anagnostaki 2016). The glycolytic process intermediates pyruvic acid and glyceraldehyde 3-phosphate are converted into 1-Deoxy-D-xylulose 5-phosphate by removing the pyruvate carboxyl. Thiamine diphosphate-mediated pyruvate decarboxylation yields an acetaldehyde molecule as an enamine, acting as a nucleophile in a reaction containing glyceraldehyde 3-phosphate. Deoxyxylulose phosphate formation follows the release from the thiamine pyrophosphate (TPP) carrier and is then converted into 2-C-methyl-D-erythritol 4-phosphate via a rearrangement reaction and reduction follows this pinacol-like rearrangement. The rearrangement stage product (i.e., predicted aldehyde) is undetectable, and without releasing any intermediates, only one enzyme catalyzes both the reduction and rearrangement events. Similar rearrangements take place during the synthesis of the amino acid leucine, isoleucine and valine (Zografos and Anagnostaki 2016). The branched-chain structure of the methylerythritol phosphate is equivalent to the isoprene unit, but the whole sequence of processes leading to the intermediate isopentenyl phosphate has yet to be determined. The reaction of methylerythritol phosphate with cytidine triphosphate (CTP) yields a cytidine diphospho derivative, which is then phosphorylated using ATP. The resulting 2-phosphate then undergoes conversion into a cyclic phosphoanhydride, with the loss of cytidine phosphate (Obiol-Pardo et al. 2011). Through as-yet-unknown mechanisms, this cyclophosphate connects the mevalonate and deoxyxylulose pathways and results in IPP. DMAPP may then be formed from IPP isomerism or synthesized independently. It remains to be determined experimentally if the mevalonate route or the deoxyxylulose phosphate route provides isoprene units for the manufacture of a certain terpenoid. Animals lack the deoxyxylulose phosphate pathway and hence rely solely on the mevalonate pathway. Plants, and many other creatures, are capable of using both pathways, usually simultaneously (Smit and Mushegian 2000). Two processes are compartmentalized in plants, with the enzymes of mevalonate pathway located in the cytoplasm and the enzymes for deoxyxylulose phosphate pathway in the chloroplasts. Accordingly, the mevalonate pathway produces triterpenoids and steroids (cytosolic products), whereas the majority of other terpenoids are produced in the chloroplasts and are obtained from deoxyxylulose phosphate (Tian et al. 2022). Undoubtedly, there are a few exceptions e.g. there are also cases where the two paths provide distinct parts of a molecule, or where common intermediates of later steps are exchanged between the two pathways. These issues will not be discussed further in this chapter, and no investigation is required to track down the isoprene unit sources. The pharmacological interest lies solely in the early route, i.e., steroid biosynthesis, which is derived by mevalonate in a great variety of species (Mohammadi-Cheraghabadi and Hazrati 2023). Thus, inhibitors of HMG-CoA reductase, the mevalonate pathway enzyme, limits production of steroids but does not alter the terpenoids formation from deoxyxylulose phosphate. Similarly, deoxyxylulose phosphate pathway inhibitors can be used to inhibit the terpenoid production while leaving steroid formation unchanged (Gershenzon and Kreis 1999).

5.1.2 The MEP Pathway

The MEP pathway provides IDP and DMADP for terpenoid biosynthesis in plastids (Nagel et al. 2019). IDP and DMADP are formed by means of condensation and reduction of D-glyceraldehyde-3-phosphate (GAP) and pyruvate over the course of seven processes, requiring three ATP and three NADPH equivalents. The committing step is catalyzed by 1-deoxy-D-xylulose 5-phosphate (DXP) synthase (DXS), which converts pyruvate and GAP into DXP and regulates flux through the MEP pathway (Bergman and Phillips 2021). Following the synthesis of DXP, DXP reductoisomerase (DXR) catalyzes its reduction to MEP via NADPH-dependent isomerization. MEP is then triggered via a cytidine diphospho linkage at C4 and the phosphorylation of the alcohol at C2. The enzyme 2C-methyl-D-erythritol-2,4-cyclodiphosphate (MEcDP) synthase (MDS) then directs the cyclization of this active intermediate to produce the unique 8-membered phosphodiester ring, MEcDP. HMBDP synthase (HDS), an iron-sulfur protein, then reduces MEcDP to 1-hydroxy-2-methyl-2-(E)-butenyl-4-diphosphate. HMBDP is further reduced by HMBDP reductase (HDR), producing IDP and DMADP in a roughly 5:1 ratio. The plastid also contains IDP isomerase, which helps to maintain appropriate IDP/DMADP ratios. Furthermore, like FDP synthase, geranyl diphosphate (GPP) and geranylgeranyl diphosphate (GGPP) synthase are the primary prenyl transferases involved in supplying plastid terpene biosynthesis, giving C10 and C20 prenyl diphosphates for the synthesis of mono- and diterpenes, respectively (Rohdich et al. 2001).

Pathways are demonstrated in Figure 5.2

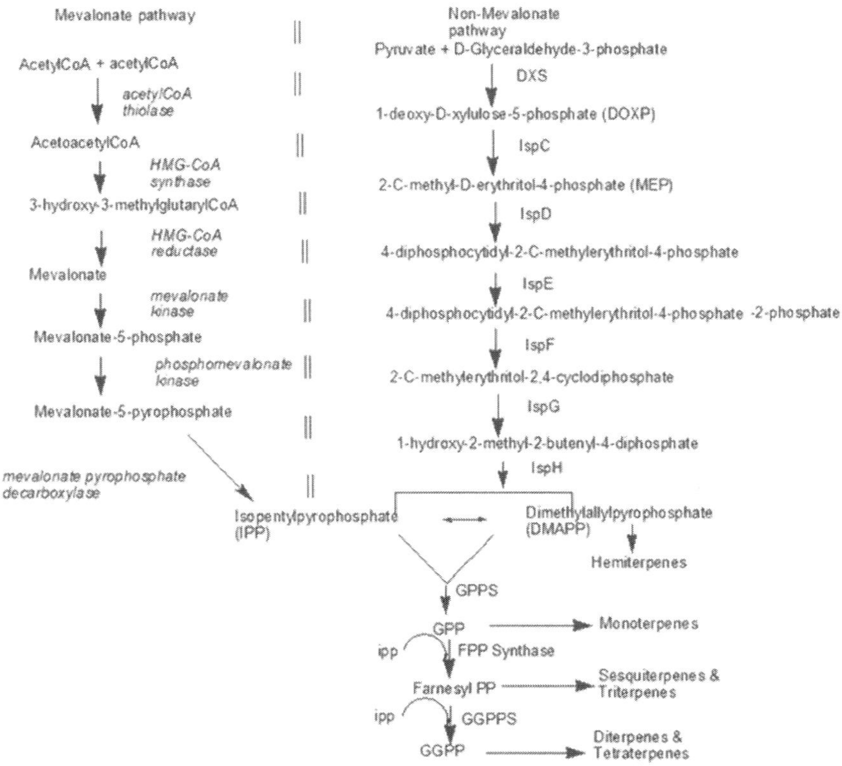

Figure 5.2 MEP pathways.

5.1.3 Secondary Metabolites

5.1.3.1 Hemiterpenes C5

DMAPP and IPP are intermediates of hemiterpene in pathways leading to the formation of more complicated terpenoid molecules. There are not many real hemiterpenes found in nature; one prominent example is isoprene, a volatile chemical emitted by many plant species, particularly trees (Gershenzon and Croteau 2018). A proton is removed from the allylic cation to generate isoprene. Once the isoprene units are formed, they can go through a variety of enzymatic modifications to yield various hemiterpenes. These transformations could include cyclization, rearrangement, or modification processes performed by particular enzymes (Gershenzon and Croteau 2018). Structure of isoprene is shown in Figure 5.3.

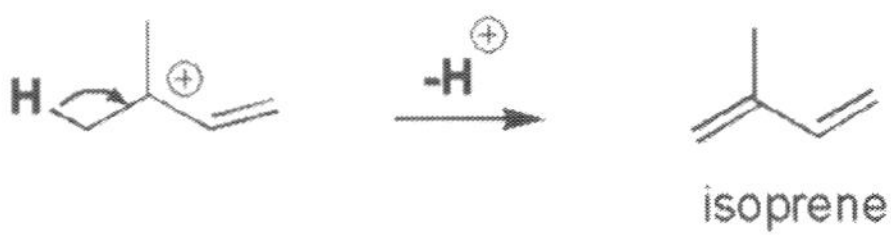

Figure 5.3 Oxidation of isoprene units.

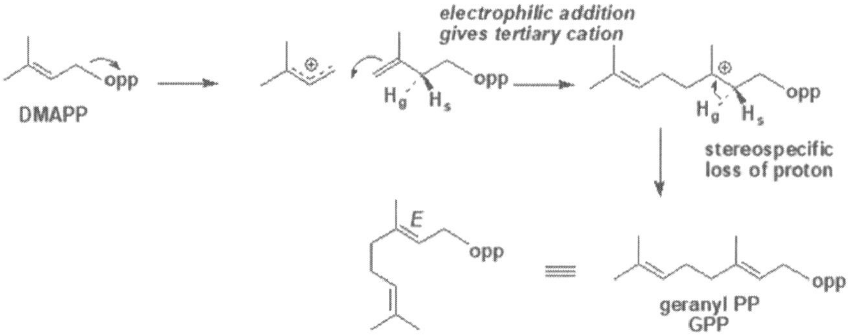

Figure 5.4 Biosynthesis of geranyl pyrophosphate.

5.1.3.2 Monoterpenes C10

The enzyme prenyl transferase converts DMAPP and IPP into geranyl diphosphate (GPP). This is thought to include the ionization of DMAPP to an allylic cation, addition to IPP's double bond, and proton loss. Stereochemically, the proton lost (HR) is similar to that lost during the isomerization of IPP to DMAPP (Chang et al. 2021) (Figure 5.4).

This yields geranyl PP, a monoterpene diphosphate with a trans double bond (E). Linalyl PP and neryl PP are isomers of geranyl PP that are likely to be formed from geranyl PP via ionization to the allylic cation, allowing for a change in attachment of the diphosphate group to the tertiary carbon in linalyl PP) or a change in stereochemistry at the double bond (to Z in neryl PP) (Zhang and Tiefenbacher 2015) (Figure 5.5).

These three molecules, with very minor alterations, can produce a variety of linear monoterpenes present in volatile oils used in flavoring and perfumery, such as geranial and neral in lemon oil, citronellal in citronella oil, and linalool in coriander oil. The monoterpene mentioned here is limonene, which is a prominent component of citrus fruit peel and caraway essential oils (Brahmi et al. 2021). An enzyme called limonene synthase cycles GPP, the common precursor of monoterpenes, into (+)-limonene during its production. The

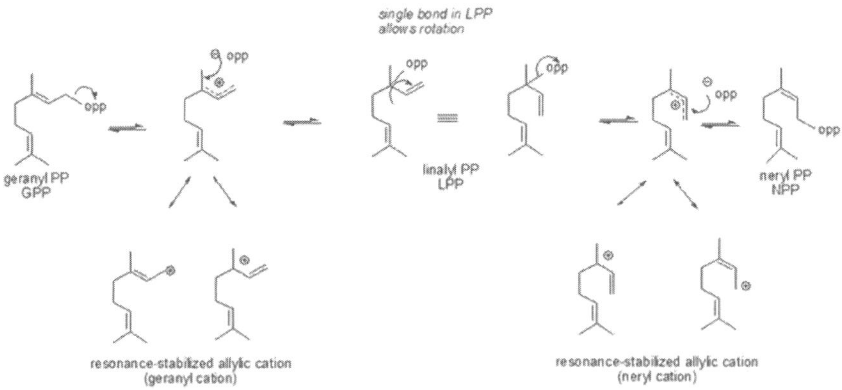

Figure 5.5 Synthesis of neryl pyrophosphate.

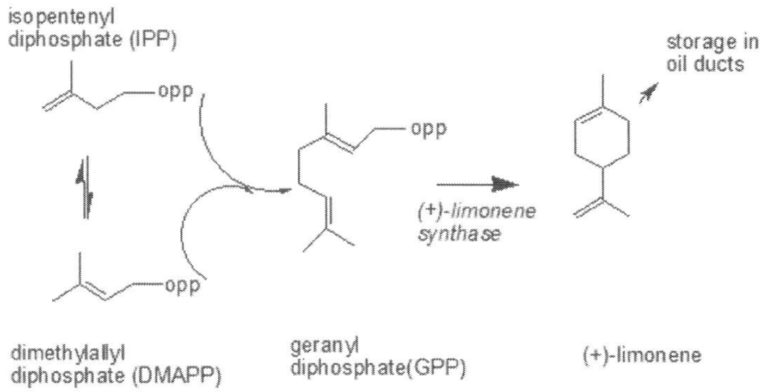

Figure 5.6 Biosynthesis of (+)- limonene.

limonene will subsequently be kept in essential oil ducts (Hyatt et al. 2007). Structure of limonene is given below in Figure 5.6.

5.1.3.3 Irregular Monoterpenes

A large variety of natural monoterpene structures containing carbon skeletons, despite being clearly formed from isoprene C5 units, do not appear to fit the standard head-to-tail coupling mechanism.

These structures are known as irregular monoterpenes and appear to be found almost exclusively in members of the plant family Compositae/ Asteraceae (Zidorn 2008). Allowing for possible rearrangements, the two isoprene units appear to have joined in a different manner, as described by data available related to their biosynthesis, yet not clear. Thus, whereas DMAPP and IPP are used in their synthesis, there is not any involvement of geranyl PP and neryl PP. The biosynthesis of irregular monoterpenes usually begins with the condensation of one or more molecules of IPP and/or DMAPP to form larger prenyl diphosphate precursors, such as geranyl diphosphate (GPP), farnesyl diphosphate (FPP), or geranylgeranyl diphosphate (GGPP), depending on the compound being synthesized. Irregular monoterpenes undergo a variety of enzymatic changes, which are performed by particular terpene synthases and other enzymes. The prenyl diphosphate precursors may undergo cyclization, rearrangement, oxidation, reduction, or other changes in order to synthesize the required irregular monoterpenoid structures (Rehman et al. 2016).

5.1.3.4 Sesquiterpenes C15

The primary precursor of sesquiterpenes, farnesyl diphosphate (FPP), is produced when geranyl diphosphate and another C5 IPP unit are added in a reaction that extends the prenyl transferase reaction. Again, an initial ionization of GPP appears likely, and the proton lost from C-2 of IPP is stereochemically similar to that lost during the preceding isoprenylation phase. FPP can then generate linear and cyclic sesquiterpenes (Stanley Fernandez et al. 2000). The increased chain length and added double bond enhance the number of potential cyclization modes, allowing for a wide range of mono-, bi-, and tri-cyclic structures to be formed. The stereochemistry of the double bond adjacent to the diphosphate can shift to an E configuration (as in FPP) or a Z configuration via ionization, as in geranyl/neryl PP (Pyun 1992) (Figures 5.7 and 5.8).

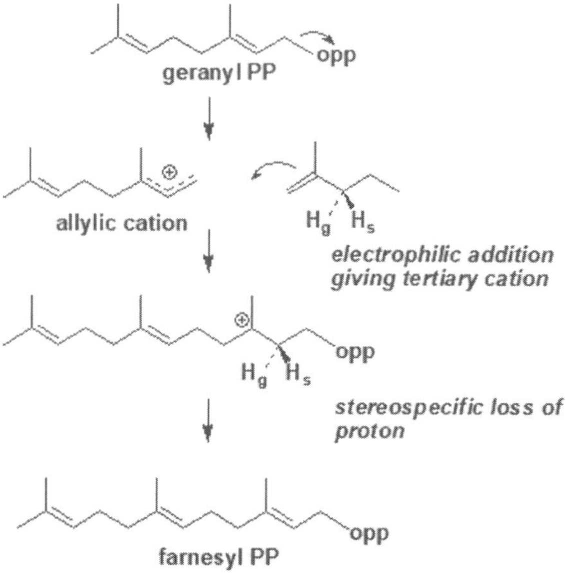

Figure 5.7 Biosynthesis of farnesyl pyrophosphate.

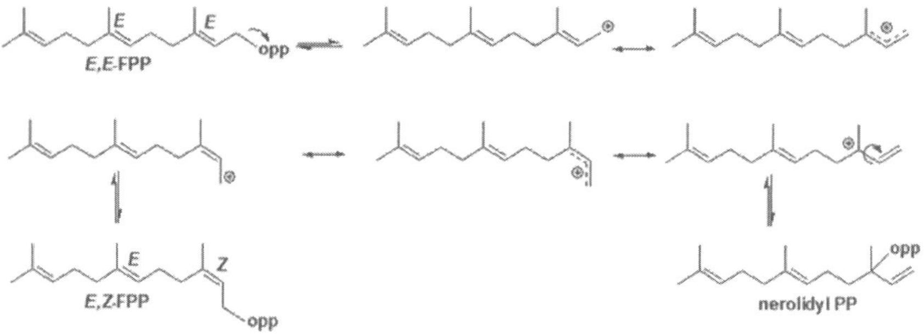

Figure 5.8 Biosynthesis of nerolidyl pyrophosphate.

Zingiberene is one example of a sesquiterpene. To begin, terpene synthase enzyme converts three isomers of farnesyl pyrophosphate, namely (2Z,6E)-FPP (1), (2Z,6Z)-FPP (2), and (2E,6E)-FPP (3), to a common FPP-derived bisabolyl cation (Figure 5.9).

This intermediate, after cyclyzation and a series of hydrogen shifts, produces zingiberene, bisabolene, and cumarene. A nucleophilic attack by X or its biogenetic equivalent on C-7 of (2Z,6E)-FPP (1) can initiate the cyclization process, eventually removing the diphosphate from C-1 to produce an enzyme-bonded bisabolyl intermediate. The X-group is eliminated, causing a series of hydrogen shifts from various positions to produce ar-Curcumene (Figure 5.10), α-Zingiberene, and β-Bisabolene (Fujisawa et al. 2010).

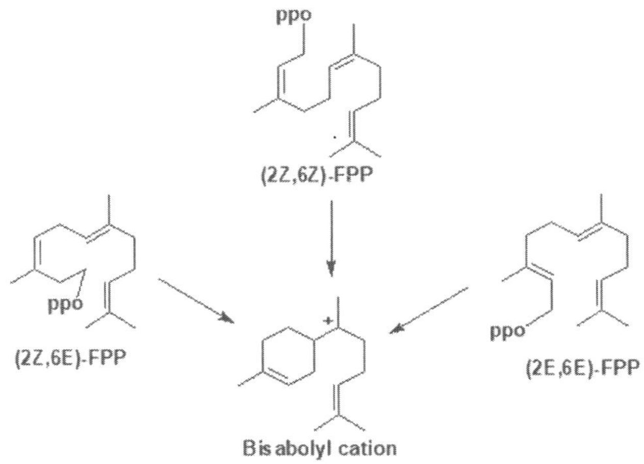

Figure 5.9 Biosynthesis of bisabolyl cation.

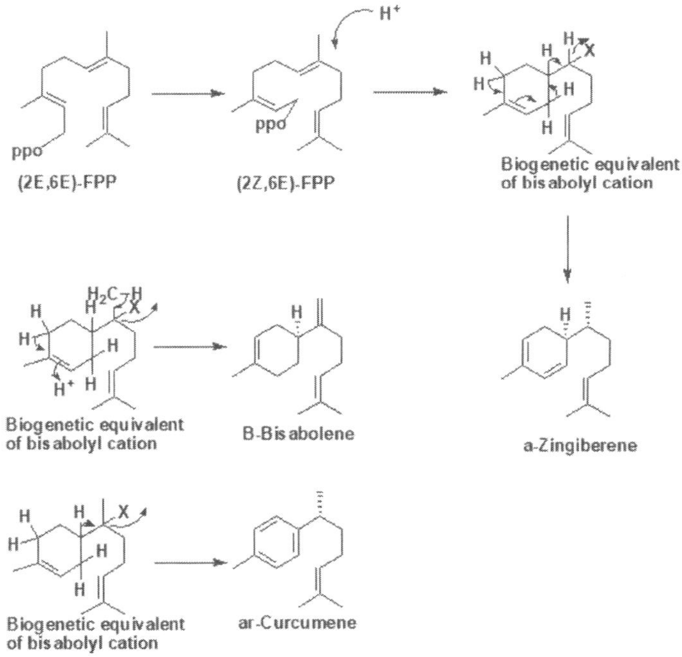

Figure 5.10 Biosynthesis of ar-curcumene.

5.1.3.5 Diterpenes-C20

Diterpenes are derived from geranylgeranyl diphosphate (GGPP), formed by addition of another IPP molecule to farnesyl diphosphate, similar to the process for lower terpenoids (Böttger et al. 2018). Formation of geranyl is demonstrated in Figure 5.11.

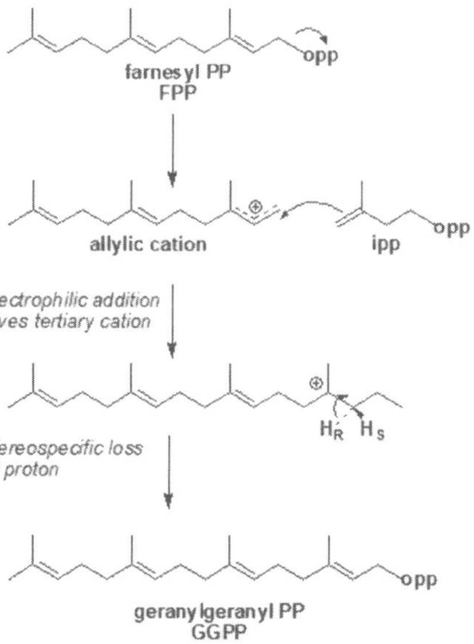

Figure 5.11 Biosynthesis of geranyl geranyl pyrophosphate.

Kaurene is synthesized from mevalonate and trans-geranylgeranyl pyrophosphate in soluble, cell-free endosperm tissue from growing *Echinocystis macrocarpa* seeds. (-)-Kaurene is the only diterpene hydrocarbon found. The particulate fraction from the same tissue catalyzes the further conversion of (-)-kaurene into a range of oxidized derivatives, including radiene (II) and (-)-kaurene (III), based on their spectral and physical properties (West and Upper 1969). The structure of kaurene is shown below in Figure 5.12.

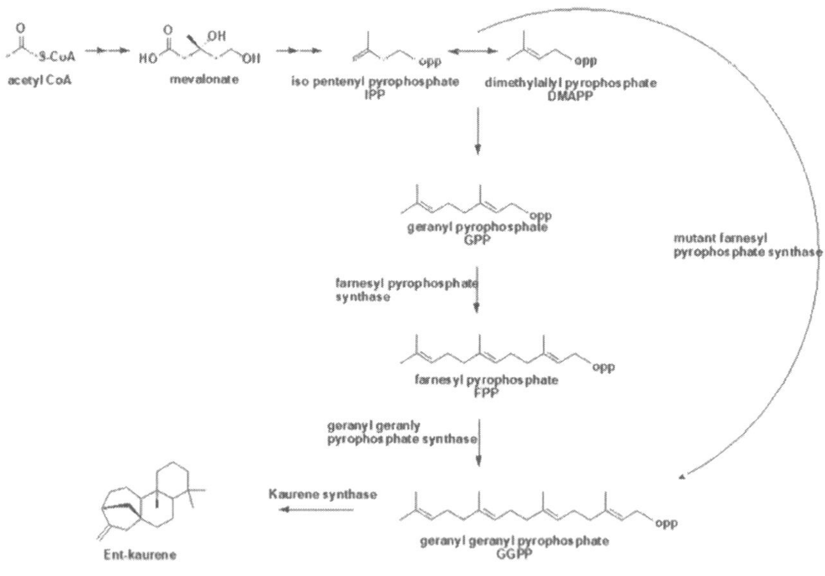

Figure 5.12 Biosynthesis of Ent-Kaurene.

5.1.3.6 *Sesquiterpenes C25*

While geranyl diphosphate (GPP) or farnesyl diphosphate (FPP) are the precursors of monoterpenes and sesquiterpenes respectively, the biosynthesis of Sesquiterpenes requires the elongation of these precursors to geranylgeranyl diphosphate (GGPP), which is composed of four isoprene units and offers no new features (apart from a 1,5-hydride shift). GFPP results from a continuation of the chain extension process, which adds to GCPP another IPP unit. Ophiobolin A, structure is shown below, for example, possesses biological activity against nematodes, bacteria, and fungi (Shen et al. 2020) (Figure 5.13).

5.1.3.7 *Triterpenes C30*

Triterpenes are not produced by extending the well-known procedure of adding IPP to the developing chain. Instead, two molecules of farnesyl PP are linked tail to tail to form the hydrocarbon squalene which was first isolated from shark liver oil (*Squalus* sp.). *Amaranthus cruentus* (Amaranthaceae) is one of several seed oils that are now known to be high in squalene (Wolosik and Markowska 2019). The coupling process, that when drawn on paper only involves removing two diphosphate groups, one molecule of FPP loses a proton from its C-1 position and gains a proton from NADPH. Once presqualene diphosphate, an intermediate in the process, was isolated from rat liver, the difficulties in developing a believable mechanism for this improbable reaction were resolved (Abe et al. 1994). All the theories that were prevailing at the time were quickly ruled out when it was identified as a cyclopropane derivative. depicts the production of presqualene PP as an attack of FPP's 2,3-double bond on the farnesyl cation, similar to chain extension with IPP. The resulting triple cation is discharged by losing a proton and forming the cyclopropane ring, yielding presqualene PP. Naturally, carbons-1 of the two FPP units must eventually couple to produce squalene, but presqualene PP synthesis has already connected one molecule's C-1 to the other's C-2. Another cyclopropane cationic intermediate is postulated to explain the ensuing shift in bonding between two units of the FPP (Mayaka 2020). The loss of diphosphate from presqualene PP results in an unfavorable cation, that, through Wagner–Meerwein

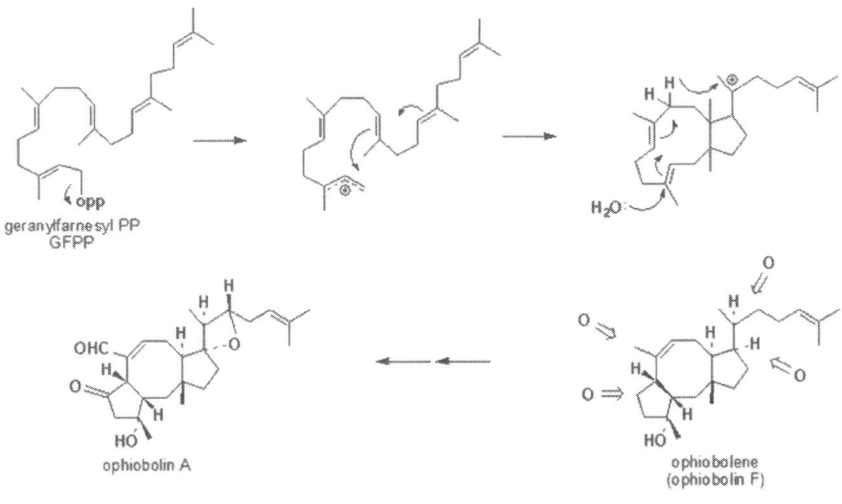

Figure 5.13 Biosynthesis of ophiobolin A.

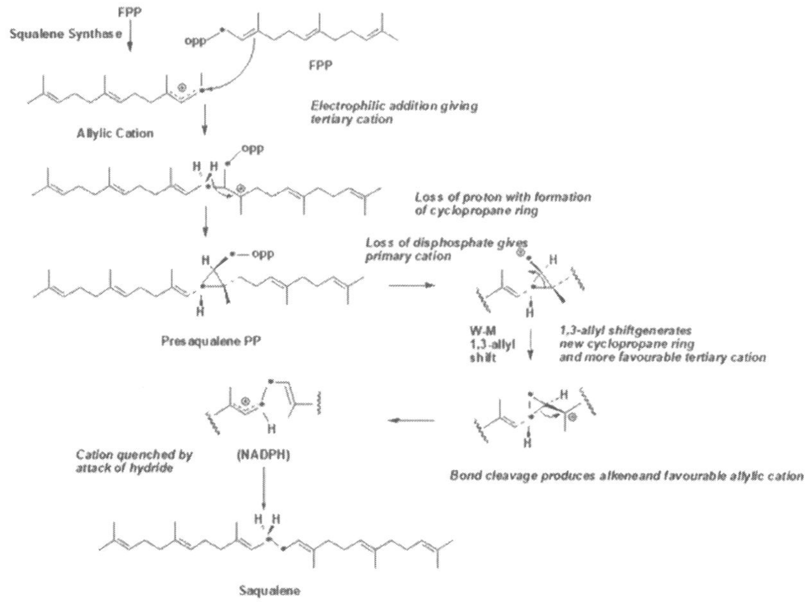

Figure 5.14 Biosynthesis of squalene.

rearrangement, creates a tertiary carbocation and forms the C-1-C-1 bond. Breakdown of the initial but now repetitive C-1-C-2 bond produces an allylic cation, and squalene production (Figure 5.14) gets completed by the addition of hydride from NADPH (Loomis and Croteau 1980).

5.1.3.7.1 Modified Triterpenoids

As previously demonstrated, triterpenoid skeletons can undergo a variety of structural changes. However, the alterations discussed here are those that give rise to the loss of numerous skeletal carbon atoms. Steroids are the most important of these degraded triterpenoids, and they are treated separately (Garg et al. 2020). The cyclization of 2,3-oxidosqualene, mediated by numerous oxidosqualene cyclases (OSCs), produces a variety of triterpenoids. Carbocation rearrangement, deprotonation, and protonation of the epoxide complete these processes (Nair and Kochupurackal 2023). A protosteryl cation converts 2,3-oxidosqualene into lanosterol, cycloartenol, and curcurbitadinol, while a dammarenyl cation converts lupeol, β-amyrin, taraxerol, and other triterpenoids. Cycloartenol synthase (CAS, EC 5.4.99.8) catalyzes the cyclization of 2,3-oxidosqualene, which yields cycloartenol (Jin et al. 2017). Biosynthesis of modified triterpenoids are given in Figure 5.15.

5.1.3.8 Tetraterpenes (C40)

Tetraterpenes, by a single category of chemicals, are represented only by carotenoids, despite the fact that several hundred natural structural variations exist (Kumari et al. 2013). Although these molecules contribute to photosynthesis, they also occur in other non-photosynthetic organisms. Phytoene (the tetraterpene skeleton) is formed by the tail-to-tail linking of two molecules of geranylgeranyl diphosphate (GGPP) in a sequence resembling that of triterpenes and squalene. A cyclopropyl molecule (prephytoene diphosphate) is temporary product in the process, and the primary distinction between both

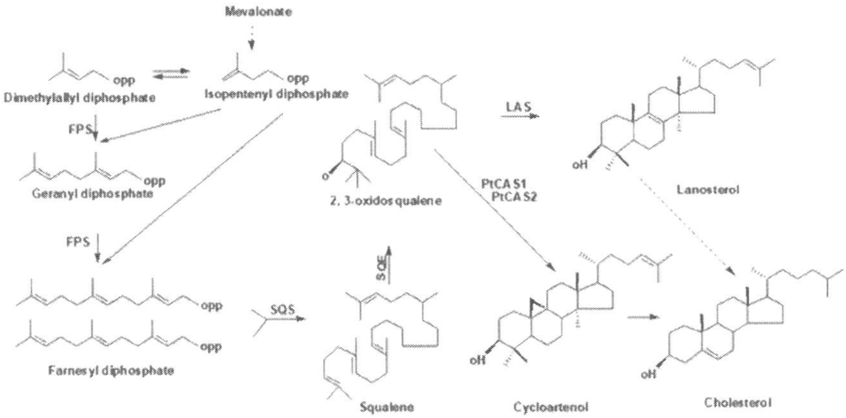

Figure 5.15 Biosynthesis of modified triterpenoids.

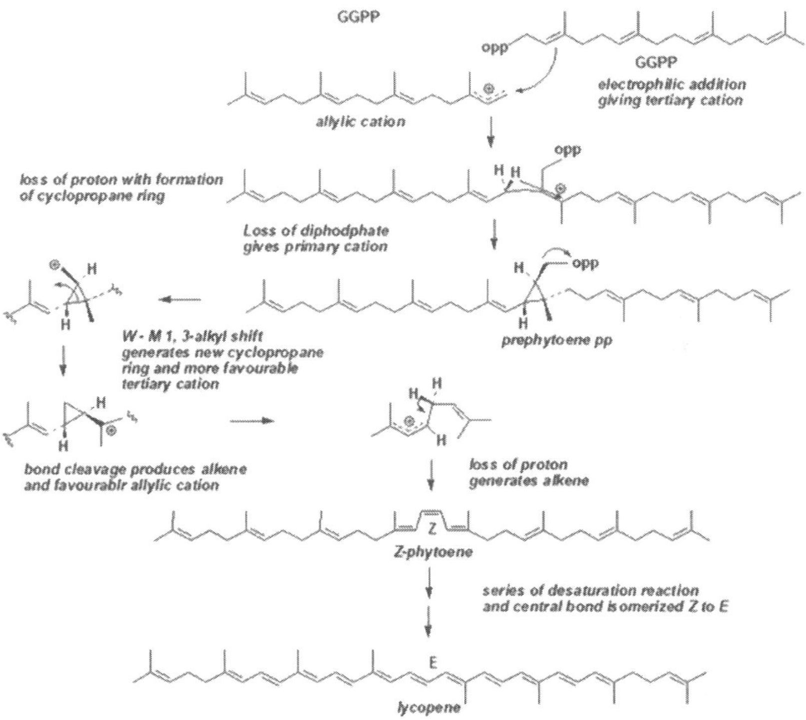

Figure 5.16 Biosynthesis of lycopene.

the triterpene and tetraterpene pathways is the release of the allylic cation (Gray et al. 1992). A hydride ion from NADPH is absorbed by the allylic cation to make squalene, but in phytoene biosynthesis, H+-ion is lost, resulting in the formation of a conjugated chain and a double bond in the middle of the compound. In bacteria, the new double bond is E (trans), whereas in fungi and plants, it is Z (cis). Lycopene structure is shown in Figure 5.16. This system of trienes avoids squalene-like cyclization. Conjugation is subsequently prolonged through a series of desaturation processes that remove pairs of hydrogens alternatively from either side of the triene system, resulting in lycopene (Martins and Ferreira 2017), which, like the most of the carotenoids, has an all-E(trans) structure. In fungi and plants, another step of isomerization is required for the central double bond modification (Zulfiqar et al. 2021). The color of carotenoids is imparted by extended π-electron system, contributing colors like orange, red and yellow pigments to the tissues of plants. Lycopene, a primary carotenoid pigment found in *Lycopersicon esculentum*; Solanaceae (ripe tomato fruit) (Soytong et al. 2021).

5.1.4 Shikimic Acid Pathway

The shikimic acid route, also known as the shikimate pathway, is the fundamental mechanism of phenolic compound production (Santos-Sánchez et al. 2019). The term "shikimic compound" comes from the 1885 isolation of the highly toxic Japanese shikimic flower (*Illicium anisatum*) (Dhaniaputri et al. 2022). It has also been reported that shikimic acid can be found in Chinese star anise (*I. verum*), sweetgum tree (*Liquidambar styraciflua*) leaves, and sweetgum bark and wood samples (Marchiosi et al. 2019). The shikimate pathway is a seven-step metabolic route used by plants for the biosynthesis of aromatic amino acids i.e, phenylalanine, tyrosine, and tryptophan (Tzin and Galili 2010). Shikimic acid, the precursor of the shikimate pathway, is formed by combining phosphoenolpyruvate (from the glycolytic process) with erythrose 4-phosphate (from the pentose phosphate pathway [PPP]) (Tzin and Galili 2010). The shikimate pathway provides an alternate route to numerous aromatic molecules, including the aromatic amino acids L-phenylalanine, L-tyrosine, and L-tryptophan (Maeda and Dudareva 2012). This pathway is used by microbes and plants but not by animals, hence aromatic amino acids are among the necessary amino acids for humans that must be supplied from their diet (Maeda and Dudareva 2012).

Seven steps are involved in the shikimate pathway to generate the end product chorismate (Weaver and Herrmann 1997). These are as follows:

- First, phosphoenolpyruvate and erythrose-4-phosphate are combined to synthesize 3-deoxy-d-arabino-heptulosonate 7-phosphate (DAHP), which is catalyzed by the enzyme, phospho-2-oxo-3-deoxyheptonate aldolase.and is technically a derivative of 2-deoxy-D-glucose-6-phosphate (Howe et al. 2003).
- In the second step, the enzyme 3-dehydroquinate synthase converts DAHP to 3-dehydroquinic acid with cobalt (II) and nicotinamide adenine dinucleotide (NAD) as cofactors (Deka 1993).
- The reduction of 3-dehydroquinic acid to quinic acid occurs in the third stage. The enzyme 3-dehydroquinate dehydratase (DHQase) catalyzes the removal of water from 3-dehydroquinic acid, which results in 3-dehydroshikimic acid (Sann et al. 2004).
- Dehydroquinic acid dehydration can be catalyzed by two types of enzymes. Type 1, linked to shikimate dehydrogenase, is not influenced by the concentration of shikimate, but type 2 is triggered by shikimate only.

- Shikimate is formed in the fourth step by the enzyme shikimate-5-dehydrogenase. It has been proposed that the two variants of DHQase regulate how dehydroquinic acid is used, resulting in the production of either shikimic acid or protocatechuic acid (Dewick 2002).
- The enzyme shikimate kinase phosphorylates the shikimate to form shikimate -3-phosphate in the fifth step. During the reaction, an ATP molecule is converted to an ADP molecule (Krell et al. 1998).
- A molecule of enol pyruvate is added to the shikimate-3-phosphate in the sixth step, forming 5-enolpyruvylshikimate-3-phosphate, a condensed product. This reaction is facilitated by enolpyruvylshikimate phosphate synthase (Bondinell et al. 1971).
- In the final step, 5-enolpyruvylshikimate-3-phosphate is transformed into the significant product chorismate by removing a phosphate group (Mir et al. 2015).

5.1.5 Pathways Leading to Aromatic Amino Acid Formation

Chorismic acid production is an important branch point in the shikimic acid pathway since it can be converted in three ways.

Chorismic acid is transformed to prephenic acid in the presence of chorismate mutase, whereas glutamine catalyzes anthranilic acid synthesis. Additionally, p-aminobenzoic acid is produced from chorismic acid (Lingens 1968). Afterwards, anthranilic acid is transformed into phosphoribosyl anthranilic acid and then into carboxyphenylaminodeoxyribulose-5-phosphate, reactions being catalyzed by phosphoribosyl anthranilate isomerase and anthranilate phosphoribosyl transferase, respectively (Doy 1961). Indolyl-glycerol phosphate synthase catalyzes the ring formation to produce indolyl-3-glycerol phosphate. Tryptophan synthase, the enzyme that catalyzes the final reaction, is made up of two components: the Alpha component catalyzes the dissociation of indolylglycerol phosphate to glyceraldehyde-3-phosphate and indole, while the Beta component catalyzes the direct condensation of indole with serine to form tryptophan (Miles 1991). Tyrosine and phenylalanine are both biosynthesized from prephanic acid via distinct routes, and they serve as precursors for the biosynthesis of phenylpropanoids. Following that, lignin, tannins, flavonoids, and coumarins are synthesized using the phenylpropanoids (Gamborg and Simpson 1964).

Prephenate dehydrogenase catalyzes the aromatization of prephanic acid to 4-hydroxyphenylpyruvic acid, which is the initial step in the synthesis of tyrosine. Tyrosine is formed as the result of transamination, which is facilitated by tyrosine aminotransferase (Schenck and Maeda 2018).

Prephenate dehydratase catalyzes the initial step in the biosynthesis of phenylalanine, which is the aromatization of prephanic acid to phenyl pyruvic acid. Phenylalanine aminotransferase next catalyzes the transamination reaction, which yields phenylalanine (Yoo et al. 2013).

Enzymes like chorismate mutase, iso-chorismate synthase, anthranilate synthase, and aminodeoxychorismate synthase regulate chorismate, which is the precursor of tryptophan, tyrosine, phenylalanine, salicylate, phylloquinone, and folate in higher plants (Svarcova et al. 2015). The shikimate pathway is shown below in Figure 5.17.

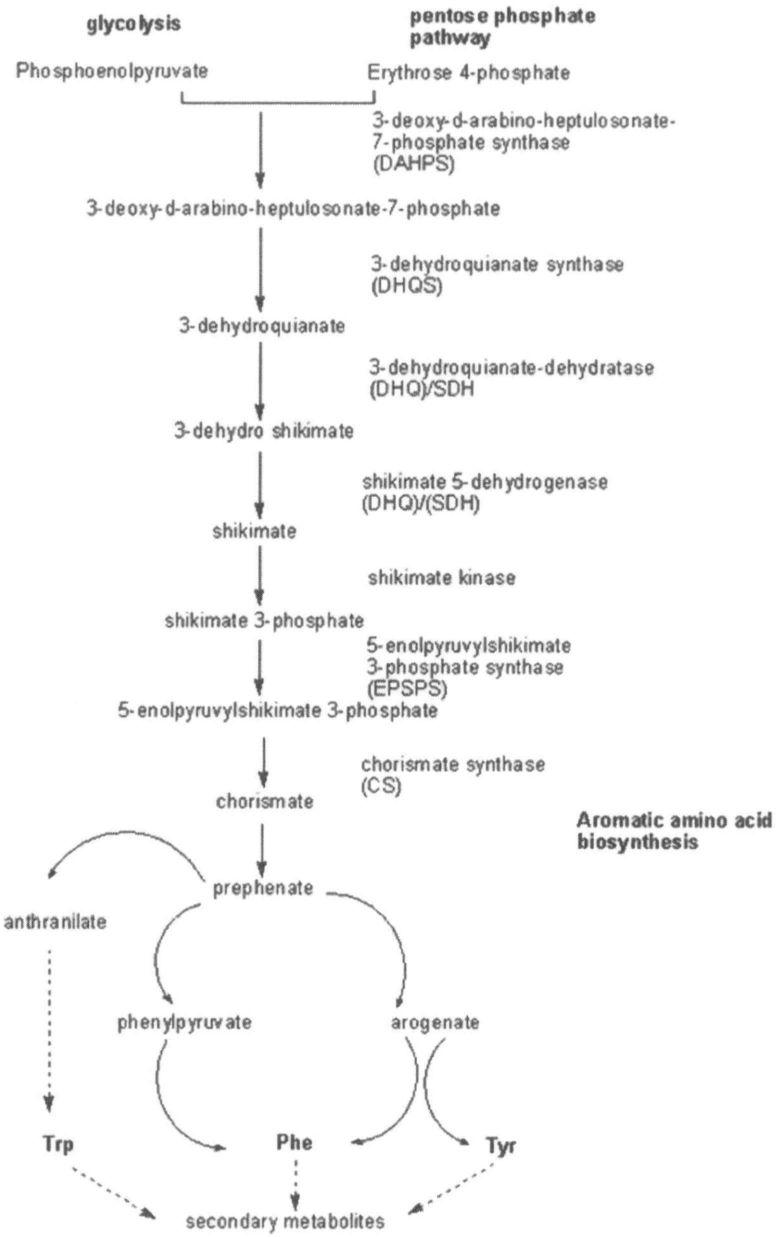

Figure 5.17 Biosynthesis of aromatic amino acid.

5.2 SECONDARY PRODUCTS

5.2.1 Cinnamic Acid

L-tyrosine and L-phenylalanine are C6C3 building blocks that are used as precursors to make a large number of naturally occurring compounds (Figure 5.18) (Dewick 2002). Elimination of the side-chain ammonia from the

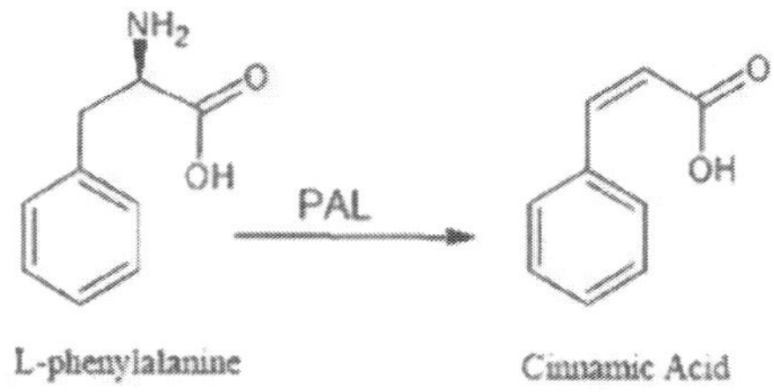

Figure 5.18 Biosynthesis of cinnamic acid from L-phenylalanine.

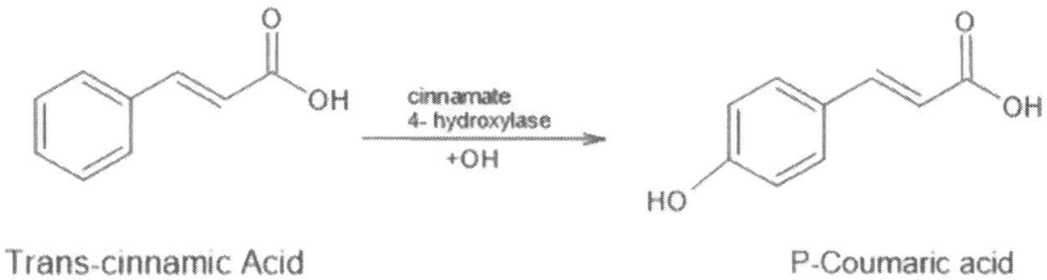

Figure 5.19 Biosynthesis of P-coumaric acid from trans-cinnamic acid.

plants is often the first step towards producing the cinnamic acid (trans-E). All plants appear to have the potential to deaminate phenylalanine through the enzyme phenylalanine ammonia lyase (PAL) (Kong 2015). Natural cinnamic acids that are more widely available include caffeic, ferulic, sinapic, and 4-coumaric acid (Otles and Nakilcioglu 2015).

Three isomeric forms of coumaric acid (ortho, meta, and para) exist in nature. It is essentially a hydroxy derivative of cinnamic acid. The 4 or p-coumaric acid has vast applications in the phytochemical and nutraceutical field. It is majorly present in cell walls of gramineous plants such as wheat, maize, and rice (Antonopoulou et al. 2022). It may also be found in vegetables like tomatoes and carrots. These are created and produced as part of a plant's defense mechanism, such as injury (Figure 5.19) (Fleuriet and Macheix 2003).

Caffeic acid (3,4 dihydroxycinnamic acid is synthesized from plant phenylpropanoid pathway by the hydroxylation of p-coumaric acid and this reaction is catalyzed by NADPH (Figure 5.20). It is a bioactive compound present in many fruits, vegetables and herbs (Poklar Ulrih et al. 2021). Caffeic acid is known for its antioxidant properties in plants. It also regulates cell expansion and turgor pressure. It also has application in potential prevention of inflammation, neurodegenerative diseases, cancer and diabetes (Alam et al. 2022).

Ferulic acid (4-hydroxy-3-methoxycinnamic acid) is synthesized by the methylation of caffeic acid (Figure 5.21). It is mainly present in wheat, flowers, fruit, seeds, peanuts and nuts (Jaganath and Crozier 2010). It forms dihydrodimers in plants and maintains its rigidity and strength. It also has activity against cancer, inflammation etc. (Zwane 2014).

It can also exist in certain dihydrodimeric forms to give strength to the plant cell. Some of the dihydrodimeric forms are shown in Figure 5.22.

Sinapic acid is also a derivative of cinnamic acid which is basically produced by hydroxylation and methylation of ferulic acid (Ruwizhi and Aderibigbe 2020) and these substitutions in ferulic acid results in formation

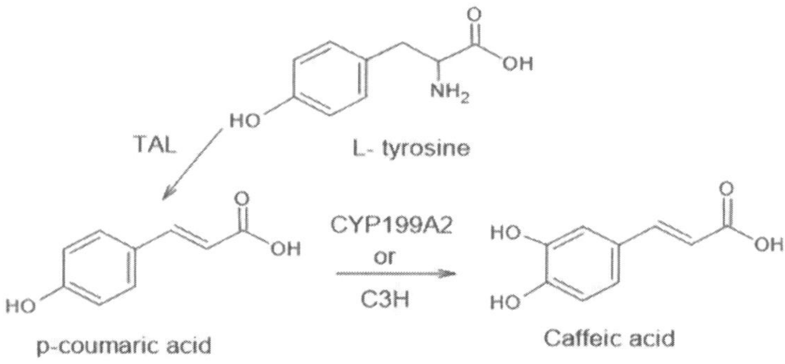

Figure 5.20 Biosynthesis of caffeic acid.

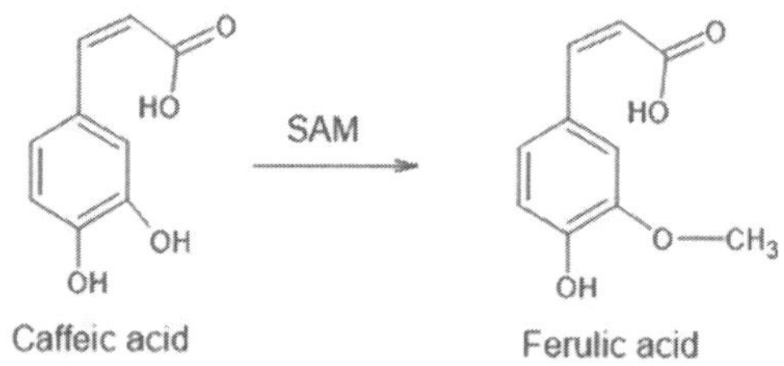

Figure 5.21 Biosynthesis of ferulic acid from caffeic acid.

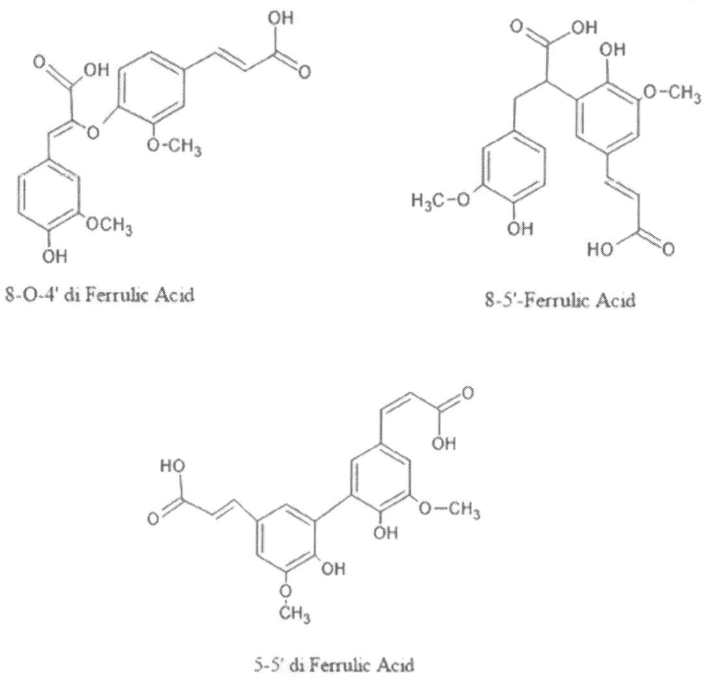

Figure 5.22 Dihydimeric forms of ferrulic acid.

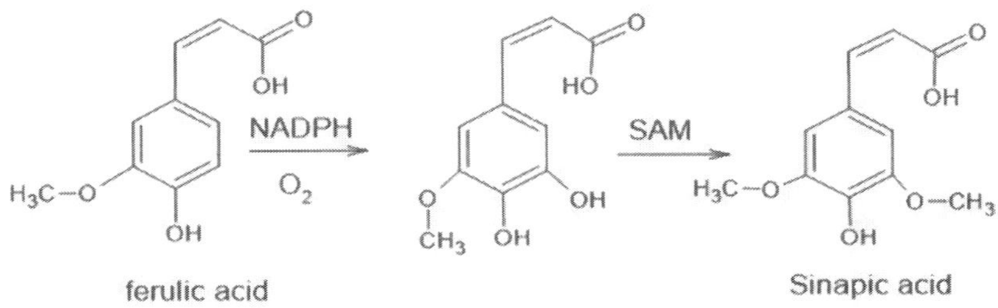

Figure 5.23 Biosynthesis of sinapic acid.

of 3,5-dimethoxy-4-hydroxycinnamic acid (sinapic acid) (Figure 5.23). It is present in many herbs such as canola, rapeseed, mustard seeds and in fruits such as oranges, grapefruits and cranberries (Pandi and Kalappan 2021). Sinapic acid and its esters serve an important function in seed germination and plant development, and also provide shield against ultraviolet radiation. It has important applications as antioxidant and as an anti-inflammatory agent (Sheahan 1996) (Figure 5.23).

5.2.2 Coumarins

Plants include coumarins in both their free form and as glycosides (Figure 5.24). They are typically found in families like the *Umbelliferae* (Apiaceae) and *Rutaceae*. Sweet clover (*Melilotus* species; Leguminosae) contains coumarin, which is responsible for the smell of freshly cut hay (Sowa-Borowiec et al. 2022).

Cinnamic acid is primarily responsible for the formation of coumarins (Figure 5.25). The process of hydroxylation of cinnamic acids is an essential stage during the synthesis of coumarins, a cinnamic acid lactone derivatives class (Wu et al. 2009).

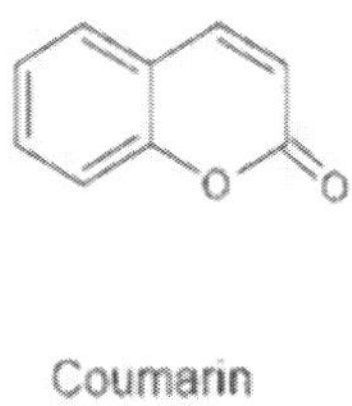

Figure 5.24 Chemical structure of coumaric acid.

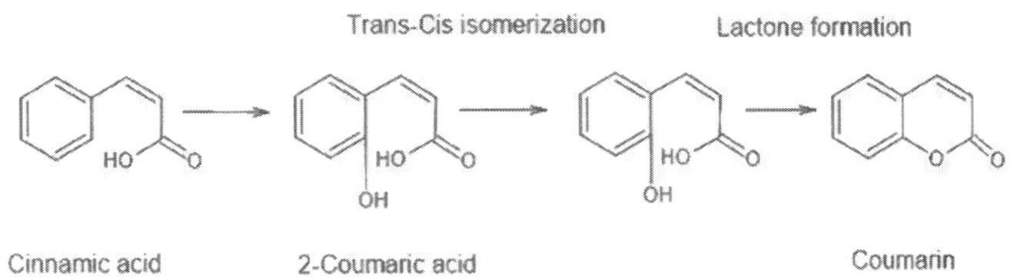

Figure 5.25 Biosynthesis of coumarin from cinnamic acid.

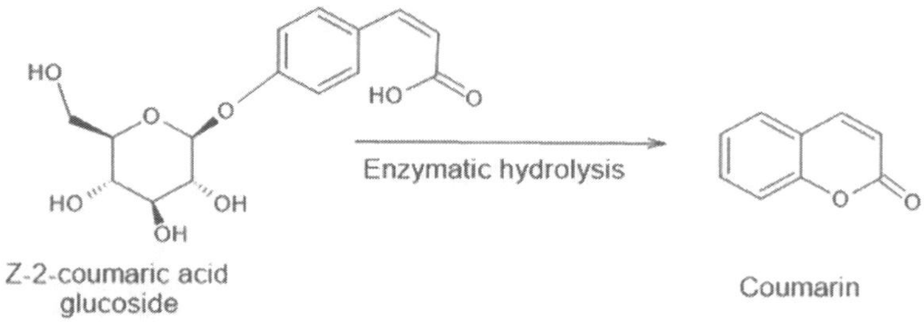

Figure 5.26 Enzymatic hydrolysis of Z-2-coumaric acid glucoside to coumaric acid.

However, there are proofs that plants have glucosides of (Z)-2-coumaric acid, and coumarin is released only when the plant tissues are damaged during harvesting and processing, causing enzymatic hydrolysis and lactonization (Figure 5.26; Porras et al. 2020).

5.2.3 Lignans and Lignins

The pathways leading to other metabolites based on C6C3 building blocks also involve cinnamic acids. The most important of them is the plant polymer lignin, which strengthens the cell wall of plants and forms cellulose microfibrils. Lignin (Figure 5.27), the substantial source of aromatic compounds, has mostly gone unexplored due to the challenges in releasing these metabolites (Kumar et al. 2019). Wood-rotting fungi, through their action, increase the availability of these beneficial compounds. Peroxidase enzymes catalyze the phenolic oxidative coupling of hydroxycinnamyl alcohol monomers to generate lignin (Singhania et al. 2022). Among these monomers, sinapyl alcohol, 4-hydroxycinnamyl alcohol (p-coumaryl alcohol), and coniferyl alcohol are the most significant; however, the specific monomers utilized depend on the kind of plant. Monocotyledons utilize all these alcohols, while monocotyledons utilize two alcohols (i.e., coniferyl, sinapyl) that are polymerized by gymnosperms (Chen et al. 2012).

Coniferyl alcohol monomers can undergo phenolic oxidative coupling, resulting in the following linkages: guaiacylglycerol β-coniferyl ether (β-aryl ether linkage), dehydrodiconiferyl alcohol (phenylcoumarin linkage), and pinoresinol (resinol linkage). These dimeric substances, known as lignans, are present in nature (Gall et al. 2017).

Pinoresinol, matairesinol, and secoisolariciresinol are a few examples of lignans. As observed in the different dimers, these dimers can react further using similar methods to form a lignin polymer with a heterogeneous series of intermolecular bondings. Although β-aryl ether type linkages occur widely, but lignin seems to be lacking organized subunits in comparison to most other naturally occurring polymeric materials (Chandra et al. 2015).

5.2.3.1 Steps Involved in Formation of Lignans

In addition to being involved in the synthesis of lignans, molecules like pinoresinol, lariciresinol, secoisolariciresinol, and matairesinol also have several pharmacological roles in plants.

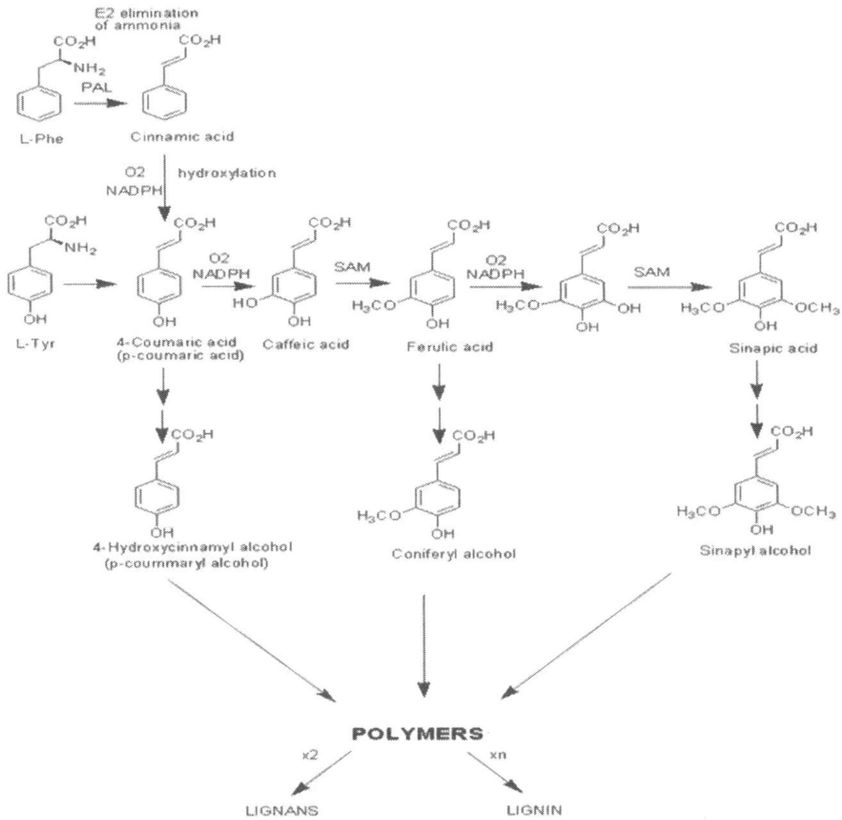

Figure 5.27 Biosynthesis of lignan and lignins.

5.2.3.1.1 Formation of Pinoresinol

Pinoresinol, a tetrahydrofuran lignan found in olives and their oils, is synthesized by catalysis of coniferyl alcohol through oxidative enzyme catalysis (peroxidases, laccase) (Murata et al. 2017). Pinoresinol inhibits low-density lipoprotein (LDL) oxidation and lipid peroxidation in plants (Kang et al. 1999); while in humans, it has pharmacological effects like acting as a hypoglycemic drug and inhibition of alpha-glucosidase (Rasouli et al. 2017). Biosynthesis of (+)- pinoresinol is shown in Figure 5.28.

5.2.3.1.2 Formation of Lariciresinol and Secoisolariciresinol

Following the initial dimerization of two monolignols, enzymes known as pinoresinol-lariciresinol reductases (PLR) catalyze two consecutive reduction processes that convert pinoresinol into lariciresinol or secoisolariciresinol (Figure 5.29; Markulin et al. 2019).

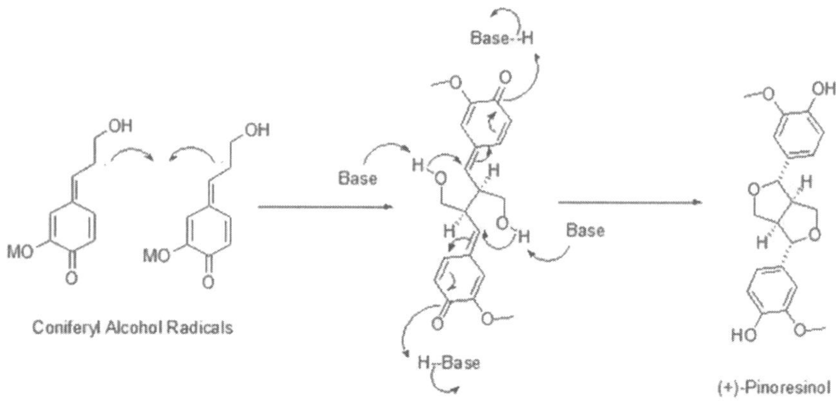

Figure 5.28 Biosynthesis of (+)- pinoresinol.

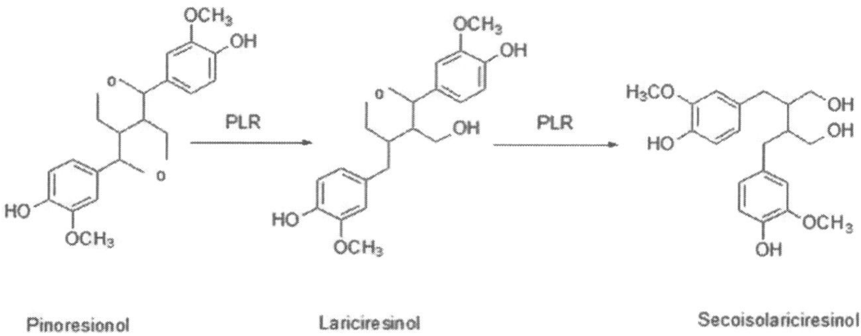

Figure 5.29 Biosynthesis of secoisolariciresinol.

5.2.3.1.3 Formation of Matairesinol

Matairesinol is synthesized by the oxidation of secoisolariciresinol in the presence NAD+-dependent secoisolariciresinol dehydrogenase (SDH; Niemeyer et al. 2003). It can be found in various grains, such as rye, and along with secoisolariciresinol, it has garnered a lot of interest due to its advantageous nutritional properties (Milder et al. 2005). Biosynthesis of (-)-matairesinol is given in Figure 5.30.

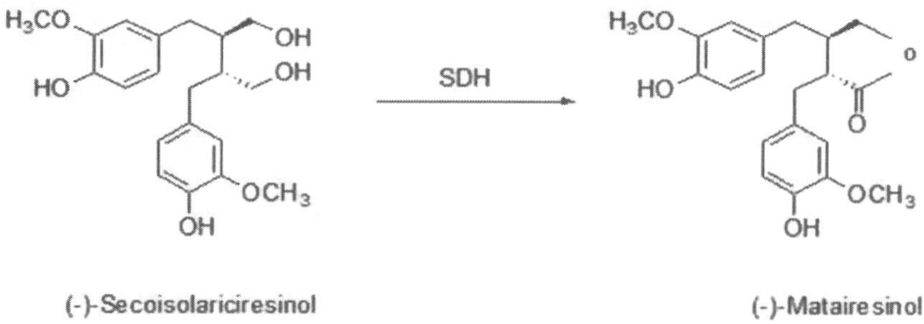

Figure 5.30 Biosynthesis of (-)-matairesinol.

5.2.4 Flavonoids and Stilbenes

A starter unit named cinnamoyl-CoA, formed by the attachment of coenzyme A (CoA) to the carboxyl group of cinnamic acid, gives flavonoids and stilbenes (Petersen et al. 2010). Cinnamate-CoA ligase, or 4-coumarate-CoA ligase, catalyzes this process, generating a polyketide which can be folded in two different ways depending on the enzyme involved (Figure 5.31). These give rise to aromatic rings by aldol- or Claisen-like reactions. A cinnamoyl-CoA unit is coupled with three malonyl-CoA units by the enzyme stilbene synthase and chalcone synthase to produce stilbenes, such as naringenin and resveratrol, respectively (Liu et al. 2021).

Resveratrol has gained popularity in recent years as an anti-inflammatory, and cancer-preventative, antioxidant, anti-platelet compound found in grapes and wine, as well as other foods (Roupe et al. 2006). The plant kingdom contains a wide variety of flavonoid derivatives, including flavones, flavanols, anthocyanidins, and catechins, which are all derived from chalcones (Karak 2019).

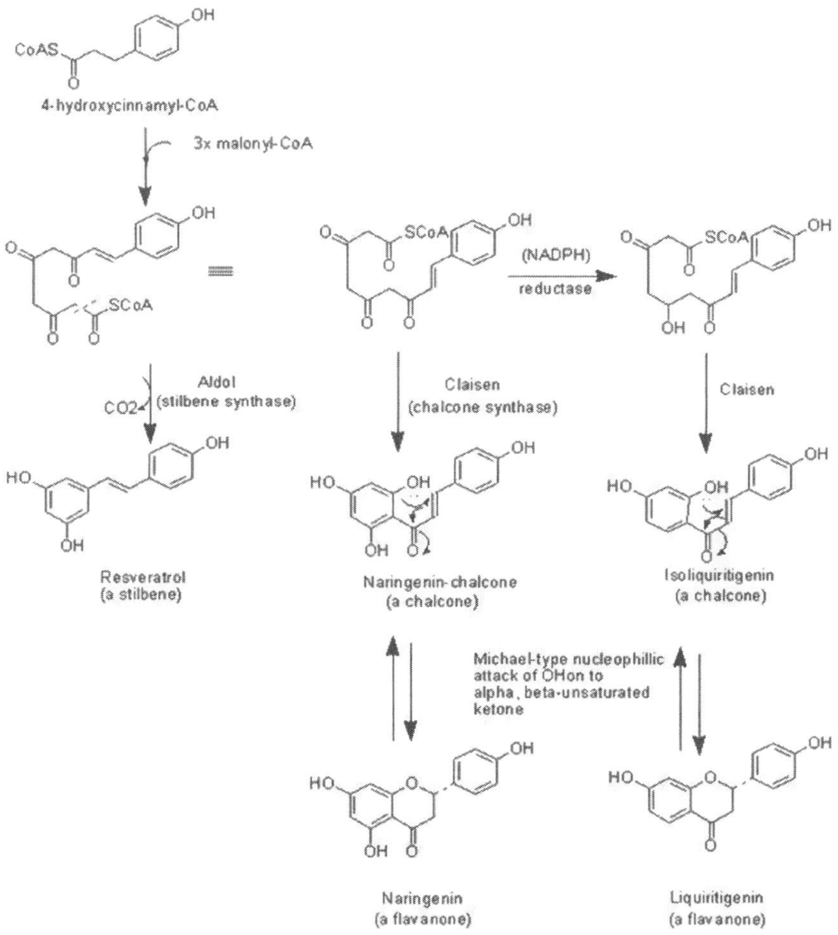

Figure 5.31 Biosynthesis of naringenin and liquiritigenin.

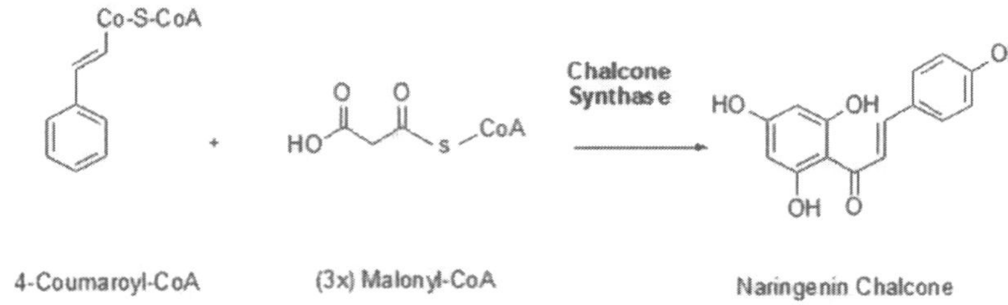

Figure 5.32 Biosynthesis of naringenin chalcone.

In addition to the modifications in the hydroxylation patterns in the two aromatic rings, which typically happen at the dihydroflavonol or flavanone stage, the variety of molecules can be further expanded by methylation, glycosylation, and dimethyl allylation (Dewick 2002).

5.2.4.1 Formation of Chalcone

Chalcone, which is essentially a starting point for the synthesis of many different products, is made by adding three molecules of malonyl-CoA to coumaroyl-CoA. Chalcone synthase catalyzes this reaction, giving naringenin chalcone (Abe et al. 2000). Biosynthesis of naringenin chalcone is given in Figure 5.32.

5.2.4.2 Formation of Dihydroflavonols

The isomerization of naringenin chalcone into naringenin yields flavanone, which is catalyzed by chalcone isomerase (Jez et al. 2000) (Figure 5.33).

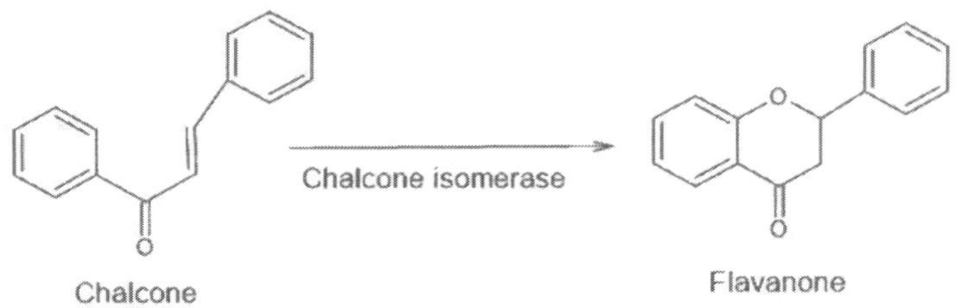

Figure 5.33 Biosynthesis of flavanone from chalcone.

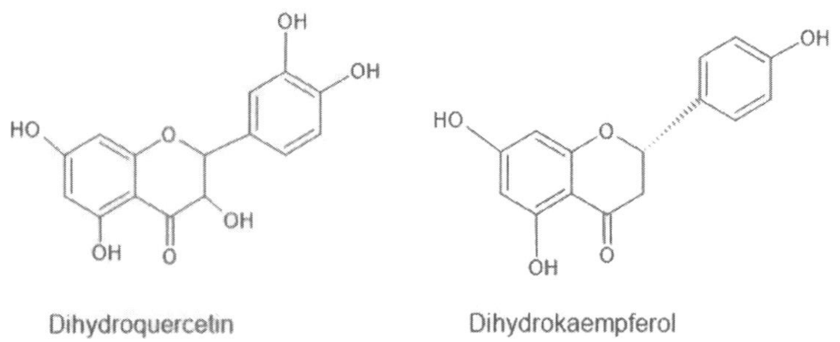

Figure 5.34 Chemical structures of dihydrokaempferol and dihydroquercetin.

Naringenin is then hydroxylated at position 3 by the enzyme flavanone 3-hydroxylase, resulting in the formation of dihydroflavonols such as dihydrokaempferol and dihydroquercetin (Figure 5.34; Amor et al. 2010).

5.2.5 Formation of Anthocyanidins

During the synthesis of anthocyanidins, first dihydroflavonols are reduced to leucoanthocyanidins, such as leucopelargonidin and leucocyanidin (Figure 5.35), in the presence of enzyme dihydroflavonol 4-reductase (DFR) (Halbwirth et al. 2006).

The next step involves the oxidation of leucoanthocyanidins to anthocyanidins such as pelargonidin, cyanidin, and delphinidin (Pervaiz et al. 2017), which is catalyzed by anthocyanidin synthase (ANS) (Figure 5.36). Anthocyanidins are further modified to produce colored compounds (Alappat and Alappat 2020).

5.2.6 Importance of Flavonoids in Plants

Plant colors are attributed to flavonoids, which include reds, blues, and violets from anthocyanidins, and yellows from chalcones and flavanols (Albert 2023). Even colorless compounds, such as flavones, absorb significantly in the ultraviolet light and are observable by insects, possibly facilitating flower pollination (Swami et al 2020).

5.2.7 Functions of Flavonoids

Flavonoids are particularly useful because they have a function of being antioxidants and protect against various types of cancers and cardiovascular diseases (Ciumărnean et al. 2020). The flavonoids like anthocyanidins, kaempferol, quercetin, catechins, and catechin gallate esters have also been shown to be potent antioxidants (Brewer 2011).

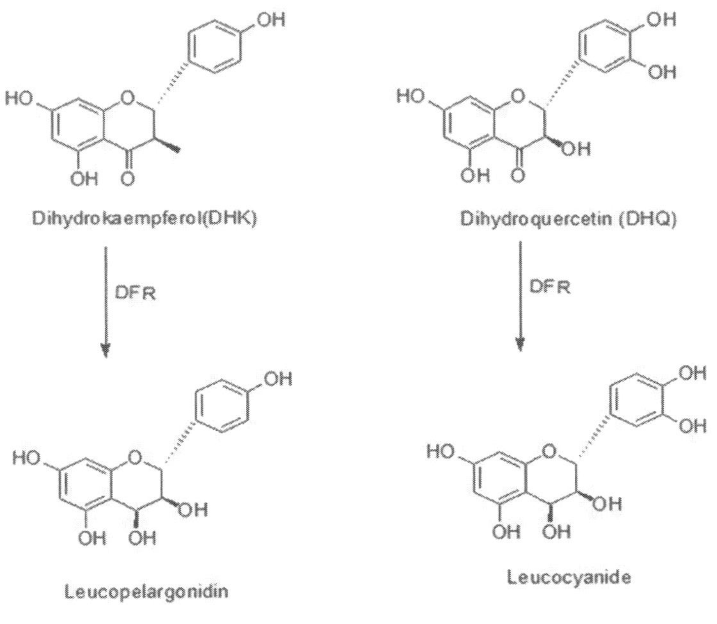

Figure 5.35 Biosynthesis of leucopelargonidin and leucocyanide.

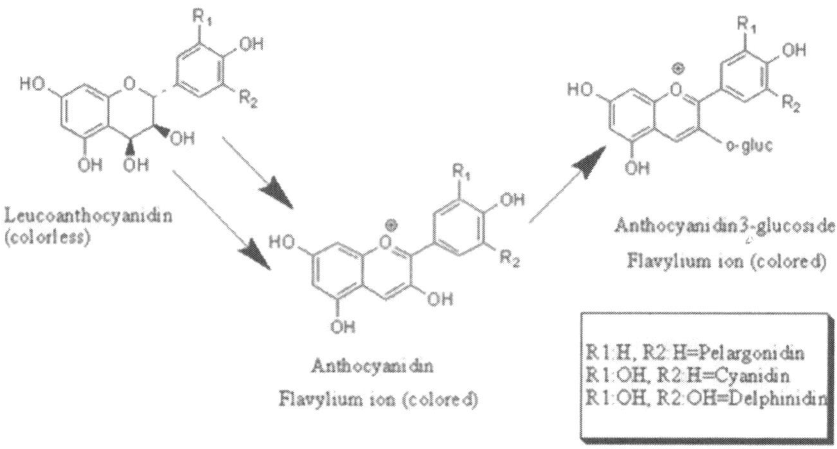

Figure 5.36 Biosynthesis of anthocyanidin-3-gluside flavylium ion.

REFERENCES

Abe I, Tomesch JC, Wattanasin S, Prestwich GD. Inhibitors of squalene biosynthesis and metabolism. *Natural Product Reports*. 1994;11:279–302.

Abe I, Morita H, Nomura A, Noguchi H. Substrate specificity of chalcone synthase: Enzymatic formation of unnatural polyketides from synthetic cinnamoyl-CoA analogues. *Journal of the American Chemical Society*. 2000;122(45):11242–3.

Alam M, Ahmed S, Elasbali AM, Adnan M, Alam S, Hassan MI, Pasupuleti VR. Therapeutic implications of caffeic acid in cancer and neurological diseases. *Frontiers in Oncology*. 2022;12:860508.

Alamgir A, Alamgir A. Medicinal, non-medicinal, biopesticides, color-and dye-yielding plants; secondary metabolites and drug principles; significance of medicinal plants; use of medicinal plants in the systems of traditional and complementary and alternative medicines (CAMs). *Therapeutic use of medicinal plants and their extracts: Volume 1: Pharmacognosy*. Springer; 2017:61–104.

Alappat B, Alappat J. Anthocyanin pigments: Beyond aesthetics. *Molecules*. 2020;25(23):5500.

Albert NW, Lafferty DJ, Moss SM, Davies KM. Flavonoids–flowers, fruit, forage and the future. *Journal of the Royal Society of New Zealand*. 2023;53(3):304–31.

Amor IL-B, Hehn A, Guedon E, Ghedira K, Engasser J-M, Chekir-Ghedrira L, Ghoul M. Biotransformation of naringenin to eriodictyol by *Saccharomyces cerevisiae* functionally expressing flavonoid 3′ hydroxylase. *Natural Product Communications*. 2010;5(12):1934578X1000501211.

Antonopoulou I, Sapountzaki E, Rova U, Christakopoulos P. Ferulic acid from plant biomass: A phytochemical with promising antiviral properties. *Frontiers in Nutrition*. 2022;8:777576.

Ashour M, Wink M, Gershenzon J. Biochemistry of terpenoids: Monoterpenes, sesquiterpenes and diterpenes. *Annual plant reviews volume 40: Biochemistry of plant secondary metabolism*. Blackwell Publishing; 2010:258–303.

Bergman ME, Phillips MA. Structural diversity and biosynthesis of plant derived p-menthane monoterpenes. *Phytochemistry Reviews*. 2021;20(2):433–59.

Bergman ME, Davis B, Phillips MA. Medically useful plant terpenoids: Biosynthesis, occurrence, and mechanism of action. *Molecules*. 2019;24(21):3961.

Bondinell W, Vnek J, Knowles P, Sprecher M, Sprinson D. On the mechanism of 5-enolpyruvylshikimate 3-phosphate synthetase. *Journal of Biological Chemistry*. 1971;246(20):6191–6.

Böttger A, Vothknecht U, Bolle C, Wolf A, Böttger A, Vothknecht U, et al. Terpenes and terpenoids. *Lessons on caffeine, cannabis & co: Plant-derived drugs and their interaction with human receptors*. 2018:153–70.

Brahmi F, Mokhtari O, Legssyer B, Hamdani I, Asehraou A, Hasnaoui I, et al. Chemical and biological characterization of essential oils extracted from citrus fruits peels. *Materials Today: Proceedings*. 2021;45:7794–9.

Brewer M. Natural antioxidants: Sources, compounds, mechanisms of action, and potential applications. *Comprehensive Reviews in Food Science and Food Safety*. 2011;10(4):221–47.

Chandra R, Yadav S, Kumar V. Microbial degradation of lignocellulosic waste and its metabolic products. *Environmental Waste Management*. 2015:249–98.

Chang HY, Cheng TH, Wang AHJ. Structure, catalysis, and inhibition mechanism of prenyltransferase. *IUBMB Life*. 2021;73(1):40–63.

Chen F, Tobimatsu Y, Havkin-Frenkel D, Dixon RA, Ralph J. A polymer of caffeyl alcohol in plant seeds. *Proceedings of the National Academy of Sciences*. 2012;109(5):1772–7.

Ciumărnean L, Milaciu MV, Runcan O, Vesa ŞC, Răchişan AL, Negrean V, et al. The effects of flavonoids in cardiovascular diseases. *Molecules*. 2020;25(18):4320.

Crozier A, Clifford MN, Ashihara H. Plant secondary metabolites: Occurrence, structure and role in the human diet. 2006.

Deka RK. Studies of the shikimate pathway enzyme, 3-dehydroquinase: University of Glasgow (United Kingdom); 1993.

Dewick PM. Medicinal natural products: A biosynthetic approach. 2002.

Dhaniaputri R, Suwono H, Amin M, Lukiati B, editors. Introduction to plant metabolism, secondary metabolites biosynthetic pathway, and in-silico molecular docking for determination of plant medicinal compounds: An overview. 7th International Conference on Biological Science (ICBS 2021); 2022: Atlantis Press.

Doy CH, Rivera Jr A, Srinivasan P. Evidence for the enzymatic synthesis of N-(5′-phosphoribosyl) anthranilic acid, a new intermediate in tryptophan biosynthesis. *Biochemical and Biophysical Research Communications*. 1961;4(1):83–8.

Fleuriet A, Macheix J-J. Phenolic acids in fruits and vegetables. *Oxidative Stress and Disease*. 2003;9:1–42.

Fujisawa M, Harada H, Kenmoku H, Mizutani S, Misawa N. Cloning and characterization of a novel gene that encodes (S)-β-bisabolene synthase from ginger, Zingiber officinale. *Planta*. 2010;232(1):121–30.

Gall DL, Ralph J, Donohue TJ, Noguera DR. Biochemical transformation of lignin for deriving valued commodities from lignocellulose. *Current Opinion in Biotechnology*. 2017;45:120–6.

Gamborg O, Simpson F. Preparation of prephenic acid and its conversion to phenylalanine and tyrosine by plant enzymes. *Canadian Journal of Biochemistry*. 1964;42(5):583–91.

Garg A, Sharma R, Dey P, Kundu A, Kim HS, Bhakta T, Kumar A. Analysis of triterpenes and triterpenoids. Recent advances in natural products analysis. 2020:393–426.

Gershenzon J, Croteau RB. Terpenoid biosynthesis: The basic pathway and formation of monoterpenes, sesquiterpenes, and diterpenes. Lipid metabolism in plants. 2018:339–88.

Gershenzon J, Kreis W. Biochemistry of terpenoids: Monoterpenes, sesquiterpenes, diterpenes, sterols, cardiac glycosides and steroid saponins. *Biochemistry of Plant Secondary Metabolism*. 1999;2:222–99.

Gray J, Picton S, Shabbeer J, Schuch W, Grierson D. Molecular biology of fruit ripening and its manipulation with antisense genes. *Plant Molecular Biology*. 1992:69–87.

Halbwirth H, Kahl S, Jäger W, Reznicek G, Forkmann G, Stich K. Synthesis of (14C)-labeled 5-deoxyflavonoids and their application in the study of dihydroflavonol/leucoanthocyanidin interconversion by dihydroflavonol 4-reductase. *Plant Science*. 2006;170(3):587–95.

Howe DL, Sundaram AK, Wu J, Gatti DL, Woodard RW. Mechanistic insight into 3-deoxy-D-manno-octulosonate-8-phosphate synthase and 3-deoxy-D-arabino-heptulosonate-7-phosphate synthase utilizing phosphorylated monosaccharide analogues. *Biochemistry*. 2003;42(17):4843–54.

Hyatt DC, Youn B, Zhao Y, Santhamma B, Coates RM, Croteau RB, Kang C. Structure of limonene synthase, a simple model for terpenoid cyclase catalysis. *Proceedings of the National Academy of Sciences*. 2007;104(13):5360–5.

Jaganath IB, Crozier A. Dietary flavonoids and phenolic compounds. *Plant Phenolics and Human Health: Biochemistry, Nutrition, and Pharmacology*. 2010;1:1–50.

Jez JM, Bowman ME, Dixon RA, Noel JP. Structure and mechanism of the evolutionarily unique plant enzyme chalcone isomerase. *Nature Structural Biology*. 2000;7(9):786–91.

Jin ML, Lee WM, Kim OT. Two cycloartenol synthases for phytosterol biosynthesis in *Polygala tenuifolia* Willd. *International Journal of Molecular Sciences*. 2017;18(11):2426.

Kang A, Lee TS. Secondary metabolism for isoprenoid-based biofuels. Biotechnology for Biofuel Production and Optimization. 2016;35–71.

Kang M-H, Naito M, Sakai K, Uchida K, Osawa T. Mode of action of sesame lignans in protecting low-density lipoprotein against oxidative damage in vitro. *Life Sciences*. 1999;66(2):161–71.

Karak P. Biological activities of flavonoids: An overview. *International Journal of Pharmaceutical Sciences and Research*. 2019;10(4):1567–74.

Kong J-Q. Phenylalanine ammonia-lyase, a key component used for phenylpropanoids production by metabolic engineering. *RSC Advances*. 2015;5(77):62587–603.

Krell T, Coggins JR, Lapthorn AJ. The three-dimensional structure of shikimate kinase. *Journal of Molecular Biology*. 1998;278(5):983–97.

Kumar P, Maharjan A, Jun HB, Kim BS. Bioconversion of lignin and its derivatives into polyhydroxyalkanoates: Challenges and opportunities. *Biotechnology and Applied Biochemistry*. 2019;66(2):153–62.

Kumari S, Priya P, Misra G, Yadav G. Structural and biochemical perspectives in plant isoprenoid biosynthesis. *Phytochemistry Reviews*. 2013;12:255–91.

Li M, Nian R, Xian M, Zhang H. Metabolic engineering for the production of isoprene and isopentenol by *Escherichia coli*. *Applied Microbiology and Biotechnology*. 2018;102:7725–38.

Liao P, Wang H, Hemmerlin A, Nagegowda DA, Bach TJ, Wang M, Chye M-L. Past achievements, current status and future perspectives of studies on 3-hydroxy-3-methylglutaryl-CoA synthase (HMGS) in the mevalonate (MVA) pathway. *Plant Cell Reports*. 2014;33:1005–22.

Lingens F. The biosynthesis of aromatic amino acids and its regulation. *Angewandte Chemie International Edition in English*. 1968;7(5):350–60.

Liu W, Feng Y, Yu S, Fan Z, Li X, Li J, Yin H. The flavonoid biosynthesis network in plants. *International Journal of Molecular Sciences*. 2021;22(23):12824.

Loomis WD, Croteau R. Biochemistry of terpenoids. Lipids: Structure and function. 1980:363–418.

Maeda H, Dudareva N. The shikimate pathway and aromatic amino acid biosynthesis in plants. *Annual Review of Plant Biology*. 2012;63:73–105.

Marchiosi R, Ferro AP, Ramos AV, Baldoqui DC, Constantin RP, Constantin RP, et al. *Calophyllum brasiliense* Cambess: An alternative and promising source of shikimic acid. *Sustainable Chemistry and Pharmacy*. 2019;14:100188.

Markulin L, Corbin C, Renouard S, Drouet S, Gutierrez L, Mateljak I, et al. Pinoresinol–lariciresinol reductases, key to the lignan synthesis in plants. *Planta*. 2019;249:1695–714.

Martins N, Ferreira IC. Wastes and by-products: Upcoming sources of carotenoids for biotechnological purposes and health-related applications. *Trends in Food Science & Technology*. 2017;62:33–48.

Mayaka RK. Chemical characterisation and antimicrobial activity of compounds from some selected medicinal Kenyan *Ganoderma* and *Trametes* species: Egerton University; 2020.

McBee DP, Hulsey ZN, Hedges MR, Baccile JA. Biological demands and toxicity of isoprenoid precursors in *Bacillus* subtilis through cell-permeant analogs of isopentenyl pyrophosphate and dimethylallyl pyrophosphate. *ChemBioChem*. 2024;25(12):e202400064.

Milder IE, Arts IC, van de Putte B, Venema DP, Hollman PC. Lignan contents of Dutch plant foods: A database including lariciresinol, pinoresinol, secoisolariciresinol and matairesinol. *British Journal of Nutrition*. 2005;93(3):393–402.

Miles EW. Structural basis for catalysis by tryptophan synthase. *Advances in Enzymology and Related Areas of Molecular Biology*. 1991;64:93–172.

Mir R, Jallu S, Singh T. The shikimate pathway: Review of amino acid sequence, function and three-dimensional structures of the enzymes. *Critical Reviews in Microbiology*. 2015;41(2):172–89.

Mohammadi-Cheraghabadi M, Hazrati S. Terpenoids, steroids, and phenolic compounds of medicinal plants. Phytochemicals in medicinal plants: Biodiversity, bioactivity and drug discovery. 2023:105.

Murata J, Ono E, Yoroizuka S, Toyonaga H, Shiraishi A, Mori S, et al. Oxidative rearrangement of (+)-sesamin by CYP92B14 co-generates twin dietary lignans in sesame. *Nature Communications*. 2017;8(1):1–10.

Nagel R, Schmidt A, Peters RJ. Isoprenyl diphosphate synthases: The chain length determining step in terpene biosynthesis. *Planta*. 2019;249:9–20.

Nair IM, Kochupurackal J. Squalene hopene cyclases and oxido squalene cyclases: Potential targets for regulating cyclisation reactions. *Biotechnology Letters*. 2023;45(5):573–88.

Niemeyer HB, Honig DM, Kulling SE, Metzler M. Studies on the metabolism of the plant lignans secoisolariciresinol and matairesinol. *Journal of Agricultural and Food Chemistry*. 2003;51(21):6317–25.

Obiol-Pardo C, Rubio-Martinez J, Imperial S. The methylerythritol phosphate (MEP) pathway for isoprenoid biosynthesis as a target for the development of new drugs against tuberculosis. *Current Medicinal Chemistry*. 2011;18(9):1325–1338.

Okamura E, Tomita T, Sawa R, Nishiyama M, Kuzuyama T. Unprecedented acetoacetyl-coenzyme A synthesizing enzyme of the thiolase superfamily involved in the mevalonate pathway. *Proceedings of the National Academy of Sciences*. 2010;107(25):11265–70.

Otles S, Nakilcioglu E. Determination of phenolic compounds (gallic, caffeic, ferulic, and p-coumaric acids). *Flow injection analysis of food additives*. CRC Press; 2015;1:393.

Pandi A, Kalappan VM. Pharmacological and therapeutic applications of sinapic acid: An updated review. *Molecular Biology Reports*. 2021;48(4):3733–45.

Pervaiz T, Songtao J, Faghihi F, Haider MS, Fang J. Naturally occurring anthocyanin, structure, functions and biosynthetic pathway in fruit plants. *Journal of Plant Biochemistry & Physiology*. 2017;5(2):1–9.

Petersen M, Hans J, Matern U. Biosynthesis of phenylpropanoids and related compounds. *Annual plant reviews volume 40: Biochemistry of plant secondary metabolism*. Wiley; 2010:182–257.

Poklar Ulrih N, Prislan I, Cigić B. Coumaric and cinnamic acids in food. *Handbook of dietary phytochemicals*. Springer; 2021:1105–43.

Porras G, Chassagne F, Lyles JT, Marquez L, Dettweiler M, Salam AM, et al. Ethnobotany and the role of plant natural products in antibiotic drug discovery. *Chemical Reviews*. 2020;121(6):3495–560.

Pyun H-J. Stereochemistry, isotope effects, and inhibition of terpene biosynthesis: I. Stereospecificity and mechanism of monoterpene cyclases. II. Inhibition by heteroatom analogs of carbocation intermediates: University of Illinois at Urbana-Champaign; 1992.

Rasouli H, Hosseini-Ghazvini SM-B, Adibi H, Khodarahmi R. Differential α-amylase/α-glucosidase inhibitory activities of plant-derived phenolic compounds: A virtual screening perspective for the treatment of obesity and diabetes. *Food & Function*. 2017;8(5):1942–54.

Rehman R, Hanif MA, Mushtaq Z, Al-Sadi AM. Biosynthesis of essential oils in aromatic plants: A review. *Food Reviews International*. 2016;32(2):117–60.

Rohdich F, Eisenreich W, Wungsintaweekul J, Hecht S, Schuhr CA, Bacher A. Biosynthesis of terpenoids: 2C-Methyl-D-erythritol 2, 4-cyclodiphosphate synthase (IspF) from *Plasmodium falciparum*. *European Journal of Biochemistry*. 2001;268(11):3190–7.

Roupe KA, Remsberg CM, Yáñez JA, Davies NM. Pharmacometrics of stilbenes: Seguing towards the clinic. *Current Clinical Pharmacology*. 2006;1(1):81–101.

Ruwizhi N, Aderibigbe BA. Cinnamic acid derivatives and their biological efficacy. *International Journal of Molecular Sciences*. 2020;21(16):5712.

Sann CL, Gower MA, Abell AD. Inhibitors of types I and II dehydroquinase. *Mini Reviews in Medicinal Chemistry*. 2004;4(7):747–56.

Santos-Sánchez NF, Salas-Coronado R, Hernández-Carlos B, Villanueva-Cañongo C. Shikimic acid pathway in biosynthesis of phenolic compounds. *Plant Physiological Aspects of Phenolic Compounds*. 2019;1:1–15.

Schenck CA, Maeda HA. Tyrosine biosynthesis, metabolism, and catabolism in plants. *Phytochemistry*. 2018;149:82–102.

Sheahan JJ. Sinapate esters provide greater UV-B attenuation than flavonoids in *Arabidopsis thaliana* (Brassicaceae). *American Journal of Botany*. 1996;83(6):679–86.

Shen L, Liu M, He Y, Al Anbari WH, Li H, Lin S, et al. Novel antimicrobial compounds as ophiobolin-type sesterterpenes and pimarane-type diterpene from *Bipolaris* species TJ403-B1. *Frontiers in Microbiology*. 2020;11:856.

Singhania RR, Patel AK, Raj T, Chen C-W, Ponnusamy VK, Tahir N, et al. Lignin valorisation via enzymes: A sustainable approach. *Fuel*. 2022;311:122608.

Smit A, Mushegian A. Biosynthesis of isoprenoids via mevalonate in Archaea: The lost pathway. *Genome Research*. 2000;10(10):1468–84.

Sowa-Borowiec P, Jarecki W, Dżugan M. The effect of sowing density and different harvesting stages on yield and some forage quality characters of the white sweet clover (*Melilotus albus*). *Agriculture*. 2022;12(5):575.

Soytong M, Guevarra P, Mateo J, Galvez H. Evaluation of tomatoes fruits flesh colour, beta-carotene and lycopene content. *International Journal of Agricultural Technology*. 2021; 17(2):727 736.

Stanley Fernandez SM, Kellogg BA, Poulter CD. Farnesyl diphosphate synthase. Altering the catalytic site to select for geranyl diphosphate activity. *Biochemistry*. 2000;39(50):15316–21.

Svarcova M, Kratky M, Vinsova J. Investigation of potential inhibitors of chorismate-utilizing enzymes. *Current Medicinal Chemistry*. 2015;22(11):1383–99.

Swami SB, Ghgare SN, Swami SS, Shinde KJ, Kalse SB, Pardeshi IL. Natural pigments from plant sources: A review. *The Pharma Innovation Journal*. 2020;9(10):566.

Swiezewska E, Danikiewicz W. Polyisoprenoids: Structure, biosynthesis and function. *Progress in Lipid Research*. 2005;44(4):235–58.

Tholl D. Biosynthesis and biological functions of terpenoids in plants. *Biotechnology of Isoprenoids*. 2015:63–106.

Tian S, Wang D, Yang L, Zhang Z, Liu Y. A systematic review of 1-Deoxy-D-xylulose-5-phosphate synthase in terpenoid biosynthesis in plants. *Plant Growth Regulation*. 2022;96(2):221–35.

Tzin V, Galili G. The biosynthetic pathways for shikimate and aromatic amino acids in Arabidopsis thaliana. *The Arabidopsis Book/American Society of Plant Biologists*. 2010;8:e0132.

Velu G, Palanichamy V, Rajan AP. Phytochemical and pharmacological importance of plant secondary metabolites in modern medicine. *Bioorganic phase in natural food: An overview*. Springer; 2018:135–56.

Vickers CE, Sabri S. Isoprene. *Biotechnology of isoprenoids*. Springer; 2015:289–317.

Weaver LM, Herrmann KM. Dynamics of the shikimate pathway in plants. *Trends in Plant Science*. 1997;2(9):346–51.

West CA, Upper CD. [17] Enzymatic synthesis of (±)-kaurene and related diterpenes. *Methods in Enzymology*. 1969;15:481–90.

Wolosik K, Markowska A. *Amaranthus cruentus* taxonomy, botanical description, and review of its seed chemical composition. *Natural Product Communications*. 2019;14(5):1934578X19844141.

Wu L, Wang X, Xu W, Farzaneh F, Xu R. The structure and pharmacological functions of coumarins and their derivatives. *Current Medicinal Chemistry*. 2009;16(32):4236–60.

Yoo H, Widhalm JR, Qian Y, Maeda H, Cooper BR, Jannasch AS, et al. An alternative pathway contributes to phenylalanine biosynthesis in plants via a cytosolic tyrosine: Phenylpyruvate aminotransferase. *Nature Communications*. 2013;4(1):2833.

Zeng L, Wang X, Tan H, Liao Y, Xu P, Kang M, et al. Alternative pathway to the formation of trans-cinnamic acid derived from L-phenylalanine in tea (*Camellia sinensis*) plants and other plants. *Journal of Agricultural and Food Chemistry*. 2020;68(11):3415–24.

Zhang Q, Tiefenbacher K. Terpene cyclization catalysed inside a self-assembled cavity. *Nature Chemistry*. 2015;7(3):197–202.

Zidorn C. *Plant chemosystematics. Chromatographic Science Series*. Springer; 2008;99:77.

Zografos AL, Anagnostaki EE. From acetate to mevalonate and deoxyxylulose phosphate biosynthetic pathways: An introduction to terpenoids. *From biosynthesis to total synthesis: Strategies and tactics for natural products*. Wiley; 2016:187–95.

Zulfiqar S, Sharif S, Saeed M, Tahir A. Role of carotenoids in photosynthesis. *Carotenoids: Structure and function in the human body*. 2021:147–87.

Zwane EN. Production, characterisation and application of a recombinant ferulic acid esterase from *Aspergillus tubingensis*: Stellenbosch: Stellenbosch University; 2014.

6 Bioactive Secondary Metabolites and Their Role in Health Management

Muhammad Fakhar-e-Alam, Muhammad Iqbal, Noor ul Amin Mohsin, Syeda Abida Ejaz, Muhammad Yasir Ali, Malik Saadullah, and Ghulam Abbas*

6.1 INTRODUCTION TO NATURAL PRODUCTS

Natural products are organic chemical substances produced within living organisms (e.g., microorganisms, animals, and plants) via primary and secondary metabolic pathways. They are pharmacologically active and, therefore, used as active components not only in traditional but in modern medicine as well. Naturally active chemical compounds are referred to as secondary metabolites and are believed to provide various health benefits to humans such as prevention of diseases. Over time, about 4,000 secondary metabolites have been reported and categorized into different classes based on their chemical structures, functions or actions, and physical characteristics [1].

6.2 IMPORTANCE OF SECONDARY METABOLITES

Organic compounds that are not directly involved in the normal growth, development, or reproduction of an organism are usually referred to as secondary metabolites [2]. These compounds are produced in a phase subsequent to growth, so these have no function in growth but they do possess some survival function. They have unusual chemical structures. In plants, they are synthesized through secondary metabolism and possess a number of biological activities, hence these are used as drugs. However, these compounds are also used as fine chemicals and in preparation of cosmetics. Secondary metabolites can be classified into three major classes of molecules namely, alkaloids, phenolics, and terpenoids [2].

6.2.1 Alkaloids

Alkaloids are defined as bitter-tasting, pharmacologically active, nitrogen-containing basic compounds of plant origin and include caffeine, morphine, cocaine, and nicotine [3]. These compounds have one or more nitrogen atoms as a part of heterocyclic system. Amino acids (i.e., tyrosine, aspartic acid, tryptophan, and lysine) are precursors of alkaloids. If consumed in high doses alkaloids are toxic for humans [4].

Upon studies various biological activities of alkaloids were screened such as anti-inflammatory, anticancer, antiviral, antihypertensive, antimicrobial, antidepressant and diuretic, anti-diarrheal activity, and anticholinergic [4–7].

6.2.2 Phenolics

Phenolics are produced though shikimic acid and malonic acid pathways in plants and contain "OH" group in their structure. Phenolics show a large range of structures and include flavonoids, furanocoumarins, tannins, phytoalexins, lignin, and anthocyanins [8]. Phenolics are also known to be responsible for organoleptic characteristics of plant-derived foods and beverages such as color, smell, and taste [9]. It has been studied that phenolics interfere with digestion and tend to slow growth. They are also believed to block enzyme activity and cell division [2].

* Corresponding author

DOI: 10.1201/9781032707273-6

6.2.3 Terpenoids

Terpenoids, being the largest class of secondary metabolites, contain more than 22,000 compounds. Isoprene (C_5H_8) is the simplest terpene and therefore, on bases of number of isoprene units present in structure, terpenoids are classified into monoterpenoids, diterpenoids, triterpenoids, and sesquiterpenoids [4]. Monoterpenoids and sesquiterpenoids are responsible for fragrance of plants, therefore, are a component of essential oils.

6.2.4 Glycosides

Glycosides are made of two units: gylcone and agylcone. The gylcone part (one or more sugars) is linked to aglycone part (a non-carbohydrate) which are joined together by a glycosidic bond. Chemical nature of the aglycone is responsible for biological activity of a glycoside. At least ten types of toxic glycosides occur in forage species [8]. Cyanogenic glycosides and glucosinolates release toxic cyanide compounds (which block aerobic cellular respiration) as defense mechanism in plants [8].

Cardiac glycosides are characterized for their positive inotropic effect and therefore, being widely used clinically for cardiovascular disorders such as for treating atrial arrhythmia and heart failure. Cardiac glycoside have been recently found biologically active against neurodegenerative diseases also. Cardiac glycosides have potential for cancer therapies due to increased susceptibility for cancer cells [10].

6.2.5 Flavonoids

A 15-carbon phenolic compound is called flavonoid. Flavonoids occur as aglycones, or as glycosides, where they are in conjugation with a sugar D-glucose or as methylated derivatives [9]. Flavonoids have been reported to possess antidiabetic, antimicrobial, anti-inflammatory, antioxidant, antigout, anti-atherosclerotic, vasorelaxant, antithrombogenic, antiulcer, anticancer, estrogenic, spasmolytic, and hepatoprotective activities [9, 11].

6.2.6 Tannins

A water soluble, bitter, astringent, and polyphenolic compound of higher-molecular weight (500–3,000) having ability to bind and inactivate various enzymes and proteins is known as a tannin. Tannins bind with many secondary metabolites and metal ions to form complexes [12].

When a tannin binds to a protein, it forms either a soluble or insoluble tannin–protein complex. They exist in the form of flavonoids polymers. Based on chemical structure and properties, tannins are classified in two groups: hydrolysable (HT) and condensed tannins (CT). CT contains a flavonoid unit (flavan-3-ol) linked by a carbon–carbon bond whereas, HT are comparatively present in less quantity in plant than CT and believed to be harmful to kidney, liver, and gastrointestinal tract. HT contains glucose molecules esterified with phenolic compounds such as gallic and ellagic acid [13].

Tannins have been reported to have many biological effects such as antimicrobial effect; they can modulate immune responses, decrease serum lipid levels, and reduce blood pressure as well as they can accelerate blood clotting [14].

6.2.7 Saponins

Amphiphilic compounds of high-molecular weight, having triterpenoid and/or steroid aglycone and sugars are known as saponins [14]. They possess glycosidic bonds and have soap-like properties. In contact with water, upon vigorous shaking, they produce frothing. Saponins are reported to be toxic as they can form complexes with cholesterol present in red blood cells, thereby, causing hemolysis. Saponins have been reported to show antidiabetic, antifungal, antitumor, antioxidative, anti-inflammatory, antiparasitic, neuroprotective, and expectorant properties [14, 15].

6.2.8 Steroids and Sterols

Plant sterols are basically steroid alcohols that are a component of plant cell membranes as they provide structural rigidity, fluidity, and integrity. Sitosterol is the most important sterol in plants, however sitossterols' 22-dehydro analogue, stigmasterol, brassica sterol, avena-sterols, and campesterol are also abundant in plants [16]. Steroids and sterols are structurally similar. However, sterols being a subgroup of steroids have a hydroxyl group at the third position of the A-ring. Sterols have been reported to have plasma cholesterol-lowering (anti-hypercholesterolemic) potential [17]. Hence, these can be used for prevention of cardiovascular diseases (CVD).

6.3 NATURAL PRODUCTS AND THEIR HEALTH MANAGEMENT ROLE

Natural products have been used by mankind for ages, and ancient folk medicine has taken advantage of their pharmacological activities processed by the secondary metabolites present in these natural products for treatment of various ailments such as fever, pain, inflammation, and so on. Subsequently, natural product active secondary metabolites are a major source of lead molecules in drug development. Current concerns about the harmful effects of synthetic chemicals used in medicine have driven research efforts globally on natural products and derived structures that might prove helpful in the development of alternative therapies for majority of prevalent ailments linked to aging process. With the potential to reduce consumption of certain harmful synthetic antioxidants, phenolic derivatives, flavonoids, and isoflavones have been shown to play preventative roles against degenerative illnesses caused by free radical activity. Little structural modifications of secondary metabolites lead to alteration of their biological activity.

6.3.1 Skin Health and Wellness

6.3.1.1 Soy

Essential amino acids, isoflavones, saponins, calcium, potassium, phytosterols, iron, and proteases soybean trypsin inhibitor (STI) and Bowman–Birk inhibitor (BBI) are only few of the most potent substances found in soy. Soy has several health benefits, making its diverse constituents suitable complements to skin care products. The phytoestrogens genistein and daidzein (Figure 6.1) are the most powerful isoflavones. A powerful antioxidant, genistein inhibits lipid peroxidation as well as chemical- and ultraviolet B (UVB)-induced carcinogenesis [18]. It has been demonstrated that genistein dramatically reduces the production of reactive oxygen species (ROS) brought on by chemical carcinogens, oxidative DNA damage, and proto-oncogene expression, as well as the start and progression of skin carcinogenesis. In human clinical studies, the proteases STI and BBI tend to contribute to skin whitening and reduce growth of unwanted body hair. In addition to its anti-hyperpigmentation properties, BBI and soy milk inhibit UV-induced pigmentation [19].

6.3.1.2 Green and Black Tea

Diverse polyphenolic components present in tea have demonstrated their influence on biochemical networks involved in inflammatory reactions, tumor-promoting responses, and cell proliferation. Hence, green tea has been found

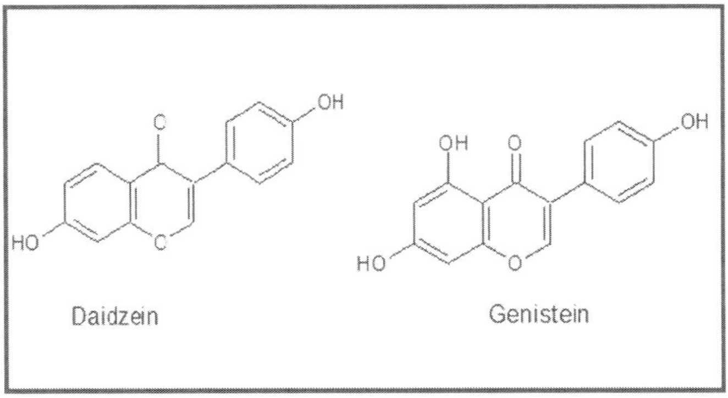

Figure 6.1 Chemical structure of daidzein and genistein.

to have anti-inflammatory and antioxidant properties. Green tea may also have antiaging properties by reducing inflammation and scavenging free radicals, which are thought to play a substantial part in the aging process. Epigallocatechin-3-gallate (EGCG) (Figure 6.2), the primary active component of green tea, has been discovered to be an effective anti-inflammatory, antioxidant, and sunscreen. Green tea extracts have photoprotective effects on human skin, decrease the number of sunburn cells, shield epidermal Langerhan cells from UV ray damage, and lessen DNA damage caused by UV radiation [20].

Compared to green tea, black tea has far fewer catechins, but more other flavonoids including quercetin, kaempferol, and theaflavin (Figure 6.2). In both human and mouse skin, it has been proven that applying black tea extracts

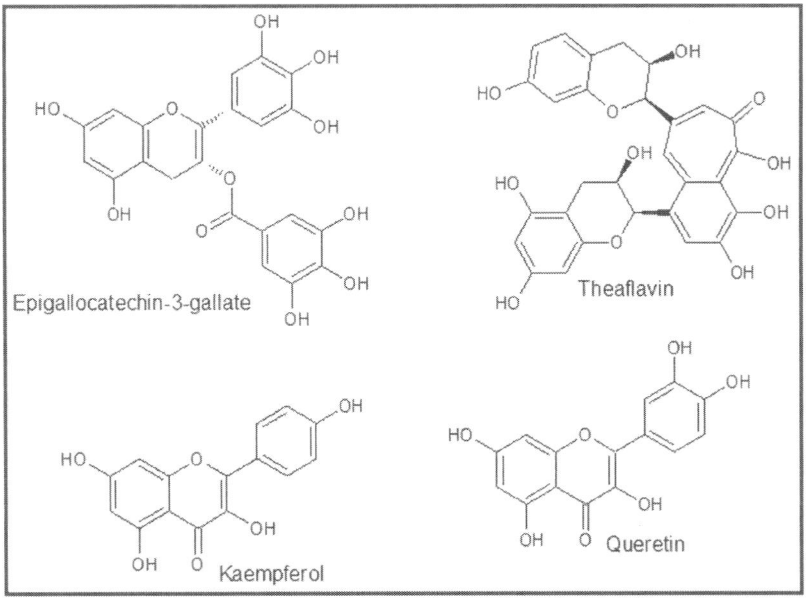

Figure 6.2 Chemical structures of epigallocatechin-3-gallate, theaflavin, kaempferol, and quercetin.

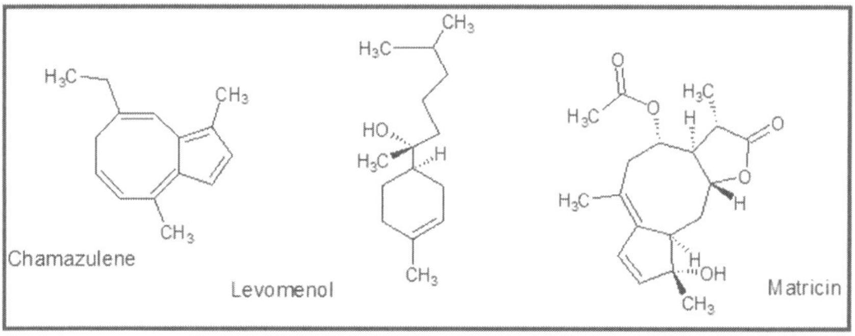

Figure 6.3 Chemical structure of chamazulene, levomenol, and matricin.

before and after exposure to UV radiation reduces the appearance of epidermal photodamage, carcinogenesis, and inflammation [21].

6.3.1.3 Chamomile

Matricaria recutita, sometimes known as German chamomile, is used traditionally as a herbal remedy for a variety of skin issues. Chamomile received approval by the German Commission E for the treatment of inflammatory mucocutaneous illnesses as well as wound and burn management. Presence of various active constitutes, such as flavonoids (apigenin, quercetin (Figure 6.2), luteolin and rutin), hydroxycoumarins, mono- and oligosaccharides, terpenoids (e.g., levomenol, bisoprolol, chamazulene, and matricine), and mucilages, provides many therapeutic effects such as antibacterial, anti-inflammatory, antioxidant, antiallergic, and analgesic [22]. Chamazulene improves wound healing and has anti-inflammatory properties. Levomenol is an anti-inflammatory and all-natural moisturizing ingredient that has been shown to improve skin texture and elasticity, it treats pruritus and reduces photodamage (Figure 6.3). Chamomile is said to contain certain antioxidant qualities in addition to claims of anti-inflammatory benefits [23].

6.3.1.4 Coffee Berry

Coffee berry, which is obtained from the fruit of the *Coffea arabica* plant, is recognized as one of the greatest sources of antioxidants and is popular for its ability to revitalize the skin. Strong polyphenols found in coffee berries include ferulic acid, quinic acid, chlorogenic acid (Figure 6.4), and condensed proanthocyanidins [24, 25].

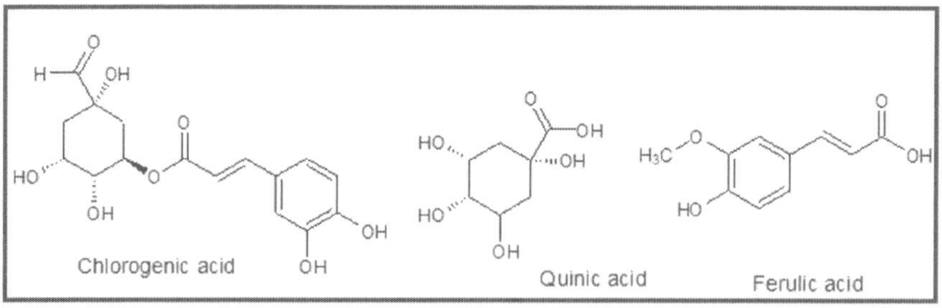

Figure 6.4 Chlorogenic acid, quinic acid, and ferulic acid.

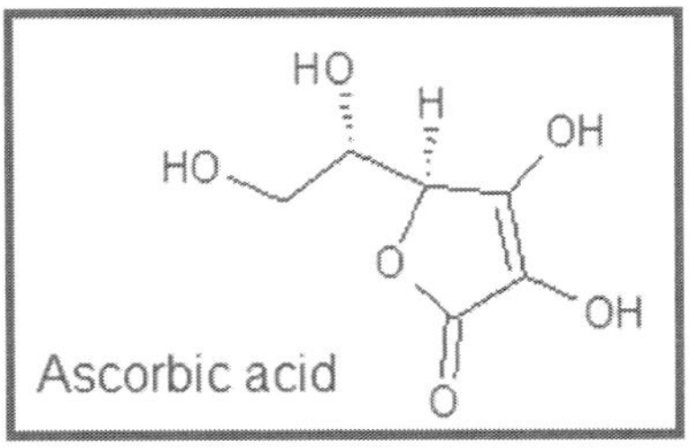

Figure 6.5 Ascorbic acid.

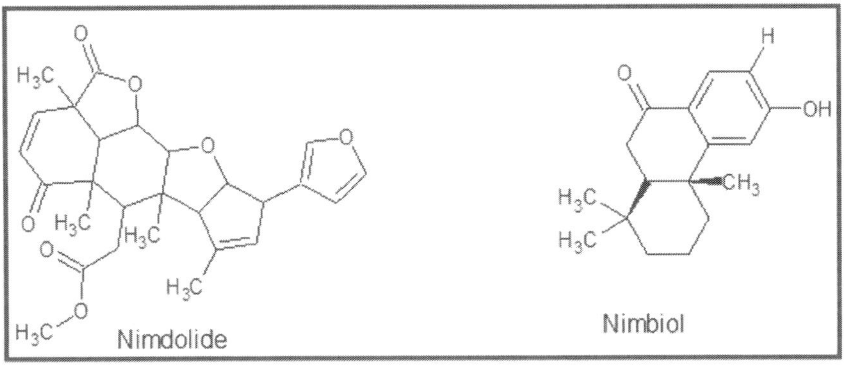

Figure 6.6 Chemical structure of nimbolide and nimbiol.

6.3.1.5 Peels of Oranges

Peel of *Citrus sinensis* provides a variety of nutrients, including calcium, potassium, magnesium, and ascorbic acid (Figure 6.5) [26, 27]. It guards against oxidative stress, skin dehydration, and free radical damage. In addition, it prevents wrinkles, aging, acne, and blemishes and has an immediate glow quality [27].

6.3.1.6 **Azadirachta indica** *(Neem)*

Azadirachta indica or neem is anti-inflammatory, antibacterial, and is very helpful for skin that is greasy and prone to acne. The antimicrobial, anti-inflammatory, and antioxidant properties of its several chemical components have an antiacne effect [28]. Ascorbic acid, n-hexacosanol, amino acids, 6-desacetylnimbinene, nimbandiol, nimbolide (Figure 6.6), 7-desacetyl-7-benzoylazadiradione, 7-desacetyl-7-benzoylgedunin, 17-hydroxyazadiradione, and nimbiol are among the active compounds found in neem leaves. Polyphenolic flavonoids, quercetin and sitosterol, present in neem are known for antibacterial and antifungal activity [29].

6.3.1.7 **Aloe barbadensis** *(Aloe Vera)*

Aloe barbadensis, that is, aloe vera is an excellent skin moisturizer. Aloe vera moisturizes, refreshes, and maintains the skin's youthful appearance and treats irritant contact dermatitis, cuts, and burns [30]. As a result of its antimicrobial properties, aloe vera is effective for treating acne and pimples. Several active constituents present in aloe vera are aloe-emodin, aloin, aloeresin A, aloveroside A, and barbaloin (Figure 6.7) [31].

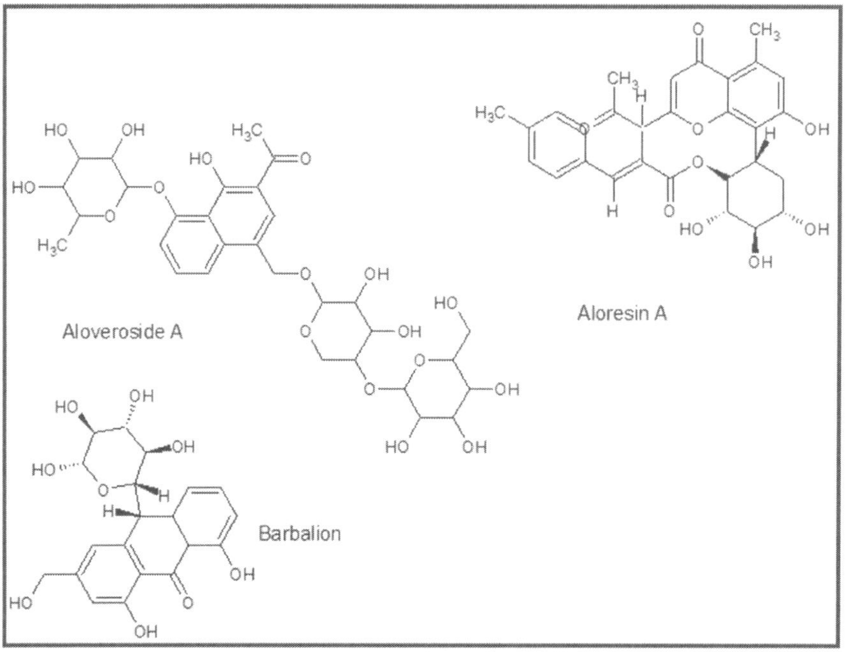

Figure 6.7 Chemical structures of aloeveroside A, aloesresin A, and barbaloin.

6.3.1.8 Fuller's Earth

Fuller's earth or *multani mitti* benefits skin in a variety of ways. It aids pore size reduction, eliminates blackheads and whiteheads, reduces freckles, relieves sunburns, cleanses skin, enhances blood circulation, improves complexion, and reduces blemishes. Magnesium chloride is abundant in *multani mitti* alongside other nutrients and hence provides skin glow [32]. Benefits of natural products for skin are described in Table 6.1.

6.3.2 Cardiovascular Health

Natural products have a significant role to play as dietary supplements and may be beneficial for treating CVDs. The antioxidative, antihypercholesteremic, antiangiogenic, anti-ischemic, inhibition of platelet aggregation, and anti-inflammatory properties of various phytochemicals may contribute to their cardioprotective effects by lowering the risk of cardiovascular disorders. Data from experimental, epidemiological, and clinical studies show that dietary nutrients have significant cardioprotective effects in the primary and secondary prevention of coronary heart disease; as a result, they are regarded as heart-healthy natural products. With respect to the mechanism of cardioprotection (Figure 6.8) produced by dietary nutritional supplements, such as flavonoids (e.g., citrus fruits, pulses, red wine, tea, and cocoa), olive oil, omega-3 fatty acids (fish oil and fish-based products), lycopene (tomato and tomato-based products), resveratrol (grapes and red wine), coffee, and soy in the prevention and treatment of cardiovascular disorders, natural products have prime importance [47]. Therefore, prophylactic and therapeutic potential of cardiovascular-friendly

Table 6.1: Benefits of Natural Products for Skin

Natural Products	Active Constituent	Benefit	Reference
Soy	Phospholipids (phosphatidylcholine and essential fatty acids), genistein, daidzein	Positive scientific evidence supports the antioxidant, antiproliferative, and anticarcinogenic properties of soy extract. Soy has been applied topically to lessen hyperpigmentation, improve skin suppleness, manage oil production, hydrate the skin, and postpone hair regrowth. Through the estrogen-like and antioxidant properties of its metabolites, soy also has the ability to reduce photoaging of the skin and prevent skin cancer.	[33–36]
Green and black tea	Epigallocatechin-3-gallate (EGCG)	Antiaging and photoprotective effects	[37, 38]
Chamomile	Levomenol and chamazulene	Improves skin texture and elasticity. Prevent skin against UV ray damage. Treats eye puffiness	[39, 40]
Coffee berry	Ferulic acid, chlorogenic acid, and quinic acid	Skin rejuvenation	[24]
Citrus sinensis peels	β-Pinene, limonene, γ-terpinene citronellol	Hydrates skin, prevents acne and blemishes	[41]
Azadirachta indica (neem)	Ferulic acid, quercetin, nimbin, nimbanene, nimbolide, ascorbic acid, n-hexacosanol, and amino acid, 6-desacetylnimbinene, nimbandiol, 7-desacetyl-7-benzoylazadiradione, 7-desacetyl-7-benzoylgedunin	Due to its antimicrobial activity, it prevents outbreak of acne and pimples	[42]
Aloe barbadensis	Emodin, aloe-emodin, aloin, aloeresin A, aloveroside, barbaloin and acemannan	Hydrates, rejuvenates the skin and acts as antiacne agent	[43–46]

natural products have been suggested based on the intriguing findings of numerous studies. Fenugreek (*Trigonella foenum-graecum* L.), onion (*Alliumcepa*), turmeric (*Curcuma longa* L.), garlic (*Allium sativum* L), tulsi (*Ocimum sanctum* L.), and arjuna (*Terminalia arjuna*) have been found to exhibit lipid-lowering and cardioprotective activities. *Commiphora mukul Engl* (Guggul), an oleo gum resin, is used to treat hypertension, atherosclerosis, hypercholesterolemia; it exhibits antioxidant activity and reduces platelet aggregation (Figure 6.9) [47].

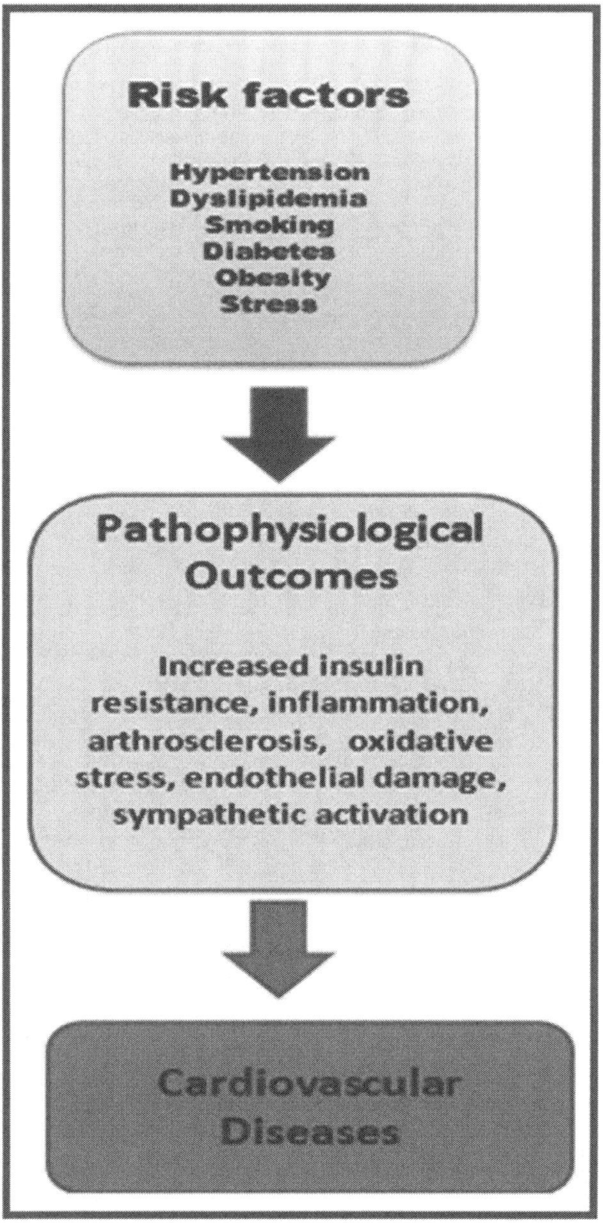

Figure 6.8 Pathophysiology of cardiovascular diseases.

6.3.2.1 Garlic

Garlic exhibits antiatherosclerotic effects due to presence of allicin (Figure 6.10), which is produced upon crushing a garlic clove. In addition to its antioxidant effects, allicin potentially influences atherosclerosis via lipoprotein modification

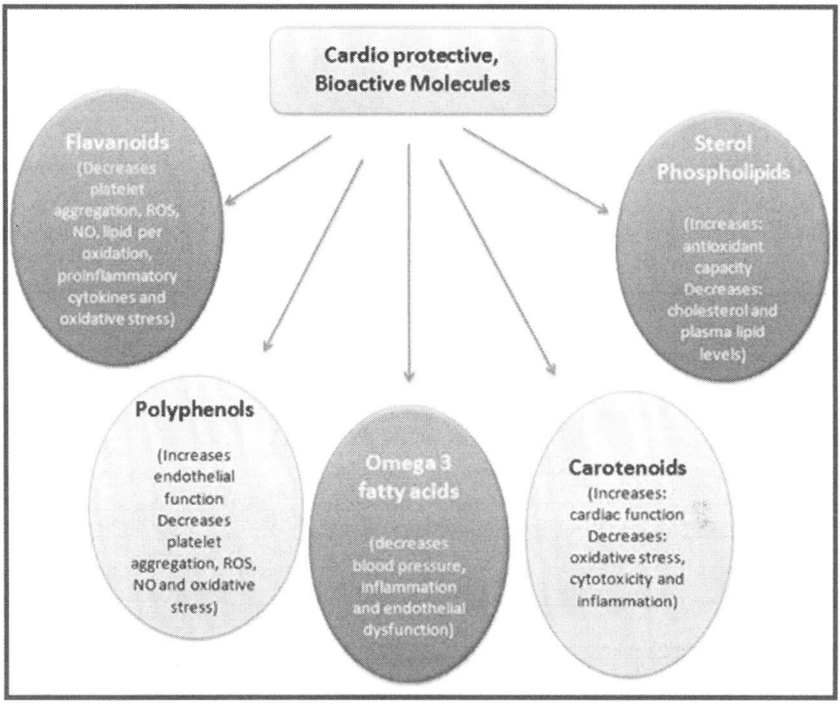

Figure 6.9 Bioactive molecules and their cardioprotective effect.

and the prevention of low-density lipoprotein (LDL) absorption and degradation by macrophage [48]. Experimental studies showed a reduction in cholesterol synthesis in rats' hepatocytes, when treated with garlic extract. Inhibition of 3-hydroxy-3-methylglutaryl (HMG)-CoA reductase and 14-alpha-demethylase by allicin was responsible for decrease in hepatic cholesterogenesis [49, 50].

6.3.2.2 Soy

CVDs are the main cause of death in postmenopausal women, and mortality from CVD is comparable in both sexes. In the postmenopausal years, there are considerable alterations in lipoprotein metabolism due to estrogen deficiency, which lead to an increase in serum cholesterol levels. The increase in blood

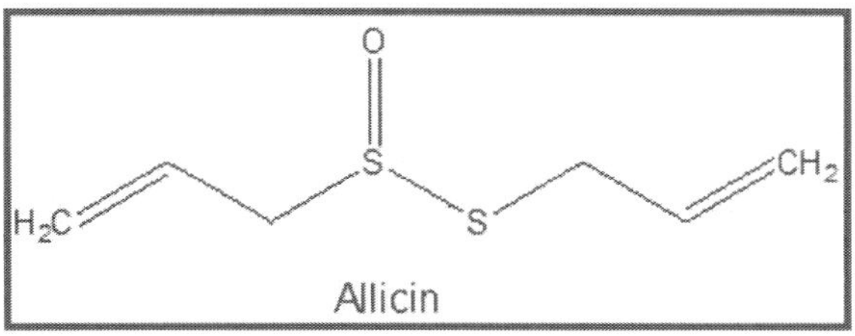

Figure 6.10 Chemical structure of allicin.

cholesterol may be stopped or reduced which can reduce the risk of CVD. It has been suggested that isoflavones may affect lipoprotein metabolism by directly affecting blood vessel walls, lowering LDL cholesterol, and raising high-density lipoprotein (HDL) cholesterol as a consequence of their estrogenic action. Most abundant sources of isoflavones, including genistein and daidzein (Figure 6.10), are soy and legume seeds (e.g., lentils, peas, and beans). Furthermore, isoflavones may also be found as traces in other plant-based foods such fruits, cereals, vegetables as well as in dairy products [51].

6.3.2.3 Nuts (Dry Fruits)

Nuts (walnuts, almonds, and pistachios, among others) contain unsaturated fatty acids like omega-3 fatty acids and dietary fiber, both of which are known to decrease blood cholesterol levels. Numerous essential bioactive elements (e.g., magnesium, potassium, and vitamin E) are also present in nuts. Nonnutritive bioactive compounds, such as ellagic acid (Figure 6.11), flavonoids, luteolin, and tocotrienols, present in nuts can alter lipids and lipoproteins thus controlling cholesterol levels in blood.

Green tea extracts are rich in EGCG, a flavan-3-ol that is a vital antioxidant for cardioprotection. Black tea has powerful antioxidant properties as well. Experimental results show that drinking green tea for 2 weeks significantly lowered the blood pressure in hypertensive rats who were at potential danger for stroke. Catechins minimize oxidative stress, which boosts nitric oxide bioavailability and endothelial function by eliminating ROS, free radicals, as well as regulating the enzymes which generate ROS. Green tea catechins also have a hypocholesterolemic impact by regulating the synthesis, absorption, and excretion of cholesterol [52].

The endothelium-derived relaxing factor (EDRF), endogenous nitric oxide (NO), is generated from L-arginine by a family of nitric oxide synthase (NOS) isoenzymes. Numerous organic substances cause the release of NO, which has a vasorelaxant effect. In a reported experimental study, on an isolated rats' aorta, ginsenosides and gallotannin, two components of *Crataegus* extract, have shown an endothelium-dependent vasodilator action [53].

6.3.2.4 **Uncaria rhynchophylla**

Calcium channel antagonism is often brought up when talking about an antihypertensive or vasorelaxant impact. In previous research, it was observed that the indole alkaloid hirsutine (Figure 6.12), which is present in *Uncaria rhynchophylla*, may lower intracellular $Ca^{2}+$ levels in isolated rats by preventing calcium release, enhancing calcium absorption into the calcium storage, and blocking the voltage-dependent calcium channel. The direct vasorelaxant effects

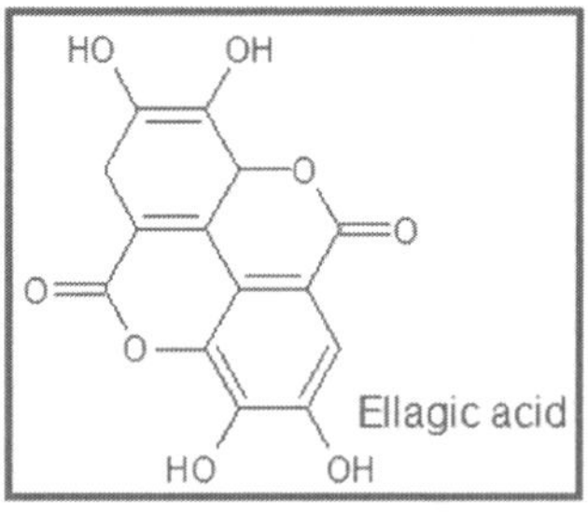

Figure 6.11 Chemical structure of ellagic acid.

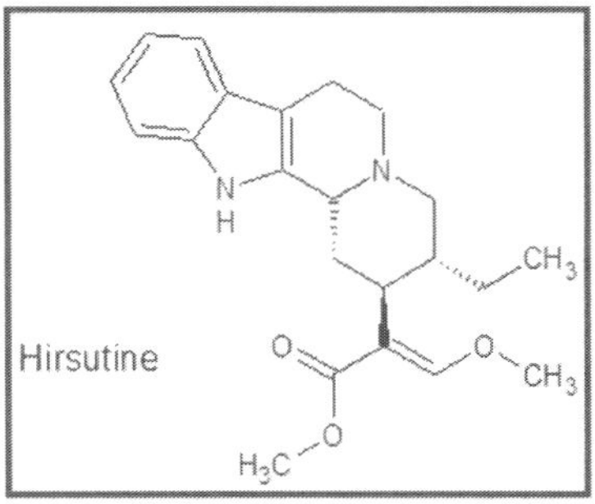

Figure 6.12 Chemical structure of hirsutine.

as well as the hypotensive effects were brought on via calcium antagonistic activities. Natural Products with Cardio-protective Effects are given in Table 6.2.

6.3.3 Liver Health

6.3.3.1 Quercetin

In the United States, nonalcoholic fatty liver disease (NAFLD) is the most prevalent form of chronic liver disease. Insulin resistance is the primary factor in this metabolic syndrome, which is the hepatic manifestation of metabolic disturbances. NAFLD is a term that refers to a group of histological conditions, such as liver steatosis, steatohepatitis, fibrosis, and hepatocarcinoma, all of which are characterized by an elevated buildup or deposition of fat in the liver and are linked to changes in the hepatic system caused by inflammation state. Hepatocytes in a primary cell culture that are exposed to high levels of glucose, insulin, and linoleic acid react by accumulating lipids, experiencing oxidative stress, and eventually dying.

Table 6.2: Natural Products with Cardio-protective Effects

Natural Products	Bioactive Principle	Mechanism of Action	Reference
Citrus fruits, pulses, red wine, berries, tea, onions, and cocoa	Flavonoids: Apigenin, glycitein, quercetin, epicatechin, hesperetin, and delphinidin	1. Antioxidant activity: prevent oxidation of cholesterol, low-density lipoprotein (LDL) 2. Reduce ischemic damage and platelet aggregation 3. Anti-inflammatory activity	[54]
Olive oil	Fatty acids (oleic acid, stearic acid, linoleic acid and palmitic acid) Phenolic acids, phenolic alcohols (HT and tyrosol) Erythrodiol and oleanolic acid Secoiridoids: 3,4-Dihydroxyphenylethanol (3,4-EDA) and oleuropein (OLE)	1. Modulates lipoprotein and carbohydrate metabolism. 2. Decreases endothelial dysfunction, oxidative damage, inflammation, and thrombosis 3. Causes endothelium-dependent vasorelaxation via erythrodiol and oleanolic acid	[55, 56]

(Continued)

Table 6.2 *(Continued)*: Natural Products with Cardio-protective Effects

Natural Products	Bioactive Principle	Mechanism of Action	Reference
Fish oil and cod liver oil	Omega-3 fatty acids: eicosapentaenoic acid (EPA) and docosahexaenoic acid (DHA)	1. Inhibits thromboxane-A2 synthesis therefore, suppresses vasoconstrictor prostanoids 2. Upregulates lipid oxidation and decreases lipid synthesis 3. Reduces the platelet aggregation, plasma norepinephrine and NO levels, hence, enhances endothelial function 4. Blocks calcium entry in vascular smooth muscles. 5. Lowers LDL levels	[57, 58]
Tomatoes, fruits (guava, papaya, watermelon, grapefruit, and apricot)	Lycopene	1. Has antioxidant role, breaks down LDL and alters endothelial functioning	[59–62]
Peanuts, seeds, and skin of grapes	Resveratrol (cis- and trans-resveratrol)	1. Plays role in myocardial ischemia-reperfusion injury due to its antiapoptotic activity 2. In endothelial cells and monocytes, it induces tissue factor expression 3. Inhibits thrombaxane synthesis 4. Upregulates high-density lipoprotein (HDL) production and downregulates LDL cholesterol and vasodilatory prostacyclin synthesis	[63–66]
Coffee	Caffeine, cafestol, kahweol, and chlorogenic acid	1. Caffeine stimulates NO production hence improves the endothelium-dependent vasodilatation 2. Chlorogenic acid decreases CVD risk by decreasing plasma homocysteine levels	[67–69]
Soybeans	Genistein, daidzein, and glycitein trypsin inhibitor known as Bowman–Birk inhibitor	1. Acts as antiesterogens and upregulates LDL-receptors 2. Decreases HMG-CoA reductase activity and improves vasodilatation 3. Increases HDL-C, inhibits tyrosine kinase activity, and stimulates bile acid excretion	[70–72]

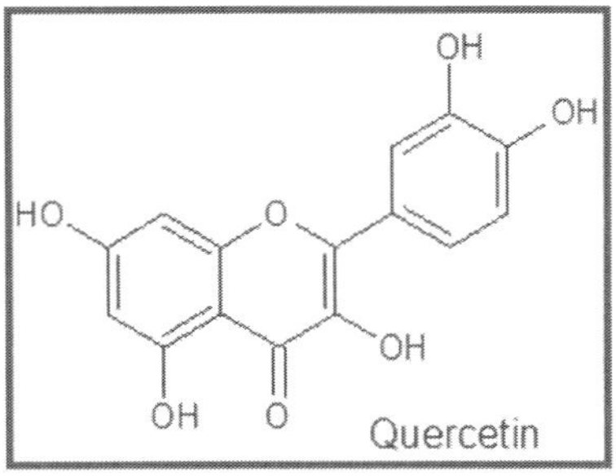

Figure 6.13 Chemical structure of quercetin.

Probably mostly via lowering oxidative stress and the expression of genes linked to steatosis (as PPAR), quercetin lowers liver fat formation and improves systemic indicators associated with metabolic syndrome. Citrus fruits, tea, onions, red grapes are rich sources of quercetin. Free radical damage and oxidative stress are linked to the pathophysiology and development of alcoholic liver disease (ALD), and these may be somewhat mitigated by antioxidants and free radical scavengers. The genesis and progression of ALD are significantly influenced by lipid metabolic disorders and oxidative stress, and the mitochondrial compartment is thought to be the primary source and vulnerable target of intracellular ROS.

Researchers have looked at the possibility that quercetin could shield hepatocytes from the oxidative damage caused by ethanol (Figure 6.13). The ethanol-induced expression of glutathione peroxidase 4 (GPX4) is often downregulated by quercetin. The effect of quercetin on the parameters traditionally linked to alcohol-induced liver injury, such as malondialdehyde (MDA), lactate dehydrogenase (LDH), superoxide dismutase, and aspartate transaminase (AST glutathione [GSH]), may also allow for an explanation of DNA demethylation associated with the upregulation of gene expression in experimental ALD [73].

Healthy liver cells' interaction with a toxic stimulus may result in either liver fibrosis or a preneoplastic cirrhotic liver. Active fibrogenic cells can be induced by a variety of cytokines, growth factors, and inflammatory mediators. Members of the epidermal growth factor (EGF) family have been postulated as having a role in this process. Amphiregulin is a particular ligand for the EGF receptor (EGFR) that is only generated in response to liver injury. Quercetin's effects on the EGFR signal and on pathways that encourage cell proliferation were recently studied. According to the findings, quercetin improved survival pathway activation while reducing the expression of genes linked to precancerous and inflammatory diseases. EGFR signaling inhibition may possibly be responsible for this effect [74].

6.3.3.2 **Curcuma longa** *(Turmeric)*

The severity of the fibrosis was noticeably reduced with the use of turmeric. The mechanism hypothesized by which curcumin exerts its positive benefits in NAFLD is the lowering of the gene transcriptions of tumor necrosis factor-alpha (TNF-α), nuclear factor (NF)-kappaB, and HMG-CoA. Water-soluble curcumin derivatives outperform parent *Curcuma longa* in terms of bioavailability (Figure 6.14), and they can significantly enhance lipid metabolism and slow the development of hepatic fibrosis.

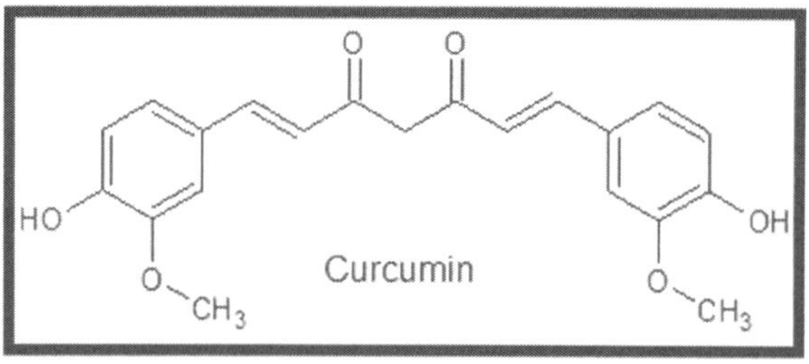

Figure 6.14 Chemical structure of curcumin.

Curcumin derivatives demonstrated a significant decrease in serum alanine transaminase and aspartate aminotransferase levels in early stages of alcohol-induced liver damage [75].

Numerous studies have documented the hypocholesterolemic action of curcumin. Curcumin appears to lower serum cholesterol levels by increasing the expression of LDL receptors in the liver, blocking LDL oxidation, increasing bile acid secretion and fecal cholesterol excretion, repressing the expression of genes involved in cholesterol biosynthesis, and preventing liver damage and fibrogenesis. Hepatic stellate cells (HSCs) undergo apoptosis induction and are inhibited from proliferating. Curcumin also inhibits the formation of extracellular matrix, downregulates the expression of LDL receptors, induces SREBP-1c, and increases HSCs' capacity to store fat. As a result, it may restore this protective functionality, demonstrating its therapeutic value in preventing liver steatosis and fibrosis [75].

Alpha-smooth muscle actin, a hallmark of fibrogenic cells, as well as the intrahepatic gene expression of monocyte chemoattractant protein-1, tissue inhibitor of metalloprotease (TIMP)-1, procollagen type I, and CD11b were all reduced by curcumin. By inhibiting hepatic protein tyrosine phosphatase 1B (PTP-1B) and subsequently increasing insulin and leptin sensitivity in the liver of rats, curcumin decreased serum insulin and leptin levels in fructose-fed rats and protected against hepatic steatosis [76].

6.3.3.3 Silybum marianum

The *Asteraceae* family member *Silybum marianum* (milk thistle) is widely known for its hepatoprotective properties. Silymarin, which is extracted from milk thistle, contains various active constituents such as silybin, silychristin (Figure 6.15), silydianin (flavonolignan), taxifolin, and quercetin (flavonoid). Silymarin has been demonstrated to have a range of anti-inflammatory actions in the liver, including stabilizing mast cells, preventing neutrophil migration, and inhibiting Kupffer cells. Cirrhosis and viral hepatitis are prominent conditions for which silymarin is recommended. Silymarin may thus be helpful in lowering hepatic inflammation in chronic liver disease, including damage from hepatitis C virus (HCV), due to its antioxidant and anti-inflammatory properties [77].

Silymarin showed antiviral properties toward HCV infection in cell culture (HCVcc), resulting in a decrease in viral entry, RNA and protein expression, and the production of infectious viruses. While silymarin did not stop HCVcc from adhering to cells, it did stop a variety of viral pseudoparticles (pp) from entering and HCVpp from incorporating with liposomes. Silymarin also stopped viral cell-to-cell transmission [78] (Figure 6.15).

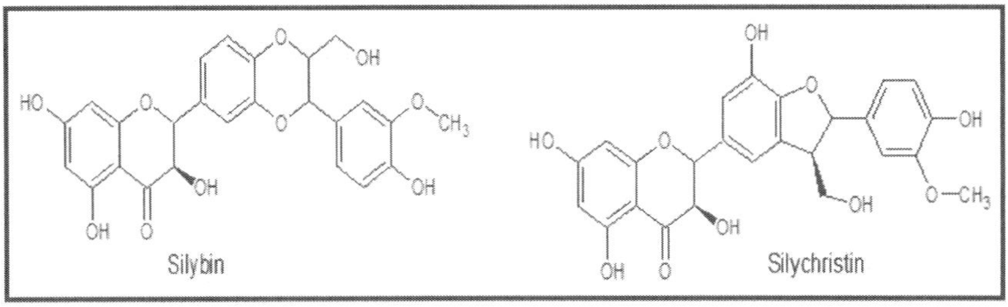

Figure 6.15 Chemical structure of silybin and silychristin.

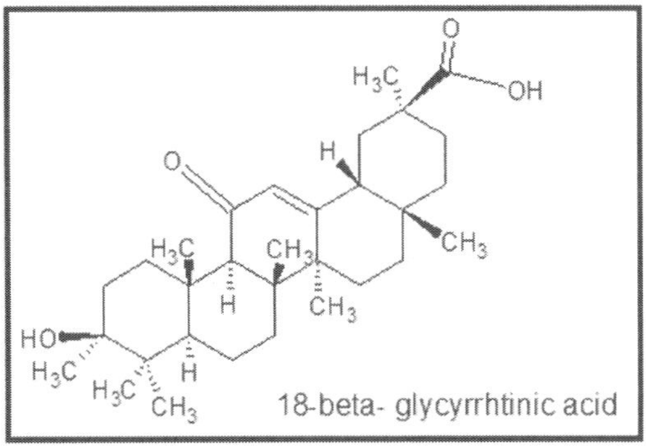

Figure 6.16 Chemical structure of 18-beta-glycyrrhetinic acid.

6.3.3.4 Glycyrrhiza glabra

The perennial plant *Glycyrrhiza glabra* (licorice root) is a member of the Leguminosae family and is endemic to the Mediterranean area as well as to central and South-Western Asia. Glycyrrhizin (GL), an active constituent of *G. glabra* is composed of two molecules of glucuronic acid and 18 beta-glycyrrhetinic acid (18β-GA). Hydroxycoumarins and beta-sitosterols are also present in *G. glabra* aqueous extract [79]. By preventing the translocation of NF-kappaB into the nucleus, 18β-GA (Figure 6.16) can inhibit the activation of HCSs and trigger their death. In addition, experimental studies suggested that 18β-GA increases the proliferation of hepatocytes in rats with CCl_4-induced liver fibrosis [80].

6.3.3.5 Schisandra chinensis *(Wuweizi)*

As per previous research, *Schisandra chinensis* seed extract can improve the liver's GSH detoxification system and thus facilitate healthy liver function processes. Schisandrin, deoxyschisandrin (schisandrin A), wuweizisu C, and gomisin A are some of the bioactive principles found in *S. chinensis* [81, 82]. These compounds have various therapeutic activities including hepatoprotective, anti-inflammatory, anticancer, anti-HIV, and immunomodulating activity [83, 84].

6.3.3.6 Coptis chinensis

Coptis chinensis contains the natural alkaloid, berberine. Although berberine is most frequently used to treat diarrhea in China, many experimental investigations have shown that it is effective at protecting against NAFLD and hyperlipidemia [85]. In one study, berberine with lifestyle intervention led to a substantial NAFLD improvement compared to lifestyle intervention alone [86].

6.3.4 Kidney Health

6.3.4.1 Cranberry

Inflammation, oxidative stress, cellular aging, and gut dysbiosis are among few risk factors that are associated with a greater likelihood of cardiovascular mortality in people with chronic renal failure (CRF). There is growing evidence that certain dietary decisions can decrease the severity of some adverse consequences.

Clinical study has shown that cranberry supplementation provides health benefits such as it causes a decline in urinary tract infections (UTIs). Cranberry polyphenols have recently been found to have anti-inflammatory and antioxidant potential, as well as the ability to modulate gut flora [87]. Previously conducted studies have explained the historic use of cranberries in treatment of chronic kidney disease. Cranberry juice contains bioactive principles including procianidin A, procianidin B, urosolic acid, and epicatechin. It is believed that cranberry extract/juice prevents microorganisms from adhering to the bladder walls [88]. Therefore, it is used to prevent the hallmarks of UTIs and its' relapse. Consuming cranberries, which have a high polyphenol content, can help prevent gastric ulcers, periodontal disease, heart disease, dental caries, and other disease conditions brought on by *Helicobacter pylori* [89].

6.3.4.2 Cordyceps Fungi

For ages, Asian nations such as China, Japan, and others have employed parasitic *Cordyceps* fungi, such *Cordyceps sinensis* (combination of a parasitic fungus and larva), for medical purposes. It contains a variety of bioactive substances, including 3′-deoxyadenosine, mannitol, nucleosides, trace elements (Zn, Mn, and Mg) [90], cordycepic acid, and cordyceps polysaccharides, ergosterol, and aminophenol [91]. Cordycepin is 3′-deoxyadenosine (3′-dA), a purine alkaloid and a derivative of the nucleoside adenosine [92]; whereas, cordycepic acid is d-mannitol. It is frequently used to cure exhaustion, restore the kidney functions, and relax the lungs [93].

Zinc provides protection to nephrons from cadmium damage and high concentrations of Zn, Mg, and Mn are crucially responsible for warming and toning of the renal yang, promoting kidney healing, and enhancing the quality of life [94]. *C. sinensis* is available as nutritional supplement. There are three types of cordyceps products in the market: capsules, oral liquid, and drinks. Table 6.3 mentions natural products and their effect on kidney health.

6.3.5 Brain Health

6.3.5.1 Ginkgo biloba

Ginkgo biloba leaf extract has shown beneficial effect in treating impairments in memory, cognitive speed, and activities of daily living. *G. biloba* extract alleviates symptoms associated with age-related cognitive impairment such as age-related amnesic condition and vascular dementia [100]. Ginkgolides A, B, and C, and bilobalide have been shown to increase circulatory perfusion, antagonize platelet-activating factor (PAF), have neuroprotective effects, and serve as cognitive activators. It stimulates choline uptake in the hippocampus, improves hypoxic tolerance, and glucose utilization [101]. Ginkgo comprises ginkgolides A and B antagonists that competitively inhibit the binding of PAF to the membrane receptor that may exert neuroprotective and antithrombotic effects [102]. In

Table 6.3: Natural Products and Their Effect on Kidney Health

Experimental Model	Natural Product	Active Constituent	Mechanism of Action	Reference
Renal ischemia/reperfusion in rats	Onions, parsley, and oranges	Apigenin	Increased the expressions of p-Akt, Bcl-2, PI3K. Caspase3 levels were reduced.	[95]
Cisplatin-induced AKI in mice	Propolis	Galangin	Inactivation of Nrf2, HO-1, and GCLC. Upregulated expression of SOD, GPx, CAT, and GSH levels. Inhibits ERK and NF-κB signaling pathways.	[96, 97]
Cisplatin-induced AKI in mice	Tea, vegetables, and berries	Myricetin	Increased GSH concentration and catalytic activity. Reduction of blood urea nitrogen, serum Cr, caspase-3, COXI, COXII, TNF-α, IL-6.	[98]
Cisplatin-induced AKI in mice	Leafy green vegetables (kale and spinach) Herbs (chives and dill)	Kaempferol	Stimulated activity of: • NF-κB, phosphorylation of IκBα • Nuclear translocation of p65. TNF-α, iNOS, IL-12 levels reduced.	[99]

addition, flavonoid glycosides and ginkgolide B may inhibit the oxidization of lipoprotein formation, platelet aggregation, and platelet adherence that may reduce the events of atherosclerosis and vascular injury [100]. Experimental data from a number of studies provides evidence for the memory enhancing effects of *G. biloba*.

6.3.5.2 Panax ginseng

Panax ginseng is sometimes called Asian or Korean ginseng. The major active components of *P. ginseng* are ginsenosides, which have been found to have a number of therapeutic properties, including anti-inflammatory, antioxidant, and anticancer effects. Results of clinical research investigations suggest that *P. ginseng* may enhance psychologic function, immunological function, and problems connected with diabetes. Experimental research demonstrated that ginseng increases immunity as the patients who took ginseng had a decreased incidence of influenza and colds, greater antibody titers, and higher natural killer cell activity levels. The importance of *P. ginseng* and its components in preventing neurodegeneration has grown dramatically [103]. The neuroprotective properties of *P. ginseng* have been ascribed principally to the antioxidative and immunomodulatory activity of its ginsenoside components. Mechanistic investigations on the neuroprotective properties of ginsenosides showed that they work not only as antioxidants but also as modulators of intracellular neuronal signaling and metabolism, cell survival/death genes, and mitochondrial function [103].

6.3.5.3 Caffeine

Caffeine is a kind of purine alkaloid. Tea and chocolate (Theobroma) are two examples of foods and extracts rich in phytochemicals, and caffeine is thought to be one of their active constituents. Caffeine's interactions with adenosine and DA D1 receptors mediate its energizing, sleep-inhibiting, and movement-promoting properties. Similar to how caffeine stimulates dopaminergic and glutamatergic activity in rodents, it acts as a competitive antagonist of inhibitory adenosine A1 and A2 receptors in mammals. Caffeine may reduce blood flow to the brain and the rest of the body by blocking adenosine A2a receptors [103]. Increases in basal cerebral blood flow, among other withdrawal symptoms, may be attributable to changes in adenosine receptor populations and/or sensitivity brought on by long-term caffeine use. Caffeine is the most extensively used psychoactive chemical in the world, and its alleged stimulant properties have made it a popular commercial ingredient. Although many studies have found that low doses of caffeine improve subjective alertness and performance on attention tasks, it has been argued that these results are simply attributable to the reduction of withdrawal symptoms experienced by deprived habitual caffeine consumers [104]. However, comparable effects have been seen in chronic caffeine users who show no signs of withdrawal. Large amounts of coffee may cause nervousness, irritability, rapid heart rate, trouble sleeping, and psychomotor agitation [105].

6.3.6 Bone Health

6.3.6.1 Osteoporosis and Natural Products

Osteoporosis is a widespread illness characterized by a loss of bone mass and an elevated risk of fractures. It is brought on by an imbalance between the body's rates of bone resorption and formation. Broken bones, incapacity to function, and ongoing agony are just a few of the clinical repercussions. Osteoporosis is caused by inadequate bone development and excessive bone resorption. Age, endocrine disorders, calcium malabsorption, limb disuse, immunological, nutritional, and hereditary factors are only a few of the complicated illnesses that can lead to osteoporosis [106]. Osteoporosis develops in two forms: The most prevalent condition in postmenopausal women is osteoporosis type 1 (postmenopausal osteoporosis), which is linked to estrogen deficiency. Senile osteoporosis, also known as type 2 osteoporosis, affects people over the age of 75 and is brought on by low calcium intake, diminishing vitamin D levels, or hyperactive parathyroid glands. Bone homeostasis depends on balance between osteoblastic bone creation and osteoclastic bone resorption. Age-related bone loss is caused by osteoblasts producing less new bone and osteoclasts producing more bone resorption. Osteoporosis related bone mass loss is generally acknowledged as a severe public health problem. Age-related bone loss can be avoided by combining dietary and pharmacological factors. The best method for treating osteoporosis is to prevent osteoclasts from resorbing bone and/or promote osteoblast activation to increase bone formation [106]. Naturally occurring bioactive principles useful for treatment are [107, 108]:

- Two examples of flavonoids that can be found in fresh food are quercetin and rutin. It was shown that both of these variables encouraged the formation of osteoblasts in bone marrow mesenchymal stem cells (BMMSCs). According to experimental data, rutin and quercetin encouraged the mineralization of ALP activity.

- Naringin, a dihydroflavonoid, is the *Drynariae rhizoma*'s main active ingredient. A new molecule named neoeriocitrin, a dihydroflavonoid, was produced by *D. rhizoma*. Neoeriocitrin had more pronounced impacts on ALP activity and proliferation.
- The bioactive herb sweroside is derived from *Fructus corni*. Sweroside significantly increased the proliferation of MG-63 cells that resemble human osteoblasts as well as rat osteoblasts.
- Sweroside increased the expression of OCN as well as ALP activity. Additionally, its antiapoptotic activities were diminished and hampered.
- Echinacoside (ECH) is a phenylethanoid glycoside that was produced by the Cistanches plant. With enhanced cell proliferation, ALP activity, COL I contents, OCN levels, and mineralization, ECH had a noticeable impact on osteoblastic cells.

6.3.6.2 Carthami flos

Safflower, Hong-Hua, or *Carthami flos* is used in Chinese medicine to promote blood circulation, treat traumatic damages, and ease muscle pains. In Korean traditional medicine, the seeds of this plant are used to promote bone formation and prevent osteoporosis [109]. The defatted seeds of *C. tinctorius* have been shown to protect ovariectomized rats from trabecular bone loss and the phenolic compounds-rich fraction stimulated the growth of osteoblast-like ROS 17/2.8 cells. The bioactive ingredients of safflower seeds have been proposed to be polyphenolic compounds such as tilianin, acacetin, matairesinol, and their derivatives [110].

6.3.6.3 Cimicifugae rhizoma

Cimicifugae rhizoma (Sheng-Man) contains four bioactive triterpene compounds isolated from the extracts, namely cimicidol-3-O-β-xyloside, cimicidanol-3-O-β-xyloside, acetylacteol-3-O-arabinoside and 7,8-didehydro-24-O-acetyl-hydroshengmanol-3-O-β-xyloside. These were shown to decrease the circulating calcium levels in ovariectomized rats. In a subsequent study, the former three triterpenes were demonstrated to suppress the formation of osteoclast-like cells and their resorbing activity. More recent results showed that an extract of *A. heracleifolia* prevented the ovariectomy-induced elevation of serum alkaline phosphatase level; it was able to preserve trabecular bone mass, bone volume, trabecular number, trabecular thickness structure model index as well as bone mineral density of the proximal tibia metaphysis and distal femur metaphysis in ovariectomized mice [111]. This plant is renowned for its heat-repelling and detoxifying effects. Clinically it is used in combination with other herbs to treat symptoms caused by excessive heat in the body, such as gum infection, mouth sores, sore throat, fever, rash, and skin diseases. More recent applications of this plant drug include the treatment of hemorrhoids, Parkinson's syndrome, and menopausal symptoms [112].

6.3.6.4 Actaea racemosa

Actaea racemosa and related species are native to North America. The roots and rhizomes, known as black cohosh, are widely used for the management of symptoms associated with menopause. It enhances osteoblastic differentiation and increases the osteoprogeterin (OPG)/RANKL ratio in human osteoblasts. A triterpene saponin fraction of the plant extract was claimed to be responsible for the osteoprotective effect. Bioactive constituent, 25-O-acetylcimigenol

3-O-β-xylopyranoside was able to block in vitro osteoclastogenesis induced by RANKL or TNF-α, and it attenuated TNF-α-induced bone loss in mice. From black cohosh, deoxyactein has also been isolated and demonstrated to promote cell growth, alkaline phosphatase activity, and collagen content. Deoxyactein was able to suppress the production of ROS and osteoclast differentiation factors such as TNF-α, interleukin (IL)-6, and RANKL [113].

6.3.6.5 Cordyceps sinensis

Cordyceps sinensis is a parasitic fungus which infects caterpillar larvae and eventually replaces the host tissue. In current dietary supplement markets, fermentation products of Cordyceps fungal filaments are also available [114]. Cordyceps is highly regarded as a kidney- and lung-tonifying agent, particularly useful for improving general physical weakness, easing respiratory discomforts, and relieving low back pain [115].

6.3.6.6 Drynaria Rhizoma

Drynaria rhizoma literally referred as healing broken bone, is a renowned liver and kidney-tonifying herb for use in bone fracture, traumatic damages. In an attempt to identify the active principles from *D. fortunei*, 11 flavonoids were obtained by bioactivity-guided isolation procedures and all were found to display proliferative activity [116]. Naringin isolated from *Drynaria* rhizomes promotes the proliferation and osteogenic differentiation of human bone mesenchymal stem cells (Figure 6.17). In a study of the potential application of naringin for regenerative treatment of inflammation-induced bone injury, co-administration of TNF-α and naringin to bone marrow mesenchymal stem cells resulted in a protection of the TNF-α-induced damages [117].

6.3.6.7 *Ginseng*

Ginsenosides are triterpene glycosides in ginseng and related *Panax* species. Early studies have reported the osteoprotective effects of ginsenosides. More recently, several ginsenosides were demonstrated to possess bone anabolic and/or anti-resorption activities [118]. Thus, ginsenoside Rb1 inhibited RANKL-induced osteoclast differentiation [119].

6.3.6.8 *Boswellia serrata*

The articular cartilage is mostly affected by osteoarthritis (OA), a degenerative condition of the joints. The knees, hips, fingers, and lower spine area are among

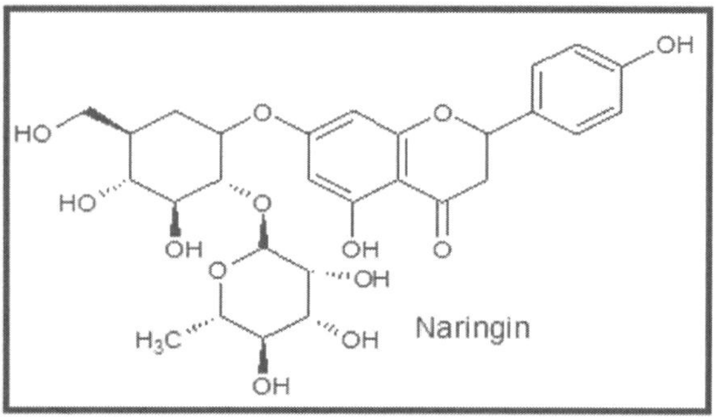

Figure 6.17 Chemical structure of naringin.

the joints most commonly affected by OA since they have been under constant stress over time. Even though hand joints have been mostly affected by OA, the knee is arguably the one that has been clinically examined the most. In cases of hip and hand OA, there were no gender differences in the prevalence. However, OA of the knee is more commonly found in women compared to males. The two most important clinical signs are pain and functional impairment, such as joint stiffness and physical incapacity [120].

One of the most popular herbal remedies for OA is *Boswellia serrata*, commonly known as lubban. A range of boswellic acids, including acetyl-ketobeta-boswellic acid (AKBA), beta-boswellic acid, and keto-beta-boswellic acid, are found in the herb *B. serrata*, which are known to been responsible for a range of therapeutic activities. According to research, AKBA dramatically lowers inflammation by inhibiting the lipoxygenase pathway. *B. serrata*-containing supplements have been shown, in several studies, to improve pain, stiffness, and physical function in individuals with knee OA. The essential oil extracted from boswellia gum resins contains boswellic acids, such as alpha-boswellic acid, 3-O-acetyl-alpha-boswellic acid, and acetyl-beta-boswellic acid. Other active ingredients found in boswellia resin include beta-ocimene, cembrene, alpha-cubebene, sabinene, beta-elemene, allo-aromandendrene, and alpha-thujene. Boswellia essential oil is therefore used as an antibacterial and antiseptic in mouthwash, as well as in medicines for treating asthma, coughs, and rheumatic and inflammatory illnesses [121].

6.3.6.9 Marine Fungus Diaporthe *Sp.*

A compound called mycoepoxydiene (MED) was found in the *Diaporthe* sp. (marine fungus). It has an alpha, beta-unsaturated gamma-lactone moiety attached to an oxygen bridged cyclooctadiene core (Figure 6.18) [122]. The biological effects of MED are diverse and include antibacterial, anticancer, and anti-inflammatory effects. Recent studies have demonstrated that MED decreases bone loss brought on by ovariectomy in rats and inhibits osteoclast growth brought on by receptor activator of NF-kappaB ligand. MED significantly lowers osteoclast production [123].

6.3.6.10 **Zoanthid Zoanthus** *Sp.*

Norzoanthamine, a member of the zoanthamine class of marine alkaloids, was found in the colonial Zoanthid (*Zoanthus* sp) (Figure 6.19). Experimental studies have shown that norzoanthamine significantly accelerates the formation of the collagen-hydroxyapatite composite, a crucial building block of bone tissue.

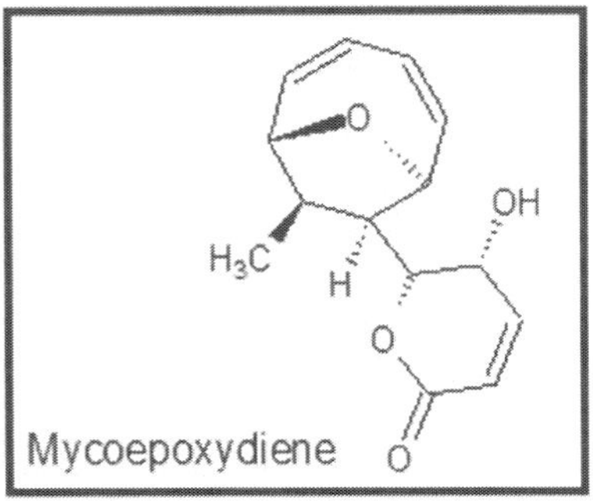

Figure 6.18 Chemical structure of mycoepoxydiene.

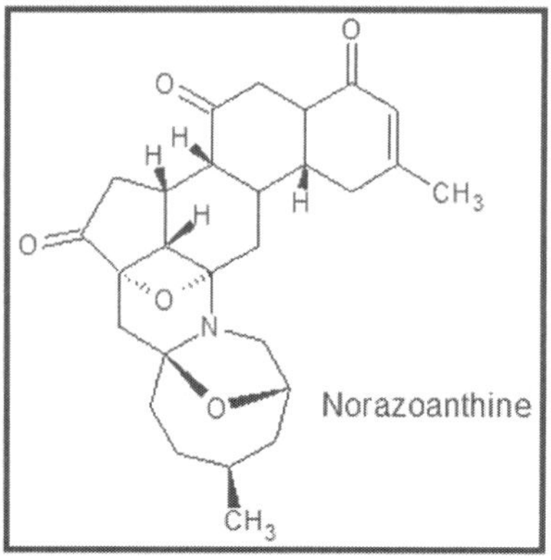

Figure 6.19 Chemical structure of norzoanthamine.

The hydroxyapatite coating on collagen fibrils on matrix vesicles promotes calcification. Norzoanthamine improves collagen release from the immobilized matrix vesicle model and raises collagen concentration, thus encouraging bone formation [124].

6.3.7 Body Weight Management

6.3.7.1 Garcinia cambogia

Obesity is marked by the enlargement of adipocytes secondary to triglyceride accumulation.

Natural extract derived from the dried fruit of the tree *Garcinia cambogia* (GC), which is native to the jungles of South India and South Asia, has been researched extensively and used as a supplement for weight reduction. The fruit rind is the source of hydroxycitric acid (HCA), also known as hydroxyl citric acid. It is an organic acid that has been shown to suppress appetite and reduce body's capacity to produce adipose tissue, making it a promising adjunct to weight loss efforts. The active component of GC extract, HCA, has been used to treat obesity and lower cholesterol levels. In humans, the decreased fatty acid production and lipogenesis may lower appetite hence lead to weight reduction. HCA was shown to block the enzyme that breaks down citrate into acetyl coenzyme A (acetyl-CoA) and oxaloacetate in the citric acid cycle (in liver). As a consequence, there is a decrease in the formation of acetyl-CoA, which is necessary for lipogenesis and the synthesis of fatty acids. Furthermore, an increase in serotonin levels mediated by GC suppresses hunger [125].

6.3.7.2 African Mango

African mango (*Irvingia gabonensis*) was studied for its weight loss potential after observing that tribes using the seed as a dietary staple had slim builds and decreased incidence of CVD. The seeds of this mango-like fruit contain high soluble fiber, which may act as a bulk forming laxative resulting in gradual absorption of dietary sugar and reduce the formation of adipocytes [126]. African mango demonstrates potential for significant weight loss of up to 12 pounds in overweight and obese subjects over a period of 10 weeks with few reported adverse events. Therefore, African mango can be added to daily diet. High performance liquid chromatograph analysis of African mango extract has revealed presence of ellagic acid, mono-, di-, and tri-O-methyl-ellagic acids, as well as their glycosides in significant amount [127].

6.3.8 Male Fertility and Health

6.3.8.1 *Saw palmetto*

Saw palmetto has been shown to stop the production of the 5-alpha reductase enzyme. This enzyme converts the androgen dihydrotestosterone from the male hormone testosterone. Although this hormone is essential for male development, it also contributes to many problems that older and middle-aged men experience. Typically, 5-alpha reductase transforms testosterone into other anabolic hormones in the testes, prostate, adrenal glands, and hair follicles. These are the same areas where males experience libido loss, prostate enlargement, and hair loss, respectively [128]. By inhibiting and reducing the production of this enzyme, saw palmetto lowers the unfavorable effects often associated with elevated dihydrotestosterone levels as men age. High testosterone levels help regulate body weight, maintain a more positive mood, and encourage better sexual function. In males with benign prostatic hyperplasia (BPH), saw palmetto has been found to help the urological system [129]. Previous studies have shown that saw palmetto bioactive constituents interact with receptors in the lower urinary tract that are linked to overactive bladder (OAB) and urinary dysfunction. Saw palmetto extract causes prostatic epithelial tissue contraction, mediated via a nonhormonal mechanism [130].

6.3.8.2 Tribulus terrestris

In traditional Chinese medicine, *Tribulus terrestris* L. (TT) also known as Gokshura is used to support health and physical performance. Mostly athletes use TT for boosting their testosterone levels and strengthening physical performance. TT has ergogenic benefits on sports performance, which includes boosting strength, recuperation after exercise, and it encourages skeletal muscle growth. Saponins present in TT exhibit an anabolic/androgenic effect by triggering the endogenous synthesis of testosterone and luteinizing hormone (LH), which may drive muscle development. Polyphenols and flavonoids present in TT serve as antioxidants and thus TT reduces inflammation by preventing oxidative and muscular damage. It strengthens endurance and boost male virility. TT supports healthy levels of male hormones while also enhancing libido and reproductive health in men. In addition, it supports a healthy endocrine system. It might help athletes gain strength and lean muscular mass [131, 132].

6.3.9 Nausea and Immunity

6.3.9.1 *Ginger* (Zingiber officinale)

Occasionally, ginger is used to cure nausea and vomiting. Even so, ginger powder seems to help with a variety of types of nausea. According to preliminary research, ginger may be helpful for treating mild to severe morning sickness and vomiting during pregnancy when taken at the suggested amount of 1 g in the form of dry ginger or its equivalent in form of ginger syrup each day. According to some reports, ginger is just as helpful as vitamin B6 in easing morning sickness symptoms. The processes by which ginger reduces motion sickness are not fully understood. Any antiemetic effects ginger may have for motion sickness are likely to be limited to effects on the gastrointestinal tract and not on the central nervous system, according to some research. According to several research, ginger may lessen the intensity and duration of nausea a patient experiences following chemotherapy. Ginger is well known for its use as a spice and herbal medicine. Ginger possesses traces of essential elements (Fe, Mg, and Ca), and it also contains vitamin C and other therapeutically active constituents such as flavonoids, sesquiterpenes, paradols, and phenolic compounds (e.g., gingerdiol, gingerol, gingerdione, and shogaols). Most prevalent uses of ginger are to treat nausea, pain, and colds. However, it also has anti-inflammatory, antiapoptotic, antitumor, antipyretic, antidiabetic, antitumorigenic, antioxidant, antiplatelet, cardiotonic, analgesic, antihyperglycemic, and cytotoxic properties [133].

6.3.10 Hair Care

6.3.10.1 Fenugreek

Seeds of *T. foenum-graecum*, commonly known as fenugreek, are traditionally used for hair fall treatment. Presence of alkaloids such as neurin, gentianine, and trigonelline in it promotes hair growth. Fenugreek seed have also reported to contain fiber, fenugrin B, vitamins, fenugreekine, and various amino acids [134]. Fresh fenugreek leaf paste is traditionally applied to the scalp, as dandruff treatment, to promote hair growth, retain natural color, and smoothen texture of hair [135].

6.3.10.2 Emblica officinalis

Amla berry increases calcium absorption, leading to the development of healthy bones, hair, teeth, and nails. Amla is the most effective natural hair conditioner. It promotes the strength of the hair follicles, which prevents early graying and delays the onset of premature graying, also preventing premature thinning of the hair as we age [136, 137]. It also offers adequate nutrition and aids in regulating blood flow. The mashed fruits promote hair development. The dried fruit is used as a component of shampoos and hair oils [137].

6.3.10.3 Acacia concinna

Shikakai (*Acacia concinna*) is a fantastic natural hair cleanser, astringent, and detangler all in one. It strengthens hair, encourages hair growth, and eliminates dandruff when used in soaps and shampoos for washing hair [138, 139].

6.3.10.4 Simmondsia chinensis

Jojoba oil contains eicosenoic, docosenoic, and oleic acids. Jojoba oil is used as a rejuvenating agent and has been traditionally used for scalp treatment. It provides nourishment, promotes hair growth, and enhances silky texture of hair [140, 141].

REFERENCES

1. Saxena M, Saxena J, Nema R, Singh D, Gupta A. Phytochemistry of medicinal plants. *Journal of Pharmacognosy and Phytochemistry*. 2013;1(6):168–82.

2. Costa T, Vieira RF, Bizzo HR, Silveira D, Gimenes MA. Secondary metabolites. *Embrapa Agroindústria de Alimentos-Capítulo em livro científico (ALICE)*. 2012;1(1):131–6.

3. Adamski Z, Blythe LL, Milella L, Bufo SA. Biological activities of alkaloids: From toxicology to pharmacology. 12th ed. 2020:210.

4. Adeyemi M. A review of secondary metabolites from plant materials for post harvest storage. *International Journal of Pure & Applied Sciences & Technology*. 2011;6(2):94–102.

5. Barbosa-Filho JM, Piuvezam MR, Moura MD, Silva MS, Lima KVB, da-Cunha EVL. Anti-inflammatory activity of alkaloids: A twenty-century review. *Revista Brasileira de Farmacognosia*. 2006;16(1):109–39.

6. Bourgaud F, Gravot A, Milesi S, Gontier E. Production of plant secondary metabolites: A historical perspective. *Plant Science*. 2001;161(5):839–51.

7. De Sousa Falcão H, Leite JA, Barbosa-Filho JM, Athayde-Filho D, Filgueiras P, De Oliveira Chaves MC. Gastric and duodenal antiulcer activity of alkaloids: A review. *Molecules*. 2008;13(12):3198–223.

8. Majak W. Review of toxic glycosides in rangeland and pasture forages. *Rangeland Ecology & Management/Journal of Range Management Archives*. 2001;54(4):494–8.

9. Tapas AR, Sakarkar D, Kakde R. Flavonoids as nutraceuticals: A review. *Tropical Journal of Pharmaceutical Research*. 2008;7(3):1089–99.

10. Prassas I, Diamandis EP. Novel therapeutic applications of cardiac glycosides. *Nature Reviews Drug discovery*. 2008;7(11):926–35.

11. Harborne JB, Williams CA. Advances in flavonoid research since 1992. *Phytochemistry*. 2000;55(6):481–504.

12. Minocha S, Kumari S, Tiwari A, Gupta AK, Gandhi S, Sharma A. An overview on tannins. *International Journal of Pharmaceutical and Biological Science Archive*. 2015;3(2):09–11.

13. Hassanpour S, MaheriSis N, Eshratkhah B. Plants and secondary metabolites (Tannins): A review. *International Journal of Forest, Soil and Erosion*. 2011;1(1):47–53.

14. Singh D, Chaudhuri PK. Structural characteristics, bioavailability and cardioprotective potential of saponins. *Integrative Medicine Research*. 2018;7(1):33–43.

15. Podolak I, Galanty A, Sobolewska D. Saponins as cytotoxic agents: A review. *Phytochemistry Reviews*. 2010;9(3):425–74.

16. Piironen V, Lindsay DG, Miettinen TA, Toivo J, Lampi AM. Plant sterols: Biosynthesis, biological function and their importance to human nutrition. *Journal of the Science of Food and Agriculture*. 2000;80(7):939–66.

17. Moghadasian MH. Pharmacological properties of plant sterols: In vivo and in vitro observations. *Life Sciences*. 2000;67(6):605–15.

18. Gopaul R, Knaggs HE, Lephart ED. Biochemical investigation and gene analysis of equol: A plant and soy-derived isoflavonoid with antiaging and antioxidant properties with potential human skin applications. *Biofactors*. 2012;38(1):44–52.

19. Leyden J, Wallo W. The mechanism of action and clinical benefits of soy for the treatment of hyperpigmentation. *International Journal of Dermatology*. 2011;50(4):470–7.

20. OyetakinWhite P, Tribout H, Baron E. Protective mechanisms of green tea polyphenols in skin. *Oxidative Medicine and Cellular Longevity*. 2012;2012(1):1–8.

21. Hsu S. Green tea and the skin. *Journal of the American Academy of Dermatology*. 2005;52(6):1049–59.

22. Abedi A, Dianati M, Gilasi H. Effect of chamomile extract on prevention of peristomal skin complications. *Medical-Surgical Nursing Journal*. 2020;9(4):50–61.

23. Srivastava JK, Shankar E, Gupta S. Chamomile: A herbal medicine of the past with a bright future. *Molecular Medicine Reports*. 2010;3(6):895–901.

24. Farris P. Idebenone, green tea, and Coffeeberry® extract: New and innovative antioxidants. *Dermatologic Therapy*. 2007;20(5):322–9.

25. Leyden J, Shergill B, Micali G, Downie J, Wallo W. Natural options for the management of hyperpigmentation. *Journal of the European Academy of Dermatology and Venereology*. 2011;25(10):1140–5.

26. Hosseini SM, Hoseinifar SH, Mazandarani M, Paknejad H, Van Doan H, El-Haroun ER. The potential benefits of orange peels derived pectin on serum and skin mucus immune parameters, antioxidant defence and growth performance in common carp (*Cyprinus carpio*). *Fish & Shellfish Immunology*. 2020;103(1):17–22.

27. Li G, Tan F, Zhang Q, Tan A, Cheng Y, Zhou Q, et al. Protective effects of polymethoxyflavone-rich cold-pressed orange peel oil against ultraviolet B-induced photoaging on mouse skin. *Journal of Functional Foods*. 2020;67(1):103834.

28. Mathur S, Kachhwaha S. Neem tree: Amazing beauty component in skin and hair care. *Advances in Pharmacology and Toxicology*. 2015;16(3):31.

29. Bhowmik D, Chiranjib YJ, Tripathi K, Kumar K. Herbal remedies of *Azadirachta* indica and its medicinal application. *Journal of Chemical and Pharmaceutical Research*. 2010;2(1):62–72.

30. Waranuch N, Phimnuan P, Yakaew S, Nakyai W, Grandmottet F, Onlom C, et al. Antiacne and antiblotch activities of a formulated combination of *Aloe barbadensis* leaf powder, *Garcinia mangostana* peel extract, and *Camellia* sinensis leaf extract. *Clinical, Cosmetic and Investigational Dermatology*. 2019;1(1):383–91.

31. Sadaf S, Jamali MS, Jan MH, Mehmood K, Mehmood Z, Anwar M, et al. Comprehension study on phytochemicals and antibacterial activity of aloe vera *Barbadensis* found in different regions of Balochistan. *Journal of Science & Technology*. 2017;6(1):186–9.

32. Bhutkar MK, Shah MM. Formulation and evolution of herbal antibacterial face pack. *Journal of Emerging Technologies and Innovative Research*. 2019;6(5):77–82.

33. Baumann LS. Cosmeceutical critique: Soy and its isoflavones. *Skin & Allergy News*. 2001;32(1):17.

34. Thornfeldt C. Cosmeceuticals containing herbs: Fact, fiction, and future. *Dermatologic Surgery*. 2005;31(1):873–81.

35. Paine C, Sharlow E, Liebel F, Eisinger M, Shapiro S, Seiberg M. An alternative approach to depigmentation by soybean extracts via inhibition of the PAR-2 pathway. *Journal of Investigative Dermatology*. 2001;116(4):587–95.

36. Wei H, Spencer J, Gelfand J, Phelps R, Lebwohl M. The isoflavone genistein: A new agent in dermatology. *Cosmetic Dermatology*. 2001;14(1):13–9.

37. Elmets CA, Singh D, Tubesing K, Matsui M, Katiyar S, Mukhtar H. Cutaneous photoprotection from ultraviolet injury by green tea polyphenols. *Journal of the American Academy of Dermatology*. 2001;44(3):425–32.

38. Zhao J, Jin X, Yaping E, Zheng ZS, Zhang YJ, Athar M, et al. Photoprotective effect of black tea extracts against UVB-induced phototoxicity in skin. *Photochemistry and Photobiology*. 1999;70(4):637–44.

39. Baumann L. Cosmeceutical critique: Chamomile. *Skin & Allergy News*. 2003;34(1):43.

40. Lee K-G, Shibamoto T. Determination of antioxidant potential of volatile extracts isolated from various herbs and spices. *Journal of Agricultural and Food Chemistry*. 2002;50(17):4947–52.

41. Himaja AK, Ashok Kumar A, Bharat Kumar B. Preparation and evaluation of poly herbal fruit face mask. *Journal of Research in Pharmaceutical Science*. 2015;2(11):07–13.

42. Susmitha S, Vidyamol K, Ranganayaki P, Vijayaragavan R. Phytochemical extraction and antimicrobial properties of *Azadirachta indica* (Neem). *Global Journal of Pharmacology*. 2013;7(3):316–20.

43. Kahramanoğlu İ, Chen C, Chen J, Wan C. Chemical constituents, antimicrobial activity, and food preservative characteristics of aloe vera gel. *Agronomy*. 2019;9(12):831.

44. Liang J, Cui L, Li J, Guan S, Zhang K, Li J. Aloe vera: A medicinal plant used in skin wound healing. *Tissue Engineering Part B: Reviews*. 2021;27(5):455–74.

45. Richardson J, Smith JE, McIntyre M, Thomas R, Pilkington K. Aloe vera for preventing radiation-induced skin reactions: A systematic literature review. *Clinical Oncology*. 2005;17(6):478–84.

46. Wahedi HM, Jeong M, Chae JK, Do SG, Yoon H, Kim SY. Aloesin from aloe vera accelerates skin wound healing by modulating MAPK/Rho and Smad signaling pathways in vitro and in vivo. *Phytomedicine*. 2017;28(1):19–26.

47. Cook NC, Samman S. Flavonoids: chemistry, metabolism, cardioprotective effects, and dietary sources. *The Journal of Nutritional Biochemistry*. 1996;7(2):66–76.

48. Varshney R, Budoff MJ. Garlic and heart disease. *The Journal of Nutrition*. 2016;146(2):416S–21S.

49. Sun Y-E, Wang W, Qin J. Anti-hyperlipidemia of garlic by reducing the level of total cholesterol and low-density lipoprotein: A meta-analysis. *Medicine*. 2018;97(18):e0255.

50. Rana S, Pal R, Vaiphei K, Sharma SK, Ola R. Garlic in health and disease. *Nutrition Research Reviews*. 2011;24(1):60–71.

51. Gacek M. Soy and legume seeds as sources of isoflavones: Selected individual determinants of their consumption in a group of perimenopausal women. *Menopause Review/Przegląd Menopauzalny*. 2014;13(1):27–31.

52. Islam MA. Cardiovascular effects of green tea catechins: Progress and promise. *Recent Patents on Cardiovascular Drug Discovery (Discontinued)*. 2012;7(2):88–99.

53. Sharifi-Rad J, Rodrigues CF, Sharopov F, Docea AO, Can Karaca A, Sharifi-Rad M, et al. Diet, lifestyle and cardiovascular diseases: Linking pathophysiology to cardioprotective effects of natural bioactive compounds. *International Journal of Environmental Research and Public Health*. 2020;17(7):2326.

54. de Whalley CV, Rankin SM, Hoult JRS, Jessup W, Leake DS. Flavonoids inhibit the oxidative modification of low density lipoproteins by macrophages. *Biochemical Pharmacology*. 1990;39(11):1743–50.

55. Visioli F, Davalos A, López de las Hazas MC, Crespo MC, Tomé-Carneiro J. An overview of the pharmacology of olive oil and its active ingredients. *British Journal of Pharmacology*. 2020;177(6):1316–30.

56. Boskou D. Olive oil: Chemistry and technology. 2006.

57. Kris-Etherton PM, Harris WS, Appel LJ. Fish consumption, fish oil, omega-3 fatty acids, and cardiovascular disease. *Circulation*. 2002;106(21):2747–57.

58. Weitz D, Weintraub H, Fisher E, Schwartzbard AZ. Fish oil for the treatment of cardiovascular disease. *Cardiology in Review*. 2010;18(5):258.

59. Bansal P, Gupta SK, Ojha SK, Nandave M, Mittal R, Kumari S, et al. Cardioprotective effect of lycopene in the experimental model of myocardial ischemia-reperfusion injury. *Molecular and Cellular Biochemistry*. 2006;289(1):1–9.

60. Friedman M. Anticarcinogenic, cardioprotective, and other health benefits of tomato compounds lycopene, α-tomatine, and tomatidine in pure form and in fresh and processed tomatoes. *Journal of Agricultural and Food Chemistry*. 2013;61(40):9534–50.

61. Das S, Otani H, Maulik N, Das DK. Lycopene, tomatoes, and coronary heart disease. *Free Radical Research*. 2005;39(4):449–55.

62. Duan L, Liang C, Li X, Huang Z, Liu S, Wu N, et al. Lycopene restores the effect of ischemic postconditioning on myocardial ischemia-reperfusion injury in hypercholesterolemic rats. *International Journal of Molecular Medicine*. 2019;43(6):2451–61.

63. Hung L-M, Chen J-K, Huang S-S, Lee R-S, Su M-J. Cardioprotective effect of resveratrol, a natural antioxidant derived from grapes. *Cardiovascular Research*. 2000;47(3):549–55.

64. Bradamante S, Barenghi L, Villa A. Cardiovascular protective effects of resveratrol. *Cardiovascular Drug Reviews*. 2004;22(3):169–88.

65. Riba A, Deres L, Sumegi B, Toth K, Szabados E, Halmosi R. Cardioprotective effect of resveratrol in a postinfarction heart failure model. *Oxidative Medicine and Cellular Longevity*. 2017;1(1):1–10.

66. Petrovski G, Gurusamy N, Das DK. Resveratrol in cardiovascular health and disease. *Annals of the New York Academy of Sciences*. 2011;1215(1):22–33.

67. Sampaio-Jorge F, Morales AP, Pereira R, Barth T, Ribeiro BG. Caffeine increases performance and leads to a cardioprotective effect during intense exercise in cyclists. *Scientific Reports*. 2021;11(1):24327.

68. Karaś A, Kawałko K, Pondel K, Bielewicz K. Cardioprotective effects of coffee. *Journal of Education, Health and Sport*. 2022;12(10):35–40.

69. Oboh G, Ojueromi OO, Ademosun AO, Omayone TP, Oyagbemi AA, Ajibade TO, et al. Effects of caffeine and caffeic acid on selected biochemical parameters in L-NAME-induced hypertensive rats. *Journal of Food Biochemistry*. 2021;45(3):e13384.

70. Zhang M, Cai S, Ma J. Evaluation of cardio-protective effect of soybean oligosaccharides. *Gene*. 2015;555(2):329–34.

71. Turner R, Baron T, Wolffram S, Minihane AM, Cassidy A, Rimbach G, et al. Effect of circulating forms of soy isoflavones on the oxidation of low density lipoprotein. *Free Radical Research*. 2004;38(2):209–16.

72. Welty FK, Lee KS, Lew NS, Zhou J-R. Effect of soy nuts on blood pressure and lipid levels in hypertensive, prehypertensive, and normotensive postmenopausal women. *Archives of Internal Medicine*. 2007;167(10):1060–7.

73. Kobori M, Masumoto S, Akimoto Y, Oike H. Chronic dietary intake of quercetin alleviates hepatic fat accumulation associated with consumption of a Western-style diet in C57/BL6J mice. *Molecular Nutrition & Food Research*. 2011;55(4):530–40.

74. Cuevas MJ, Tieppo J, Marroni NP, Tuñón MJ, González-Gallego J. Suppression of amphiregulin/epidermal growth factor receptor signals contributes to the protective effects of quercetin in cirrhotic rats. *The Journal of Nutrition*. 2011;141(7):1299–305.

75. Graham A. Curcumin adds spice to the debate: Lipid metabolism in liver disease. *British Journal of Pharmacology*. 2009;157(8):1352–3.

76. Li JM, Li YC, Kong LD, Hu QH. Curcumin inhibits hepatic protein-tyrosine phosphatase 1B and prevents hypertriglyceridemia and hepatic steatosis in fructose-fed rats. *Hepatology*. 2010;51(5):1555–66.

77. Matveev A, Koniaeva E, Kurchenko V, Shchekatikhina A. Hepatoprotective properties of silymarin. *Eksperimental'naia i Klinicheskaia Gastroenterologiia= Experimental & Clinical Gastroenterology*. 2011(2):130–5.

78. Wagoner J, Negash A, Kane OJ, Martinez LE, Nahmias Y, Bourne N, et al. Multiple effects of silymarin on the hepatitis C virus lifecycle. *Hepatology*. 2010;51(6):1912–21.

79. Stickel F, Schuppan D. Herbal medicine in the treatment of liver diseases. *Digestive and Liver Disease*. 2007;39(4):293–304.

80. Moro T, Shimoyama Y, Kushida M, Hong YY, Nakao S, Higashiyama R, et al. Glycyrrhizin and its metabolite inhibit Smad3-mediated type I collagen gene transcription and suppress experimental murine liver fibrosis. *Life Sciences*. 2008;83(15–16):531–9.

81. Wang R, Kong J, Wang D, Lien LL-m, Lien EJ-c. A survey of Chinese herbal ingredients with liver protection activities. *Chinese Medicine*. 2007;2(1):1–8.

82. Zhu M, Yeung RY, Lin KF, Li RC. Improvement of phase I drug metabolism with Schisandra chinensis against CCl4 hepatotoxicity in a rat model. *Planta Medica*. 2000;66(06):521–5.

83. Ip S-P, Ko K-M. The crucial antioxidant action of schisandrin B in protecting against carbon tetrachloride hepatotoxicity in mice: A comparative study with butylated hydroxytoluene. *Biochemical Pharmacology*. 1996;52(11):1687–93.

84. Wang JP, Raung SL, Hsu MF, Chen CC. Inhibition by gomisin C (a lignan from *Schizandra chinensis*) of the respiratory burst of rat neutrophils. *British Journal of Pharmacology*. 1994;113(3):945–53.

85. Yuan X, Wang J, Tang X, Li Y, Xia P, Gao X. Berberine ameliorates nonalcoholic fatty liver disease by a global modulation of hepatic mRNA and lncRNA expression profiles. *Journal of Translational Medicine*. 2015;13(1):1–11.

86. Liu Y, Zhang L, Song H, Ji G. Update on berberine in nonalcoholic fatty liver disease. *Evidence-Based Complementary and Alternative Medicine*. 2013;1(1):1–8.

87. de Almeida Alvarenga L, Borges NA, Moreira LdSG, Teixeira KTR, Carraro-Eduardo JC, Dai L, et al. Cranberries: Potential benefits in patients with chronic kidney disease. *Food & Function*. 2019;10(6):3103–12.

88. Madden E, McLachlan C, Oketch-Rabah H, Calderón AI. Safety of cranberry: Evaluation of evidence of kidney stone formation and botanical drug-interactions. *Planta Medica*. 2021;87(10/11):803–17.

89. Amin R, Thalluri C, Docea AO, Sharifi-Rad J, Calina D. Therapeutic potential of cranberry for kidney health and diseases. *eFood*. 2022;3(5):e33.

90. Zhu J, Wang Z. Several metal tests in *Cordyceps sinesis* and its mycelium. *Journal of Jining Medical College*. 1993;12:52–4.

91. Wei L. Research advanced in functional composition content of *Cordyceps*. *West China Journal of Pharmaceutical Sciences*. 2003;18:359–60.

92. Li SP, Li P, Ji H, Zhang P, Dong TT, Tsim K. The contents and their change of nucleosides from natural *Cordyceps sinensis* and cultured *Cordyceps mycelia*. *Yao xue xue bao= Acta Pharmaceutica Sinica*. 2001;36(6):436–9.

93. Zhou X, Gong Z, Su Y, Lin J, Tang K. Cordyceps fungi: Natural products, pharmacological functions and developmental products. *Journal of Pharmacy and Pharmacology*. 2009;61(3):279–91.

94. Lynch RJ, Duckworth RM. Microelements: Part I: Zn, Sn, Cu, Fe and I. *The Impact of Nutrition and Diet on Oral Health*. 2020;28:32–47.

95. Wang X, Wang W, Wang J-Z, Yang C, Liang C-Z. Effect of apigenin on apoptosis induced by renal ischemia/reperfusion injury in vivo and in vitro. *Renal Failure*. 2018;40(1):498–505.

96. Huang Y-C, Tsai M-S, Hsieh P-C, Shih J-H, Wang T-S, Wang Y-C, et al. Galangin ameliorates cisplatin-induced nephrotoxicity by attenuating oxidative stress, inflammation and cell death in mice through inhibition of ERK and NF-kappaB signaling. *Toxicology and Applied Pharmacology*. 2017;329:128–39.

97. Kang HG, Lee HK, Cho KB, Park SI. A review of natural products for prevention of acute kidney injury. *Medicina*. 2021;57(11):1266.

98. Hassan SM, Khalaf MM, Sadek SA, Abo-Youssef AM. Protective effects of apigenin and myricetin against cisplatin-induced nephrotoxicity in mice. *Pharmaceutical Biology*. 2017;55(1):766–74.

99. Wang Z, Sun W, Sun X, Wang Y, Zhou M. Kaempferol ameliorates *Cisplatin* induced nephrotoxicity by modulating oxidative stress, inflammation and apoptosis via ERK and NF-κB pathways. *AMB Express*. 2020;10(1):1–11.

100. Silberstein R, Pipingas A, Song J, Camfield D, Nathan P, Stough C. Examining brain-cognition effects of ginkgo biloba extract: Brain activation in the left temporal and left prefrontal cortex in an object working memory task. *Evidence-Based Complementary and Alternative Medicine*. 2011, p.164139.

101. Mullaicharam A. A review on evidence based practice of Ginkgo biloba in brain health. *International Journal of Pharmaceutical Chemistry and Analysis*. 2013;1:24–30.

102. Moulton PL, Boyko LN, Fitzpatrick JL, Petros TV. The effect of *Ginkgo biloba* on memory in healthy male volunteers. *Physiology & Behavior*. 2001;73(4):659–65.

103. Reay JL, Scholey AB, Kennedy DO. Panax ginseng (G115) improves aspects of working memory performance and subjective ratings of calmness in healthy young adults. *Human Psychopharmacology: Clinical and Experimental*. 2010;25(6):462–71.

104. Acquas E, Tanda G, Di Chiara G. Differential effects of caffeine on dopamine and acetylcholine transmission in brain areas of drug-naive and caffeine-pretreated rats. *Neuropsychopharmacology*. 2002;27(2):182–93.

105. Cao C, Cirrito JR, Lin X, Wang L, Verges DK, Dickson A, et al. Caffeine suppresses amyloid-β levels in plasma and brain of Alzheimer's disease transgenic mice. *Journal of Alzheimer's Disease*. 2009;17(3):681–97.

106. Lin JT, Lane JM. Osteoporosis: A review. *Clinical Orthopaedics and Related Research (1976–2007)*. 2004;425:126–34.

107. Tu KN, Lie JD, Wan CKV, Cameron M, Austel AG, Nguyen JK, et al. Osteoporosis: A review of treatment options. *Pharmacy and Therapeutics*. 2018;43(2):92.

108. An J, Yang H, Zhang Q, Liu C, Zhao J, Zhang L, et al. Natural products for treatment of osteoporosis: The effects and mechanisms on promoting osteoblast-mediated bone formation. *Life Sciences*. 2016;147:46–58.

109. Zheng G, Hu C, Guo H, Lv C, editors. Healing functions of carthami flos against contusion and other potential damages. 2019 IEEE International Conference on Bioinformatics and Biomedicine (BIBM); 2019: IEEE.

110. Tu Y, Xue Y, Guo D, Sun L, Guo M. *Carthami flos*: A review of its ethnopharmacology, pharmacology and clinical applications. *Revista Brasileira de Farmacognosia*. 2015;25:553–66.

111. Li J, Liu J, He C, Yu Z, Du Y, Kadota S, et al. Triterpenoids from *Cimicifugae rhizoma*, a novel class of inhibitors on bone resorption and ovariectomy-induced bone loss. *Maturitas*. 2007;58(1):59–69.

112. Li J-X, Yu Z-Y. *Cimicifugae rhizoma*: From origins, bioactive constituents to clinical outcomes. *Current Medicinal Chemistry*. 2006;13(24):2927–51.

113. Kolios L, Schumann J, Sehmisch S, Rack T, Tezval M, Seidlova-Wuttke D, et al. Effects of black cohosh (*Cimicifuga racemosa*) and estrogen on metaphyseal fracture healing in the early stage of osteoporosis in ovariectomized rats. *Planta Medica*. 2010;76(09):850–7.

114. Zhang D-W, Wang Z-L, Qi W, Zhao G-Y. The effects of *Cordyceps sinensis* phytoestrogen on estrogen deficiency-induced osteoporosis in ovariectomized rats. *BMC Complementary and Alternative Medicine*. 2014;14(1):1–7.

115. Qi W, Wang P-J, Guo W-J, Yan Y-B, Zhang Y, Lei W. The mechanism of *Cordyceps* sinensis and strontium in prevention of osteoporosis in rats. *Biological Trace Element Research*. 2011;143:302–9.

116. Chen L, Tao Z-S, Chen H, Zhou K, Zhou D-S. Combined treatment with alendronate and *Drynaria rhizome* extracts. *Zeitschrift für Gerontologie und Geriatrie*. 2018;51(8).

117. Yu F, Lv Q, Tong Z, Song W, Zhao Z, Yuan B, et al. Effect of total flavonoids from *Drynaria rhizome* on bone loss in ovariectomized rats. *Tropical Journal of Pharmaceutical Research*. 2019;18(6):1285–9.

118. Gong Y-S, Chen J, Zhang Q-Z, Zhang J-T. Effect of 17β-oestradiol and ginsenoside on osteoporosis in ovariectomised rats. *Journal of Asian Natural Products Research*. 2006;8(7):649–56.

119. Che C-T, Wong MS, Lam CWK. Natural products from Chinese medicines with potential benefits to bone health. *Molecules*. 2016;21(3):239.

120. Rashan L, Hakkim FL, Idrees M, Essa MM, Velusamy T, Al-Baloshi M, et al. Boswellia gum resin and essential oils: Potential health benefits – An evidence based review. *International Journal of Nutrition, Pharmacology, Neurological Diseases*. 2019;9(2):53–71.

121. Italiano G, Raimondo M, Giannetti G, Gargiulo A. Benefits of a food supplement containing *Boswellia serrata* and bromelain for improving the quality of life in patients with osteoarthritis: A pilot study. *The Journal of Alternative and Complementary Medicine*. 2020;26(2):123–9.

122. Senthilkumar K, Venkatesan J, Kim S-K. Marine derived natural products for osteoporosis. *Biomedicine & Preventive Nutrition*. 2014;4(1):1–7.

123. Kuramoto M, Hayashi K, Yamaguchi K, Yada M, Tsuji T, Uemura D. Structure-activity relationship of norzoanthamine exhibiting significant inhibition of osteoporosis. *Bulletin of the Chemical Society of Japan*. 1998;71(4):771–9.

124. García-García P, Reyes R, Évora C, Delgado A, Fernández JJ, Daranas AH. Osteoprotective effect of the marine alkaloid norzoanthamine on an osteoporosis model in ovariectomized rat. *Biomedicine & Pharmacotherapy*. 2022;147:112631.

125. Fassina P, Adami FS, Zani VT, Machado ICK, Garavaglia J, Grave MTQ, et al. The effect of *Garcinia cambogia* as coadjuvant in the weight loss process. *Nutricion Hospitalaria*. 2015;32(6):2400–8.

126. McLendon AN, Spivey J, Woodis CB. African mango (*Irvingia gabonensis*) extract for weight loss: A systematic review. *The Journal of Nutrition Therapy*. 2013;2(1):53–8.

127. Sun J, Chen P. Ultra high-performance liquid chromatography with high-resolution mass spectrometry analysis of African mango (*Irvingia gabonensis*) seeds, extract, and related dietary supplements. *Journal of Agricultural and Food Chemistry*. 2012;60(35):8703–9.

128. Bent S, Kane C, Shinohara K, Neuhaus J, Hudes ES, Goldberg H, et al. Saw palmetto for benign prostatic hyperplasia. *New England Journal of Medicine*. 2006;354(6):557–66.

129. Gerber GS, Kuznetsov D, Johnson BC, Burstein JD. Randomized, double-blind, placebo-controlled trial of saw palmetto in men with lower urinary tract symptoms. *Urology*. 2001;58(6):960–3.

130. Marks LS, Partin AW, Epstein JI, Tyler VE, Simon I, Macairan ML, et al. Effects of a saw palmetto herbal blend in men with symptomatic benign prostatic hyperplasia. *The Journal of Urology*. 2000;163(5):1451–6.

131. Zheleva-Dimitrova D, Obreshkova D, Nedialkov P. Antioxidant activity of *Tribulus terrestris*: A natural product in infertility therapy. *The International Journal of Pharmacy and Pharmaceutical Sciences*. 2012;4(4):508–11.

132. Fernández-Lázaro D, Fernandez-Lazaro CI, Seco-Calvo J, Garrosa E, Adams DP, Mielgo-Ayuso J. Effects of *Tribulus terrestris L.* on sport and health biomarkers in physically active adult males: A systematic review. *International Journal of Environmental Research and Public Health*. 2022;19(15):9533.

133. Shahrajabian MH, Wenli S, Cheng Q. Pharmacological uses and health benefits of ginger (*Zingiber officinale*) in traditional Asian and ancient Chinese medicine, and modern practice. *Notulae Scientia Biologicae*. 2019;11(3):309–19.

134. Bhargavi MN, Devang PJ. A review on some Indian medicinal plants useful in hair care. *Current Traditional Medicine*. 2023;9(2):141–5.

135. Shahi Z, Mehrizi MK, Hadizadeh M. A review of the natural resources used to hair color and hair care products. *Journal of Pharmaceutical Sciences and Research*. 2017;9(7):1026.

136. Nikmanesh N, Mohammadi-Motamed S. Amla oil, a pharmaceutical product based on traditional knowledge for hair loss treatment. *Research Journal of Pharmacognosy*. 2019;6(1):57–61.

137. Khosla S, Sharma S. A short description on pharmacogenetic properties of *Emblica officinalis*. *Spatula DD*. 2012;2(3):187–93.

138. Sang S-H, Akuwoah GA, Liew K-B, Lee SK, Keng J-W, Lee S-K, et al. Natural alternatives from your garden for hair care: Revisiting the benefits of tropical herbs. *Heliyon*. 2023;9(11):e21876. Available at SSRN 4462088.

139. Sharma L, Agarwal G, Kumar A. Medicinal plants for skin and hair care. *Indian Journal of Traditional Knowledge*. 2003;2:62–68.

140. Ashour ML, Ayoub NA, Singab ANB, Al Azizi MM. *Simmondsia chinensis* (Jojoba): A comprehensive pharmacognostic study. *Journal of Pharmacognosy and Phytochemistry*. 2013;2(2):97–120.

141. Bharti M, Shrivastav A, Abid M, Khan NA. A review on hair growth regulator. *Journal of Drug Delivery and Therapeutics*. 2020;10(5):368–75.

7 Green Synthesis and Its Role in Therapeutic Intervention

Humaira Gul, Maryam Farrukh, Minahil Fatima, Maria Manan, Muhammad Yasir Ali, and Hassan Murtaza Naqvi*

7.1 INTRODUCTION

Advancement of technology at an unprecedented pace has led to global urbanization which has raised the living standards of life. With such development, the pressure of conserving our ecosystem has also intensified (Ivanković et al., 2017). In particular, the pharmaceutical industries are utilizing processes that generate excessive energy and hazardous waste, further exacerbating the strain on our ecosystem (Patneedi and Prasadu, 2015). This growing concern provokes an urge to adopt methods that allow us to meet our needs along with minimizing the environmental impact. One promising approach is green synthesis, which focuses on creating chemical products and processes that reduce the use and production of hazardous substances and by-products. One of the key things about this method is that it does not create any harm during the synthetic process (Li and Anastas, 2012).

From the term "green" one can quickly infer that it focuses on the use of plant materials, biodegradable compounds, and renewable resources. The scope of this process is not limited to concern for the ecosystem but it also aims to improve the safety of human health (Asif, 2021).

One of the major contributors to the world economy is pharmaceutical sector with an estimated revenue of 1.27 trillion US dollars. Despite this fact, it is directly or indirectly involved in climate change by producing approximately 1.9 million tons of CO_2 annually (Mishra et al., 2021). The active drug residues, intermediates, and their metabolites due to their incomplete degradation are the main cause of aquatic contamination and their levels can be significantly toxic to aquatic life (Leder et al., 2015). It highlights the need to integrate green synthesis into drug development and therapeutic practices to lessen this rising pollution. This approach extends beyond the environmental concern to include the safety, effectiveness, and sustainability of pharmaceuticals.

7.2 ROLE OF GREEN SYNTHESIS IN THE ADVANCEMENT OF NANOMEDICINE

The intersection of nanomedicine with green synthesis brings innovation to drug delivery systems and therapeutic intervention. Green nanomedicines (Jahangirian et al., 2017) offer more safety and biocompatibility by minimizing the risk ratio. They are best employed in targeting drug delivery to the cells as they behave differently from their conventional forms due to their nano size (Rafique et al., 2017). One such innovation is the green synthesis of silver nanoparticles (NPs) which has enhanced the efficiency of silver sulfadiazine in treating burn infections (Adhya et al., 2014). While silver sulfadiazine is a gold standard in treating topical burn infections, studies have shown that silver can be toxic to some host cells and may also delay wound healing. Therefore, it is crucial to achieve a minimum yet effective concentration of silver to reduce the risk of cytotoxicity while maintaining the therapeutic benefits. To overcome this problem, researchers have designed its nano-cream for better effects by using green synthesis (Jadhav et al., 2016).

* Corresponding author

 DOI: 10.1201/9781032707273-7

It also brings a revolution in cancer treatments by contributing significant advancements to this field. For example, thiazole, one of the heterocyclic compounds, serves as a scaffold in more than 18 anticancer drugs and is well known for its multitargeted effects on the cell (Sharma et al., 2020). However, the traditional methods used to synthesize thiazole were not free from environmental hazards. Typically, it was produced from haloketones, thioamides, and some other reactants through reactions that required high temperatures, toxic solvents, and high reaction time and often resulted in unwanted by-products. These safety and environmental concerns have led researchers to turn to green synthesis for a better approach to the production of thiazole (Sharma et al., 2021b).

7.3 PRINCIPLES OF GREEN SYNTHESIS

Green synthesis is based on 12 principles that provide a comprehensive framework of the processes involved, all directed toward the minimization of substances that harm both the environment and human health (Doble et al., 2010). The following principles were presented by two scientists Paul Anastas and John Warner in their book *Green Chemistry Theory and Practice*.

7.3.1 Prevention of Waste

The first principle in the principles of green synthesis is the prevention of waste. It means that preventing the production of waste is far better than treating or disposing as it is more cheaper and favorable for the environment (Sheldon et al., 2022). Waste refers to any substance that has lost their utilization and can come in a variety of different forms according to their nature and composition. In 1992, Roger Sheldon presented a term, now commonly known as E-factor. It provides a measure to assess the quantity of waste generated in a manufacturing process. Likewise, 80% of the waste in pharmaceutical industries is linked with the solvents use (Cue and Zhang, 2009). The use of solvents must be reduced to prevent waste. One prime example is the green synthesis of sertraline by the use of ethanol as a sole solvent eliminates the need for other four solvents including dichloromethane, tetrahydrofuran, toluene, and hexane. This resulted in a significant decrease in solvent consumption from 250 L to just 25 L/Kg of sertraline. So, pharmaceutical industries must devise methods and procedures that have minimal impact on the environment along with researching new therapies (Anastas and Eghbali, 2010).

7.3.2 Atom Economy

Atom economy or atom efficiency implies that the maximum reactant should be incorporated into the final product to give the maximum output of the reaction (Sheldon, 2016). This concept was introduced by Barry Trost in 1991 because he believed that increasing the utility of atoms could prevent waste formation at the molecular level. This principle ensures that the maximum raw materials should be utilized to form the product to prevent leftover waste. It is calculated by the ratio of the molecular weight of the desired products to the molecular weight of the reactants used (Anastas and Beach, 2007). One example of this principle is ibuprofen. The primary issue with the older boots method is that it yields low economic costs as it utilizes only 40% of the raw materials (MUREŞAN, 2018). In 1990 a new green method for the synthesis of ibuprofen was introduced that converted up to 99% of raw materials and intermediates into the final product. The new process has significantly higher atomic efficiency with minimal production of waste hence acting as a safe process for the environment.

7.3.3 Designing Safer Chemicals and Products

Chemical products should be designed to be less toxic yet effective (Garrett, 1996). This has always been the goal and challenge for pharmaceutical industries and the basis of the research of newer drugs, that should be safer for human use while maintaining its efficacy. The objective of producing safer drugs is a balance between the optimal performance of procedure and drug product performance thereby minimizing the toxicity to a safer level (Karagölge and Gür, 2016). There are various pharmaceutical examples following this principle. Procaine, now an obsolete drug is a local anesthetic that has the potential to cause severe allergic reactions. Now this drug is improved to a relatively safer drug named lidocaine by replacing the ester group with amide. By, redesigning the drug product we have achieved a safer and more effective therapy (Gordh et al., 2010).

7.3.4 Utilization of Safe Solvents

The use of solvents and other related substances should be minimized. They should only be used when required and only safer ones should be preferred. Mostly the traditional solvents are toxic and inflammable. They are not completely recycled even by energy-efficient distillation and offer considerable losses. So, their use must be replaced with environmentally friendly solvents. The choice of solvent depends on the safety of workers, the environment, and the process (Kerton and Marriott, 2013). Green synthesis also prefers to use solventless systems, water, and supercritical fluid when possible. Water is the universal solvent and can drive most reactions. It is safe, cheap, and non-toxic and can be used to run large-scale processes. One example of this principle is the traditional synthesis of ibuprofen involving hazardous organic solvents like dichloromethane and toluene which are carcinogens and neurotoxic. This led researchers to the use of green solvent, ethanol as an alternative. Ethanol is much safer to use and its loss is much easier to recover than others. This principle has replaced the older process with a much safer process (MUREŞAN, 2018).

7.3.5 Use of Renewable Raw Materials

Renewable feedstock should be used where it is possible because it can be used again rather than to use non-renewable resources. Non-renewable and non-biodegradable materials contribute to significant waste production and product loss (Zimmerman et al., 2020). Artemisinin, an anti-malarial drug is extracted from the plant *Artemisia annua* of Chinese origin. However, obtaining a high yield of it requires spacious land, water, adequate time, and other resources that are difficult to manage. To meet the need for artemisinin, scientists genetically engineered yeast and bacteria to produce it from glucose which is present in excess.

7.3.6 Reduce Derivatives

Derivate is a compound that is derived from the parent molecule. During a process, the use of unnecessary derivatization like blocking and unblocking groups, and temporary steps modification should be minimized. Minimizing the use of chemical derivatives in a reaction is essential because the formation of derivatives requires more steps and reagents (Kottappara et al., 2020). For instance, Aspirin is traditionally prepared by the reaction of salicylic acid with acetic anhydride. Although this process looks like a linear reaction the reactant acetic anhydride is produced from acetic acid first and then utilized in reaction. Following this principle, scientists substituted this with the direct reaction of salicylic acid with acetic acid to reduce the need for acetic anhydride. Another example is the production of other antibiotics from penicillin. The traditional process of production of 6-aminopenicillic acid is now modified to use an immobile catalyst penicillin amidase which eliminates the need for −60°C temperature for solvents.

7.3.7 Biocatalyst

By the use of a catalyst, any process can be accelerated by minimizing its activation energy and time. Such an approach eliminates the need for other requirements like high temperature, pressure and after the process recovers in unchanged form. Catalysts which are specially of natural origin should be used. This principle is illustrated with example of traditional use of rhodium, which is a heavy metal, as a catalyst in the synthesis of sitagliptin, an oral glucose-lowering agent. This step leads to a significant amount of waste. So, scientists replace this rhodium catalyst with biocatalyst transaminase that catalyzes ketone intermediate to desired chiral amine. By replacing, it improves the reaction selectivity, milder the required conditions, and reduces waste (Patel, 2016).

7.3.8 Degradation Design

The product should be synthesized in such a way that then when it degrades, an innocuous end product must be product. It ensures that the degraded product does not cause any harm to the environment (Beach et al., 2009). One such example is the use of halothane, an inhalational anesthetic which not completely degrade and can cause ozone depletion due to the presence of halogens. To address this, concern newer anesthetics are produced like sevoflurane and desflurane which are more biodegradable than halothane while maintaining the required therapeutic effect.

7.3.9 Analytical Procedures

Green synthesis implies that analytical procedures should be further developed so that they can minimize the use of solvents and also produce less waste. Traditional methods require a large amount of sample with excess solvents and other requirements. The goal of green chemistry is to prevent waste and make the analytical process more sustainable. For example, chromatographic techniques are widely used in pharmaceutical industries in drug development, quality control, and separation and purification of raw materials but the main drawback is that large quantities of solvents are used. These solvents produce large amounts of waste also so they must be modified and prefer environmentally friendly solvents (Keith et al., 2007).

7.4 METHODS FOLLOWING GREEN SYNTHESIS

Green synthesis has led to the development of several methods and techniques that allow the synthesis processes to be conducted more effectively and safely without causing harm to the environment. There are broad categories of methods but here we cover only a few of them.

7.4.1 Microwave-assisted Synthesis

Microwave-assisted synthesis is a technique that utilizes microwaves for the conduction of a reaction. It is a type of electromagnetic radiation that carries lower energy and frequency usually about 300 to 300,000 megahertz (Belwal, 2013). In the electromagnetic spectrum, it lies between radio and visible regions. It usually carries an energy of 0.037 Kcal and has better interaction with ionic and polar molecules. It has a unique ability to synthesize organic molecules, especially the heterocyclic compounds that play an important role in medicinal chemistry. Furthermore, it has the ability to provide modifications in selective regions and also requires solvent and catalyst-free conditions. This has led to certain advantages like short reaction time, less economic burden, and more safety (Anastas and Zimmerman, 2012).

The basic principle in this technique is the use of microwaves as the heating source as we use in microwave ovens. It involves the interaction between the charged particle of material and the photons of energy either by collision or conduction by which they generate heat. These processes also contribute to the drug discovery and the synthesis of pharmaceuticals (Larhed and Hallberg, 2001).

Microwave-assisted synthesis helps in the synthesis of novel formulations like novel pyrazolone derivatives. Pyrazolone is famous for its numerous effects like anti-inflammatory, anticancer, and antibiotic effects. It leads to the synthesis of more efficacious derivatives. One of which is pyrazolone attached to aminopyridine is synthesized by a microwave-assisted synthesis process which is a solvent-free process having an average reaction time of no more than 10 minutes (Antre et al., 2011).

Another example is the synthesis of imidazole by microwave-assisted synthesis (Henary et al., 2020). Imidazole is structural constituent of many antifungal drugs like ketoconazole, miconazole and clotrimazole. Traditionally it was prepared by several methods which required a reaction time of almost 4 hours. Microwave-assisted technique enable us to complete the reaction in 10 minutes without the need for solvent. Likewise, triazole which is the structural component of sitagliptin, fluconazole, and voriconazole is also synthesized by microwave-assisted synthesis (Mahato et al., 2018).

Phenytoin which is an antipsychotic drug is also synthesized by this method. The reaction occurs between benzil and urea in a solvent medium at 160 watts for 30 minutes. But in the case of conventional methods it takes about 3 hours for equimolar synthesis (Nerkar et al., 2013).

7.4.2 Ultrasound Wave-Assisted Synthesis

Ultrasound wave-assisted synthesis is an environmentally friendly process that utilizes ultrasonic energy for synthetic procedures or chemical reactions. Ultrasonic energy is high-speed sound energy whose frequency lies between 20 kilohertz to several megahertz (Mohammadi Ziarani et al., 2020). Such reactions can't proceed without the presence of solvents. So, according to the principle of green synthesis universal solvent, water is utilized in most of the cases. The principle behind this process is the formation, growth, and collapse of micro-sized bubbles when ultrasonic waves of high intensity are irradiated on them. The explosion of the bubbles releases high energy in the form of heat and pressure creating the required conditions for the reaction (Sharma et al., 2021a).

This method is important in the synthesis of various heterocyclic compounds which are the basis of the drug products. Benzodiazepines which are known for their sedative, hypnotic, and anxiolytic properties are synthesized efficiently by this method without the need for a catalyst in approximately 15 minutes (Draye et al., 2020).

Benzofuran which is present in citalopram and griseogulvin is also synthesized by this reaction. Other examples are isoxazole found in antibiotics like dicloxacillin, cloxacillin, and cycloserine (Borah and Chowhan, 2022).

7.5 GREEN SYNTHESIS OF THERAPEUTIC COMPOUNDS

Different therapeutic compounds, either of synthetic or natural origin, in various forms are synthesized by utilizing the green synthesis technique.

7.5.1 Antioxidants

Moringa oliefera which belongs to Moringaceae family is well recognized for its antioxidant properties (Shalaby et al., 2022) The bioactive constituent present in the seeds leaves and pods of this plant is flavonoids. Flavonoids are a class of active phytoconstituents that have a variety of chemical structures. They are well known for their scavenging action of free radicals and reactive oxygen species (Agati et al., 2012). When it comes to designing this active constituent in drug product form it's not easy to make a suitable pharmaceutical form. Because most of the phytoconstituents are not easily soluble and also easily deteriorate which affects their bioavailability and efficacy. Because we are considering the green approach for the synthesis, we will direct to the method which is reliable and environment friendly. To improve the drug stability and efficacy scientists have developed the nano-formulation of the leaf extract of moringa (Kiran et al., 2021). For this purpose, moringa leaves are washed and dried and then the leaf powder is mixed with solvent ethanol. Due to their insolubility in water, it is not used as the solvent. Then, by using ultrasound-assisted synthesis the NPs are made, and their antioxidant potential is tested by various tests. These tests confirmed that antioxidant activity of nano-formulation is enhances rather than classical dosage form (Virk et al., 2023).

7.5.2 Antimicrobials

Citrus medica commonly known as finger citron has been previously used for headaches, stomachs, and infectious diseases due to its potential antimicrobial activity (Panara et al., 2012). Its activity lies in its essential oil present in the mesocarp and epicarp of the fruit. However, due to its hydrophobic nature, it is not soluble in water which limits its availability to systemic circulation. Practically it has been made in emulsion, but droplet size is still a problem. So now they are designed for nanoemulsion. Nanosize of the drug enhances the ability of penetration and effectiveness (Li et al., 2018).

Sweet orange, *C. sinensis*, essential oil is also prepared in microemulsion form for better efficacy by using the principles of green synthesis (Souza et al., 2021).

Cancer is a leading cause of death worldwide with an annual mortality rate of 9.6 million. It has led scientists to develop advancements in its therapies. Curcumin is a bioactive compound present in *Curcuma longa* known for its anticancer and anti-mutagenic action. It inhibits several growth factors, enzymes, receptors, and transcription factors of cancerous cells. Although it has anticancer properties, its potential is not fully acquired due to its poor solubility, crystalline nature, and low bioavailability. To improve its bioavailability and solubility, the scientists have reduced its particle size, and introducing NPs of curcumin has improved the targeted drug therapy and its half-life along with minimizing side effects. Curcumin-capped NPs are another strategy that is therapeutically equivalent and requires a smaller number of active constituents. It shows better efficacy with comparison to curcumin NPs or powdered curcumin (Venkatas et al., 2022). It is also better therapy than chemotherapy which destroys cancer cells along with healthy cells and also causes severe side effects like severe nausea, vomiting, weight-loss, and hair loss. NPs of curcumin are made by various procedures of green synthesis like ultrasonic assisted synthesis (Singh et al., 2022)

7.5.3 Wound Healing

Many medications are already available that are frequently used to treat wound healing processes, including inflammatory reactions, the production of blood clots, cell proliferation, and platelet function. These medications include glucocorticoid steroids, nonsteroidal anti-inflammatory medications, and chemotherapy treatments (Bartlett and Hartle, 2013, Berretta et al., 2016, Dhatariya, 2013, Franz, 2007, Varga et al., 2017). The available treatments for treating wounds are costly, hazardous to nearby healthy tissues, and harmful. Therefore, the development of alternative techniques is required for focused treatment and cost-effective wound care, which indicates very high decreasing and good scavenging activities (Nethi et al., 2019, Ovais et al., 2018). In such a case, as a burgeoning field that bridges chemistry, physics, and biology, innovative nanotechnology research can offer another platform for the development of emerging therapeutic compounds employed for wound treatment. As was previously indicated, chemical routes are frequently used in the broad synthetic pathways used for NP production. The chemical processes used are typically highly expensive and involve the use of hazardous and poisonous compounds that pose several environmental concerns (Nath and Banerjee, 2013). Green synthesis, on the other hand, focuses on replacing dangerous chemicals with safer alternatives, and as such, it creates new processes to reduce or even get rid of materials that, when employed to make environmentally friendly NPs, have negative impacts on the environment and human health. Due to the need for green technology, nanoscience is now investigating the biological domain and presenting a unique platform for NP synthesis under aseptic environmental circumstances. The green pathway is a biocompatible, secure, and environmentally benign method of synthesizing NPs for use in biomedical applications by using fungi, plants, bacteria, algae, and microorganisms (Thongtra and Sapronova, 2017). It's crucial to remember that the different synthesis methods employed to produce NPs dictate their physicochemical features and attributes (morphology, surface area, changing size, etc.) (Ghotekar et al., 2021, Korde et al., 2020, Pagar et al., 2019). Reaction parameters including temperature, duration, concentration of precursor, and pH of reaction mixture are carefully set to provide favorable physicochemical qualities (Gill et al., 2020, Pagar et al., 2019, Singh et al., 2020, Yen et al., 2020). Controlling these variables will therefore produce the finest NPs, which will have exceptional qualities and useful applications. It will also shorten reaction times and provide easy scaling. It further notes that plant phytochemicals, which control the size and form of the particles as well as bioreducing the metal precursor, are essential to the ecologically benign synthesis of metal-based NPs (Barwant et al., 2022, Ghotekar et al., 2018, Nagore et al., 2021).

7.5.4 Infectious Disease

Essential oils, organic acids, and bacteriocins are among the organic substances that have been employed as antimicrobial packaging materials (Suffredini et al., 2014). The food sector uses nanosensors to detect pathogens while processing food (Helmke and Minerick, 2006). Numerous NPs, including metal oxide, chitosan, copper, and ag NPs with antibacterial qualities include titanium oxide (TiO_2) and zinc oxide (ZnO) (Manjula et al., 2022). Conventional techniques for diagnosing infectious diseases are laborious, time-consuming, and slow, especially in an emergency. Microbial biosensors have set a new standard for fast, affordable, and precise diagnosis of infectious pathogens, hormone abnormalities, and DNA. The amount of epinephrine was measured by immobilizing the white rot fungus *Phanerochaete chrysosporium* ME446 on a Pt electrode using glutaraldehyde crosslinked with gelatin for NPs with sizes ranging from 5 to 100 μm and 1.04 μM detection limits. Lactase, an enzyme, catalyzes redox processes in fungal cells that convert adrenaline to epinephrine quinone, increasing the current flow (Liu and Yan, 2012). When antigens for viral infections are added, as in the case of rat basophilic leukemia, mast cells initiate an exocytotic reaction minutes later. This response has been utilized to create biosensors for pathogen detection (Curtis et al., 2008). To investigate the SOS response triggered by toxic substances in DNA, which was observed in DNA damage identified by the fluorescent protein regulated by a recA gene promoter, *Escherichia coli* SOS-EGFP was analyzed (Curtis et al., 2008). Lab-on-a-chip is a microfluid device that enables the detection of the analyte even at very low concentrations by energetically transferring the targeted analyte to the surface of the microbial biosensors. Lab-on-a-chip operates on the electrochemical system by integrating microbes onto a microfluid chip (De Leo et al., 2012). Forensic identification for recognition of body fluidic traces containing DNA and miRNA as evidential proof has been analyzed using microbial biosensors (Wang et al., 2013).

7.5.5 Antioxidant Therapy

Neurodegenerative diseases affect millions of people worldwide, and their incidence is rising dramatically as life expectancy increases (Erkkinen et al., 2018). The diseases in this category represent a collection of chronic, progressive disorders that result in loss of neurons in the central nervous system, resulting in deficits in specific mental functions. Among the most common neurodegenerative diseases are multiple sclerosis, Alzheimer's disease (AD), Huntington's disease, amyotrophic lateral sclerosis, and Parkinson's disease (PD). There is no clear etiology for most neurodegenerative diseases, but these conditions are thought to share similar molecular and cellular characteristics that contribute to their progression. Oxidative stress, mitochondrial function abnormalities, protein misfolding, excitotoxicity, and calcium homeostasis dysregulation are a few of the contributing factors (Ilieva et al., 2009, Magalingam et al., 2018, Taylor et al., 2016, Zeng et al., 2018).

A very common example of neurodegeneration is the use of amphetamine-type stimulants. Globally, about 25 million people use amphetamine-type stimulants, especially methamphetamine (Meth), which is more than the combined amount of heroin and cocaine addicts (Han et al., 2021). Its long duration of action, low production cost, and low cost make it highly sought-after by abusers (German et al., 2014), despite the fact that Meth abuse can cause serious health complications in humans (Sepehr et al., 2020). In addition to causing neurodegenerative changes in addicts' brains, this drug can also damage other body organs (Mozaffari et al., 2019). There are a number of symptoms associated with this substance, including agitation, anxiety, aggression, paranoia, hypertension, hyperthermia, and psychosis (Lineberry and Bostwick, 2006). Many drugs cause apoptosis in the nervus system, which leads to neuronal body death (Cadet and Krasnova, 2009). Meth's chemical resemblance to dopamine, which enables it to cross dopamine axons, liberate dopamine from cytoplasmic vesicles, and transfer it into synaptic clefs, is the mechanism underlying its harmful effects (Sulzer et al., 2005). Meth neurotoxicity is caused by the interaction of superoxide radicals in nerve terminals with dopamine quinones. Dopamine metabolism can produce harmful hydroxyl radicals in addition to hydrogen peroxide, particularly when it interacts with metal ions (Cadet and Brannock, 1998). According to the mounting data, meth also contributes to oxidative stress by creating an imbalance between the generation of reactive oxygen species (ROS) and antioxidant enzymes' capacity to scavenge them. Overproduction of ROS can harm a cell's proteins, lipids, and nuclear and mitochondrial DNA (Chen et al., 2022, Peirouvi and Razi, 2022). It has recently been discovered that substances that inactivate free radicals, such as antioxidants, lipid peroxidation inhibitors, and spin-trapping chemicals lessen the toxicity linked to methamphetamine use (Tata and Yamamoto, 2007). Thus, preventing methamphetamine-induced neurotoxicity may benefit from increasing the expression of antioxidant enzymes as well as using exogenous antioxidant substances.

Many phytoconstituents show potent antioxidant potential (Fereidooni et al., 2020, Khaleghi and Khorrami, 2021, Saghafi et al., 2019, Zhang et al., 2023), which have the great capability to initiate the cellular antioxidant process with decrease probability of adverse effects. In addition to extracts from plants, green- or biosynthesized NPs—metal NPs made by encasing these extracts—may also provide a great deal of promise for protecting cells. Because the reducing agents in the extracts lower the metal salts throughout the synthesis process and coat the NPs with their bioactive compounds, the nanomaterials generated are green, ecologically friendly, and cheaply priced. This makes the NPs much more flexible. As a result, they take on the extracts' therapeutic properties and nanomaterial characteristics, such as their large surface areas, tiny diameters, and high surface energy (Khorrami et al., 2020). Silver NPs (AgNPs) are among the most often employed metal NPs in the world due to their unique electrical and optical properties as well as their biological activities (Bayrami et al., 2020, Desireddy et al., 2013, Ravindran et al., 2013). AgNPs made from the extract of black peel pomegranate act as neuroprotective negotiator in the SH-SY5Y (human neuroblastoma cell line) cells against the cell death caused by Meth. Pretreated cells with the extract and NPs, followed by exposing them to Meth showed that these green synthesized compounds could significantly defend the cells by regulating the genes involved in cell death (necrosis and apoptosis). These protective mediators also play their role in inhibiting the cells' DNA fragmentation, lessening the MDA levels, decreasing the level of intercellular ROS, and maintaining the mitochondrial membrane potential at normal levels (Khorrami et al., 2023).

7.5.6 Cardiovascular Health

Currently used anticoagulant medications based on heparin are mostly used in bloodlines, thromboembolic event treatment, and blood clot control and preventive treatments. Despite their effectiveness, these anticoagulants need to be continuously monitored and counteracted using antidotes (Crowther and Warkentin, 2008) because of the potential for bleeding during their administration, which is more common in individuals who are at a high risk of bleeding (Meijers et al., 2014). These risks are mitigated by the use of NPs as anticoagulants. There are numerous potential applications for magnetic NPs, including magnetic resonance imaging, medicine delivery, gene transfer, and magnetic drug targeting (Ali et al., 2021). An active nano-anticoagulant, Hep-SPIONs (heparin-stabilized ferrimagnetic iron oxide NPs), was found to be effective when used in hemodialysis. This Hep-SPION showed similar anticoagulant activity to the commercial heparin commonly used in medical procedures. It also has promising magnetic properties. Hep-SPIONs are also known to be suitable for animal hemodialysis due to their significant blood compatibility. These properties of Hep-SPIONs make them potentially useful for a range of purifying blood procedures in the future due to their considerable effectiveness and safety (Zhao et al., 2018). An important component of thrombolytic therapy is the ability to dissolve thrombus. Coronary heart disease and venous thromboembolism are two common types of thrombotic occlusions that cause severe morbidity and mortality (González-Palomo et al., 2021, Benjamin et al., 2019). It is currently possible to treat these with thrombolytic drugs/therapy using plasminogen activators (PAs) injections. In addition to its major limitations, PAs can have a number of side effects, including rapid drug elimination from the body, a restricted therapeutic window, and the possibility of hemorrhaging. A combination of medical applications and nanotechnology is called nanomedicine and it can reduce the risk of deadly side effects associated with current thrombolytic drugs (Zenych et al., 2020).

7.5.7 Anti-inflammatory Therapy

7.5.7.1 Arthritis

Certain autoimmune diseases, such as rheumatoid arthritis (RA), result in rapid progression, swelling, inflammation, and eventually joint destruction. The currently accessible therapies for the inflammation cure, give beneficial results initially but show adversarial side effects during prolonged use. The drugs could be administered directly at the site of inflammation, which would be an improved method of administration. It would take a lot of time and money to complete this project. Anti-inflammatory drugs can be delivered directly onto inflammation sites using green synthesized NPs as vectors (Stevenson et al., 2011). NPs can deliver drugs, such as glucocorticoids and clodronate, to animal models of arthritis when they are encapsulated with liposomes (Metselaar et al., 2003). Clodronate liposomes administered through the IV route suppressed RA symptoms by targeting macrophages directly (Richards et al., 1999). Similarly, glucocorticoid prednisolone contained in polyethylene glycol (PEG)-coated liposomes showed tenfold greater efficacy than the usual method of treatment. Within two days, the symptoms will disappear, and after six days, complete remission usually occurs (Allen et al., 1995). These findings indicate that NPs may be suitable candidates for treating autoimmune disorders, where they have anti-inflammatory properties (Mayegowda et al., 2023).

7.5.7.2 Inflammatory Bowel Disease

More than 4.9 million people globally suffer from inflammatory bowel diseases (IBDs), which include Crohn's disease and ulcerative colitis (Wang et al., 2023). These are chronic idiopathic diseases that arise from a complex interplay of factors such as individual susceptibility, interactions with gut microflora, and the immune system's role (Lee and Chang, 2021, Shan et al., 2022). The dysfunction between effector and regulatory cells leads to an uncontrolled release of inflammatory molecules (Liyanage et al., 2023). Oxidative stress and inflammation are closely related. Oxidative stress is made worse by IBD because activated immune cells produce more ROS and RNS and because antioxidant defenses are compromised. Through a number of processes, including the activation of nitric oxide synthases, NADPH oxidases, and inflammatory pathways like NF-kB, inflammation can cause oxidative stress. On the other hand, oxidative stress can worsen inflammation by encouraging the release of cytokines that promote inflammation, activating immune cells, and impairing the intestinal barrier (Bourgonje et al., 2020). By creating inflammatory cytokines and soluble mediators like reactive oxygen species (ROS), neutrophil-myeloperoxidase (MPO) catalyzes the creation of powerful ROS. Neutrophils and monocytes concentrate in the gastrointestinal wall and contribute to the pathophysiology of inflammatory bowel disease (IBD) (Chami et al., 2018).

Current treatments include immunosuppressants and pharmacotherapy; tumor necrosis factor (TNF) inhibitors have shown particularly encouraging results. However, issues including lack of response and declining efficacy have spurred research into cutting-edge therapies like cell therapy, apheresis, exosomes, tiny molecules, and improved gut ecology. Mucosal healing which is linked to better long-term outcomes is the ultimate objective (Hossein-Khannazer et al., 2021). Since the ENS may be able to regulate immune responses, maintain the integrity of the mucosal barrier, alter gastrointestinal motility, and reestablish neuroimmune communication, approaches that center on the ENS have potential. Nanotechnology holds the potential to give the most effective means of distributing drugs or directly treating diseases in a more controlled and safe manner by enabling the targeted distribution of molecules and the aggregation of NPs and bioactive chemicals on the appropriate tissue (Patra et al., 2018). Additionally, it facilitates the passage over barriers like as membranes and extremely high pH (Zhang et al., 2012), which is particularly significant for IBD. NPs offer a lot of promise for application in IBD therapy because of their small size, large surface area, and stability in the gastrointestinal tract. Many types of NPs are being investigated for their potential in targeted medication delivery, including those derived from plants, polymeric materials, lipid-based materials, metallic NPs, and exosomes (Zhang and Merlin, 2018).

Gold nanoparticles (AuNPs) have great potential as an IBD treatment. Because of its well-known antioxidant qualities and great biocompatibility, gold has a long history of use in therapy. AuNPs have found uses in therapy as well as imaging thanks to developments in nanotechnology.

Their potential to cure a wide range of inflammatory disorders, including neuroinflammation (Fujita et al., 2021) (Di Bella et al., 2021), skin inflammation (Danscher and Rasmussen, 2023), and autoimmune inflammation (Dhandapani et al., 2023). Their effectiveness in treating inflammatory bowel illnesses has been demonstrated in numerous trials (Abdelmegid et al., 2019, Dhandapani et al., 2023, Hussein and Saleh, 2018, Zhu et al., 2018). One notable advantage of AuNPs is their exceptional stability, especially at very low pH values. Owing to their stability, they can penetrate the intestines and begin treating the illness linked to IBD when taken orally. This characteristic makes them particularly suitable for targeted therapy in the treatment of IBD (Wang et al., 2022).

Because algae have strong anti-inflammatory qualities, extracts from them have already been utilized to treat IBD. These are brought about by a high polysaccharide content (Besednova et al., 2020) which can protect the intestine's biological, chemical, immunological, and physical barriers (Yang et al., 2022). Since brown algae have a polysaccharide content greater than 50% of their dry weight with some species reaching values over 70% they are particularly promising for these applications (Furness, 2006, Li et al., 2021). Previously known as Cystoseira tamariscifolia, Ericaria selaginoides (ES) has a 50% polysaccharide content, which is lower than some other closely related species (Vizetto-Duarte et al., 2016), However, it has been reported that its anti-inflammatory potential is higher than that of most algal species (Pedro et al., 2022). Since the accumulation and reduction of gold ions is essential for the obtention of these NPs, the high polysaccharide content also enables the effective generation of AuNPs by green synthesis (Mukherjee et al., 2021). AuNPs in ES extract were previously synthesized in an environmentally friendly manner, and their cytocompatibility was confirmed (Machado et al., 2021).

7.5.7.3 Anticancer Therapeutics

NPs have long been investigated for their possible anticancer effects, and some studies have found that particular NPs have these effects. Selenium NPs were shown to be coated on titanium oxide nanotubes and to inhibit the proliferation of MG-63 cancer cells, according to a 2019 study. In addition to other advantageous qualities like antibacterial capabilities, the TiO_2 tubes coated with these Se-NPs demonstrated anticancer capabilities (Bilek et al., 2019). As predicted, the current cancer treatment approaches, such as cancer chemotherapy and cancer biology, are not being utilized to their full potential (Llinàs et al., 2018). Because mesoporous silica NPs dissolve pharmaceuticals more quickly than other materials, they have proven to be an excellent drug delivery vector. Improved stability and availability in in vivo settings provide more evidence for this (Barkat et al., 2021). While graphite g-C3H4 has been used for various activities like remediation, energy production, microbial disinfection, and drug delivery, especially in cancer therapy has been a potential resource due to its effectiveness, surface activity, stability, and biocompatibility (Pourmadadi et al., 2023). Green-synthesized CeO2 NPs have also shown promising and potential activity against cancers and also for diabetes due to their stability (Pansambal et al., 2023).

In the clinical domain, NPs are now employed for cancer theranostics. AuNPs are one type of NP that is gaining a lot of attention because of their strong performance in imaging techniques and ease of use. The development of large-scale production in a more environmentally friendly way is made possible by the widespread availability of biological precursors employed in plant-based synthesized AuNPs. Traditional cancer treatments, such as chemotherapy and surgery, have many drawbacks and often don't yield positive outcomes. A longer half-life, ease of modification with ligands identified by cancer cell surface receptors, and enhanced absorption via receptor-mediated endocytosis are all features of AuNPs. Studies on the use of AuNPs as contrast agents for X-ray-based imaging techniques (i.e., computed tomography) have been conducted in order to take use of their special properties. The use of AuNPs, which are produced by nontoxic and biocompatible plants, as nanocarriers for the delivery of therapeutic biomolecules has the potential to significantly advance the successful treatment of a variety of malignancies. Cancers have been identified utilizing fluorescent plant-based markers, such as AuNPs, that were created using Medicago sativa, Olax Scandens, H. ambavilla, and H. lanceolatum. Additionally, different extracts have been used to create green synthetic AuNPs, which have been used to treat a variety of solid tumor types. However, the size, surface reactivity, and surface area of AuNPs are the main factors that determine their cytotoxicity. The advantages of plant-based materials in cancer treatment are first discussed in this article. Next, taking into account AuNPs' significant role in medicine, the use of AuNPs in cancer therapy and detection is emphasized, with a focus on the challenges associated with using such NPs in drug delivery systems (Sargazi et al., 2022).

7.6 CHALLENGES AND FUTURE DIRECTIONS

7.6.1 Strategies for Scaling up of Green-Synthesized Nanomaterials

Large-scale manufacturing of nanomaterials is required to replace current materials with NPs that have superior properties. Understanding physiochemical phenomena at the most fundamental level is made possible by the unique potential that nanoscience presents. Theoretically, this data could help optimize and enhance the structure that is being studied. Even with these fundamental advancements, there is still a big obstacle in nanotechnology. This is due to the ongoing difficulty in translating scientific discoveries published in research journals into applications for industrial technology. There are several facets to the problem. Firstly, materials behave differently at higher and lower scales, just as they do at the nanoscale. More precisely, at the meso- and macro scales, the degree of control at the nanometer scale tends to decrease. Second, unless a substantial return is guaranteed, the industry is reluctant to invest extensively in developing new large-scale procedures for the fabrication of nanomaterials. The problem is especially significant in the applied sciences since there is a divide between research conducted in labs and on an industrial scale (Sukweenadhi et al., 2022).

7.6.2 Regulatory Frameworks for Green Synthesis

Green chemistry offers unique opportunities for innovations through product substitution, the development of new feedstock, catalysis in aqueous media, microwave utilization, the possibility of alternatives, or the use of natural solvents. The current state of affairs regarding the impending global crises has an impact on green chemistry. This paper provides an overview of the environmental protocols used in the synthesis of several FDA-approved pharmaceuticals with high volume demand, high chemical and optical purity requirements, and green chemistry utilization. Safer chemicals and auxiliaries, the employment of catalysts in research, and the development of degradation with more efficient synthetic formulations are just a few of the many advancements in green chemistry. This conceptual paper demonstrates how metrics and principles impact the life cycle of chemicals, from design to disposal, as well as the steps taken to advance sustainability. It is driven by the desire to reduce expenses and improve the sustainability of the production process. The implementation of green chemistry in the pharmaceutical sector has been hindered by regulatory limitations and self-imposed requirements. The objective of the new statute was to adopt a technique that creates no pollutants at all, instead of addressing pollutants after products. It asserts that the regulatory framework's economic modernization will concurrently protect the environment and public health(SHAILAJA and PRIYANKA).

7.6.3 Integration of Green Synthesis with Traditional Medicines

From the beginning of human civilization to the present, the value of medical plants has been demonstrated by the extensive research done on the ever-expanding and pharmaceutically significant naturally occurring plant products. The active ingredients found in plants are highly biodynamic, which means that the world will always have access to a ready supply of medicinally useful herbal remedies. Nanomedicine is the term used recently to describe the uses of nanotechnology for monitoring and controlling biological systems, as well as for treating and diagnosing disorders. For this reason, it is essential to incorporate nanocarriers into conventional medical procedures for the treatment of chronic illnesses like cancer, diabetes, and asthma. The goal of this study is to present the most recent research on the diverse types of NPs produced by herbal plants, their biological activity, and their varied applications. From February 2021 to June 2021, the databases of PubMed, Science Direct, Medline, Scopus, Google Scholar, Iran Medex, and Web of Science were searched to gather pertinent information for this review study(Sathyanarayanan et al., 2022).

7.7 FUTURE RESEARCH DIRECTIONS

Nanotechnology has been expanding quickly and has many uses in various fields. NPs have an impact on both humans and animals and can lead to a variety of lung, renal, and other health issues. Ongoing efforts are necessary to address the dearth of information regarding the dangers of extended use. Green synthesis provides a secure method for creating nontoxic NPs with extra advantageous properties. Because natural substitutes are being used, green synthesis has already been used in several industries. One of the most interesting study fields at the moment is nanomedicine, which covers drug delivery, cancer diagnosis, and other therapeutic alternatives. It is necessary to produce NPs with consistent sizes, consistent characteristics, biocompatibility with drug loading, and limited release to the intended cells. While NPs have made great strides in diagnosis and treatment, their potential still has to be expanded to include treating parasite infections and responding to cancer therapy, two neglected and underutilized areas. The primary focus of basic research should be on the basic markers of NPs that enable targeted delivery of these particles to diseased tissues without compromising normal cellular functioning. Molecular genetics of disease is a significant indicator of advanced nanomedicine, as is modeling drug–cell interactions and animal model studies. Research on the effectiveness and toxicity of these NPs is necessary. To properly manage the near future with care, customized healthcare must be developed on the principles and mechanisms by incorporating nanodevices and nanorobots; these green-synthesized NPs have the potential to be a safe and viable biomaterial.

7.8 CONCLUSION

Demand for green technology, based on microorganisms and plants, that is ecologically benign has increased significantly in recent years. A wide range of uses for NPs made from biological sources have been made possible by research, in response to the growing demand for these compounds. Because of the antibacterial qualities of its metal components, NPs are commonly used to combat bacteria, fungi, and some viruses.

However, various biomolecules are essential components of green-synthesized NPs that have achieved tremendous appeal due to their low cost and eco-friendliness. Green-synthesized NPs' antibacterial qualities are used in a variety of industries, including food packaging, skin care product active ingredients, illness treatment, and medication delivery.

Thanks to its antibacterial, anticancer, antioxidant, anticoagulant, antiparasitic, diagnostic, and nanoimaging characteristics, this new field of nanomedicine has received a lot of attention. While it should be emphasized that excessive use and widespread deployment of NPs may result in toxicity as a result of the build-up of released metals and ions that can have fatal consequences for people, concentrated utilization has not yet been identified. In conclusion, the utilization of biological and biodegradable resources in the synthesis of NPs will surely usher in an eco-friendly era, making green synthesis a generally good and very significant endeavor in all sectors of science. However, there are a few issues that still need to be researched, including the extraction method, irregular particle sizes, environmental factors, the use of biological resources, and most importantly the application without toxicity. In light of the aforementioned difficulties, enhancing green-synthesized NPs will unquestionably be a wise decision in a straightforward, affordable, energy-efficient, and significant procedure. Investigating the difficulties and problems that may arise in translating these to clinical trials and providing support for the advancement of improvised healthcare, particularly in the area of tailored medications, may be beneficial going forward.

REFERENCES

Abdelmegid, A. M., Abdo, F. K., Ahmed, F. E. & Kattaia, A. A. 2019. Therapeutic effect of gold nanoparticles on DSS-induced ulcerative colitis in mice with reference to interleukin-17 expression. *Scientific Reports*, 9, 10176.

Adhya, A., Bain, J., Ray, O., Hazra, A., Adhikari, S., Dutta, G., Ray, S. & Majumdar, B. K. 2014. Healing of burn wounds by topical treatment: A randomized controlled comparison between silver sulfadiazine and nano-crystalline silver. *Journal of Basic and Clinical Pharmacy*, 6, 29.

Agati, G., Azzarello, E., Pollastri, S. & Tattini, M. 2012. Flavonoids as antioxidants in plants: Location and functional significance. *Plant Science*, 196, 67–76.

Ali, A., Shah, T., Ullah, R., Zhou, P., Guo, M., Ovais, M., Tan, Z. & Rui, Y. 2021. Review on recent progress in magnetic nanoparticles: Synthesis, characterization, and diverse applications. *Frontiers in Chemistry*, 9, 629054.

Allen, T. M., Brandeis, E., Hansen, C. B., Kao, G. Y. & Zalipsky, S. 1995. A new strategy for attachment of antibodies to sterically stabilized liposomes resulting in efficient targeting to cancer cells. *Biochimica et Biophysica Acta (BBA)-Biomembranes*, 1237, 99–108.

Anastas, P. & Eghbali, N. 2010. Green chemistry: Principles and practice. *Chemical Society Reviews*, 39, 301–312.

Anastas, P. T. & Beach, E. S. 2007. Green chemistry: The emergence of a transformative framework. *Green Chemistry Letters and Reviews*, 1, 9–24.

Anastas, P. T. & Zimmerman, J. B. 2012. *Innovations in green chemistry and green engineering: Selected entries from the encyclopedia of sustainability science and technology*. Springer.

Antre, R. V., Cendilkumar, A., Goli, D., Andhale, G. S. & Oswal, R. J. 2011. Microwave assisted synthesis of novel pyrazolone derivatives attached to a pyrimidine moiety and evaluation of their anti-inflammatory, analgesic and antipyretic activities. *Saudi Pharmaceutical Journal*, 19, 233–243.

Asif, M. 2021. Green synthesis, green chemistry, and environmental sustainability: An overview on recent and future perspectives of green chemistry in pharmaceuticals. *Green Chemistry & Technology Letters*, 7, 18–27.

Barkat, A., Beg, S., Panda, S. K., Alharbi, K. S., Rahman, M. & Ahmed, F. J. 2021. Functionalized mesoporous silica nanoparticles in anticancer therapeutics. *Seminars in Cancer Biology*, 69, 365–375.

Bartlett, R. & Hartle, A. 2013. Routine use of dexamethasone for postoperative nausea and vomiting: The case against. *Anaesthesia*, 68(9), 892–896.

Barwant, M., Ugale, Y., Ghotekar, S., Basnet, P., Nguyen, V.-H., Pansambal, S., Ananda Murthy, H., Sillanpaa, M., Bilal, M. & Oza, R. 2022. Eco-friendly synthesis and characterizations of Ag/AgO/Ag2O nanoparticles using leaf extracts of *Solanum elaeagnifolium* for antioxidant, anticancer, and DNA cleavage activities. *Chemical Papers*, 76, 4309–4321.

Bayrami, M., Bayrami, A., Habibi-Yangjeh, A., Shafeeyan, M. S., Feizpoor, S., Arvanagh, F. M., Nourani, M. R. & Taheri, R. A. 2020. Biologically synthesised ZnO/CuO/Ag nanocomposite using propolis extract and coated on the gauze for wound healing applications. *Iet Nanobiotechnology*, 14, 548–554.

Beach, E. S., Cui, Z. & Anastas, P. T. 2009. Green chemistry: A design framework for sustainability. *Energy & Environmental Science*, 2, 1038–1049.

Belwal, S. 2013. Green revolution in chemistry by microwave assisted synthesis: A review. *Modern Chemistry*, 1, 22–25.

Benjamin, E. J., Muntner, P., Alonso, A., Bittencourt, M. S., Callaway, C. W., Carson, A. P., Chamberlain, A. M., Chang, A. R., Cheng, S. & Das, S. R. 2019. Heart disease and stroke statistics—2019 update: A report from the American Heart Association. *Circulation*, 139, e56–e528.

Berretta, M., Rinaldi, L., Di Benedetto, F., Lleshi, A., De Re, V., Facchini, G., De Paoli, P. & Di Francia, R. 2016. Angiogenesis inhibitors for the treatment of hepatocellular carcinoma. *Frontiers in Pharmacology*, 7, 428.

Besednova, N. N., Zaporozhets, T. S., Kuznetsova, T. A., Makarenkova, I. D., Kryzhanovsky, S. P., Fedyanina, L. N. & Ermakova, S. P. 2020. Extracts and marine algae polysaccharides in therapy and prevention of inflammatory diseases of the intestine. *Marine Drugs*, 18, 289.

Bilek, O., Fohlerova, Z. & Hubalek, J. 2019. Enhanced antibacterial and anticancer properties of Se-NPs decorated TiO2 nanotube film. *PLoS One*, 14, e0214066.

Borah, B. & Chowhan, L. R. 2022. Ultrasound-assisted transition-metal-free catalysis: A sustainable route towards the synthesis of bioactive heterocycles. *RSC Advances*, 12, 14022–14051.

Bourgonje, A. R., Feelisch, M., Faber, K. N., Pasch, A., Dijkstra, G. & Van Goor, H. 2020. Oxidative stress and redox-modulating therapeutics in inflammatory bowel disease. *Trends in Molecular Medicine*, 26, 1034–1046.

Cadet, J. L. & Brannock, C. 1998. Invited review free radicals and the pathobiology of brain dopamine systems. *Neurochemistry International*, 32, 117–131.

Cadet, J. L. & Krasnova, I. N. 2009. Molecular bases of methamphetamine-induced neurodegeneration. *International Review of Neurobiology*, 88, 101–119.

Chami, B., Martin, N. J., Dennis, J. M. & Witting, P. K. 2018. Myeloperoxidase in the inflamed colon: A novel target for treating inflammatory bowel disease. *Archives of Biochemistry and Biophysics*, 645, 61–71.

Chen, L., Ru, Q., Xiong, Q., Yang, J., Xu, G. & Wu, Y. 2022. Potential effects of Nrf2 in exercise intervention of neurotoxicity caused by methamphetamine oxidative stress. *Oxidative Medicine and Cellular Longevity*, 2022, 4445734.

Crowther, M. A. & Warkentin, T. E. 2008. Bleeding risk and the management of bleeding complications in patients undergoing anticoagulant therapy: Focus on new anticoagulant agents. *Blood, The Journal of the American Society of Hematology*, 111, 4871–4879.

Cue, B. W. & Zhang, J. 2009. Green process chemistry in the pharmaceutical industry. *Green Chemistry Letters and Reviews*, 2, 193–211.

Curtis, T., Naal, R. M. Z., Batt, C., Tabb, J. & Holowka, D. 2008. Development of a mast cell-based biosensor. *Biosensors and Bioelectronics*, 23, 1024–1031.

Danscher, G. & Rasmussen, S. 2023. nanoGold and μGold inhibit autoimmune inflammation: A review. *Histochemistry and Cell Biology*, 159, 225–232.

De Leo, E., Galluccio, L., Lombardo, A. & Morabito, G. 2012. Networked labs-on-a-chip (NLoC): Introducing networking technologies in microfluidic systems. *Nano Communication Networks*, 3, 217–228.

Desireddy, A., Conn, B. E., Guo, J., Yoon, B., Barnett, R. N., Monahan, B. M., Kirschbaum, K., Griffith, W. P., Whetten, R. L. & Landman, U. 2013. Ultrastable silver nanoparticles. *Nature*, 501, 399–402.

Dhandapani, S., Wang, R., Cheol Hwang, K., Kim, H. & Kim, Y.-J. 2023. Enhanced skin anti-inflammatory and moisturizing action of gold nanoparticles produced utilizing *Diospyros* kaki fruit extracts. *Arabian Journal of Chemistry*, 16, 104551.

Dhatariya, K. 2013. II. Does dexamethasone-induced hyperglycaemia contribute to postoperative morbidity and mortality?. *British Journal of Anaesthesia*, 110(5), 674–675.

Di Bella, D., Ferreira, J. P., SIlva, R. D. N. O., Echem, C., Milan, A., Akamine, E. H., Carvalho, M. H. & Rodrigues, S. F. 2021. Gold nanoparticles reduce inflammation in cerebral microvessels of mice with sepsis. *Journal of Nanobiotechnology*, 19, 1–15.

Doble, M., Rollins, K. & Kumar, A. 2010. *Green chemistry and engineering.*

Draye, M., Chatel, G. & Duwald, R. 2020. Ultrasound for drug synthesis: A green approach. *Pharmaceuticals*, 13, 23.

Erkkinen, M. G., Kim, M.-O. & Geschwind, M. D. 2018. Clinical neurology and epidemiology of the major neurodegenerative diseases. *Cold Spring Harbor Perspectives in Biology*, 10, a033118.

Fereidooni, F., Komeili, G., Fanaei, H., Safari, T., Khorrami, S. & Feizabad, A. K. 2020. ☆ Protective effects of ginseng on memory and learning and prevention of hippocampal oxidative damage in streptozotocin-induced Alzheimer's in a rat model. *Neurology, Psychiatry and Brain Research*, 37, 116–122.

Franz, M. G. 2007. Optimizing healing of the acute wound by minimizing complications. *Current Problems in Surgery*, 44, 679–766.

Fujita, T., Zysman, M., Elgrabli, D., Murayama, T., Haruta, M., Lanone, S., Ishida, T. & Boczkowski, J. 2021. Anti-inflammatory effect of gold nanoparticles supported on metal oxides. *Scientific Reports*, 11, 23129.

Furness, J. B. 2006. The organisation of the autonomic nervous system: Peripheral connections. *Autonomic Neuroscience: Basic and Clinical*, 130, 1–5.

Garrett, R. L. 1996. *Pollution prevention, green chemistry, and the design of safer chemicals.* ACS Publications.

German, C. L., Fleckenstein, A. E. & Hanson, G. R. 2014. Bath salts and synthetic cathinones: An emerging designer drug phenomenon. *Life Sciences*, 97, 2–8.

Ghotekar, S., Pansambal, S., Bilal, M., Pingale, S. S. & Oza, R. 2021. Environmentally friendly synthesis of Cr2O3 nanoparticles: Characterization, applications and future perspective: A review. *Case Studies in Chemical and Environmental Engineering*, 3, 100089.

Ghotekar, S., Savale, A. & Pansambal, S. 2018. Phytofabrication of fluorescent silver nanoparticles from *Leucaena leucocephala L.* leaves and their biological activities. *Journal of Water and Environmental Nanotechnology*, 3, 95–105.

Gill, R., Ghosh, S., Sharma, A., Kumar, D., Nguyen, V.-H., Vo, D.-V. N., Pham, T.-D. & Kumar, P. 2020. Vertically aligned ZnO nanorods for photoelectrochemical water splitting application. *Materials Letters*, 277, 128295.

González-Palomo, A. K., Fernández-Macias, J. C., Rangel-Ramírez, V. V., Jimenez-Avalos, J. A. & Cortés-Garcia, J. D. 2021. Transcriptomics applied in research of non-communicable diseases. *Transcriptomics from aquatic organisms to humans*.

Gordh, T., Gordh, T. E., Lindqvist, K. & Warner, D. S. 2010. Lidocaine: The origin of a modern local anesthetic. *The Journal of the American Society of Anesthesiologists*, 113, 1433–1437.

Han, B., Compton, W. M., Jones, C. M., Einstein, E. B. & Volkow, N. D. 2021. Methamphetamine use, methamphetamine use disorder, and associated overdose deaths among US adults. *JAMA Psychiatry*, 78, 1329–1342.

Helmke, B. P. & Minerick, A. R. 2006. Designing a nano-interface in a microfluidic chip to probe living cells: Challenges and perspectives. *Proceedings of the National Academy of Sciences*, 103, 6419–6424.

Henary, M., Kananda, C., Rotolo, L., Savino, B., Owens, E. A. & Cravotto, G. 2020. Benefits and applications of microwave-assisted synthesis of nitrogen containing heterocycles in medicinal chemistry. *RSC Advances*, 10, 14170–14197.

Hossein-Khannazer, N., Torabi, S., Hosseinzadeh, R., Shahrokh, S., Asadzadeh Aghdaei, H., Memarnejadian, A., Kadri, N. & Vosough, M. 2021. Novel cell-based therapies in inflammatory bowel diseases: The established concept, promising results. *Human Cell*, 34, 1289–1300.

Hussein, R. M. & Saleh, H. 2018. Promising therapeutic effect of gold nanoparticles against dinitrobenzene sulfonic acid-induced colitis in rats. *Nanomedicine*, 13, 1657–1679.

Ilieva, H., Polymenidou, M. & Cleveland, D. W. 2009. Non–cell autonomous toxicity in neurodegenerative disorders: ALS and beyond. *Journal of Cell Biology*, 187, 761–772.

Ivanković, A., Dronjić, A., Bevanda, A. M. & Talić, S. 2017. Review of 12 principles of green chemistry in practice. *International Journal of Sustainable and Green Energy*, 6, 39–48.

Jadhav, K., Dhamecha, D., Bhattacharya, D. & Patil, M. 2016. Green and ecofriendly synthesis of silver nanoparticles: Characterization, biocompatibility studies and gel formulation for treatment of infections in burns. *Journal of Photochemistry and Photobiology B: Biology*, 155, 109–115.

Jahangirian, H., Lemraski, E. G., Webster, T. J., Rafiee-Moghaddam, R. & Abdollahi, Y. 2017. A review of drug delivery systems based on nanotechnology and green chemistry: Green nanomedicine. *International Journal of Nanomedicine*, 2957–2978.

Karagölge, Z. & Gür, B. 2016. Sustainable chemistry: Green chemistry. *Journal of the Institute of Science and Technology*, 6, 89–96.

Keith, L. H., Gron, L. U. & Young, J. L. 2007. Green analytical methodologies. *Chemical Reviews*, 107, 2695–2708.

Kerton, F. M. & Marriott, R. 2013. *Alternative solvents for green chemistry.*

Khaleghi, M. & Khorrami, S. 2021. Down-regulation of biofilm-associated genes in mecA-positive methicillin-resistant *S. aureus* treated with *M. communis* extract and its antibacterial activity. *AMB Express*, 11, 85.

Khorrami, S., Dogani, M., Mahani, S. E., Moghaddam, M. M. & Taheri, R. A. 2023. Neuroprotective activity of green synthesized silver nanoparticles against methamphetamine-induced cell death in human neuroblastoma SH-SY5Y cells. *Scientific Reports*, 13, 11867.

Khorrami, S., Kamali, F. & Zarrabi, A. 2020. Bacteriostatic activity of aquatic extract of black peel pomegranate and silver nanoparticles biosynthesized by using the extract. *Biocatalysis and Agricultural Biotechnology*, 25, 101620.

Kiran, M., Kumar, C. R., Shwetha, U., Onkarappa, H., Betageri, V. & Latha, M. 2021. Green synthesis and characterization of gold nanoparticles from Moringa oleifera leaves and assessment of antioxidant, antidiabetic and anticancer properties. *Chemical Data Collections*, 33, 110714.

Korde, P., Ghotekar, S., Pagar, T., Pansambal, S., Oza, R. & Mane, D. 2020. Plant extract assisted eco-benevolent synthesis of selenium nanoparticles-a review on plant parts involved, characterization and their recent applications. *Journal of Chemical Reviews*, 2, 157–168.

Kottappara, R., Palantavida, S. & Vijayan, B. K. 2020. Reduce derivatives. *Green chemistry and applications.*

Larhed, M. & Hallberg, A. 2001. Microwave-assisted high-speed chemistry: A new technique in drug discovery. *Drug Discovery Today*, 6, 406–416.

Leder, C., Rastogi, T. & Kümmerer, K. 2015. Putting benign by design into practice-novel concepts for green and sustainable pharmacy: Designing green drug derivatives by non-targeted synthesis and screening for biodegradability. *Sustainable Chemistry and Pharmacy*, 2, 31–36.

Lee, M. & Chang, E. B. 2021. Inflammatory bowel diseases (IBD) and the microbiome—searching the crime scene for clues. *Gastroenterology*, 160, 524–537.

Li, C.-J. & Anastas, P. T. 2012. Green chemistry: Present and future. *Chemical Society Reviews*, 41, 1413–1414.

Li, Y., Zheng, Y., Zhang, Y., Yang, Y., Wang, P., Imre, B., Wong, A. C., Hsieh, Y. S. & Wang, D. 2021. Brown algae carbohydrates: Structures, pharmaceutical properties, and research challenges. *Marine Drugs*, 19, 620.

Li, Z.-H., Cai, M., Liu, Y.-S. & Sun, P.-L. 2018. Development of finger citron (*Citrus medica L.* var. *sarcodactylis*) essential oil loaded nanoemulsion and its antimicrobial activity. *Food Control*, 94, 317–323.

Lineberry, T. W. & Bostwick, J. M. 2006. Methamphetamine abuse: A perfect storm of complications. Mayo Clinic Proceedings. 77–84.

Liu, J.-M. & Yan, X.-P. 2012. Competitive aptamer bioassay for selective detection of adenosine triphosphate based on metal-paired molecular conformational switch and fluorescent gold nanoclusters. *Biosensors and Bioelectronics*, 36, 135–141.

Liyanage, N., Nagahawatta, D., Jayawardena, T. U. & Jeon, Y.-J. 2023. The role of seaweed polysaccharides in gastrointestinal health: Protective effect against inflammatory bowel disease. *Life*, 13, 1026.

Llinàs, M. C., Martínez-Edo, G., Cascante, A., Porcar, I., Borrós, S. & Sánchez-GARCÍA, D. 2018. Preparation of a mesoporous silica-based nano-vehicle for dual DOX/CPT pH-triggered delivery. *Drug Delivery*, 25, 1137–1146.

Machado, S., González-Ballesteros, N., Gonçalves, A., Magalhães, L., Sárria Pereira De Passos, M., Rodríguez-Argüelles, M. C. & Castro Gomes, A. 2021. Toxicity in vitro and in zebrafish embryonic development of gold nanoparticles biosynthesized using *Cystoseira* macroalgae extracts. *International Journal of Nanomedicine*, 5017–5036.

Magalingam, K. B., Radhakrishnan, A., Ping, N. S. & Haleagrahara, N. 2018. Current concepts of neurodegenerative mechanisms in Alzheimer's disease. *BioMed Research International*, 2018, 3740461.

Mahato, A. K., Sahoo, B. M., Banik, B. K. & Mohanta, B. C. 2018. Microwave-assisted synthesis: Paradigm of green chemistry. *Journal of the Indian Chemical Society*, 95, 1327–1339.

Manjula, N., Sarma, G., Shilpa, B. M. & Suresh Kumar, K. 2022. Environmental applications of green engineered copper nanoparticles. *Phytonanotechnology*.

Mayegowda, S. B., Sarma, G., Gadilingappa, M. N., Alghamdi, S., Aslam, A., Refaat, B., Almehmadi, M., Allahyani, M., Alsaiari, A. A. & Aljuaid, A. 2023. Green-synthesized nanoparticles and their therapeutic applications: A review. *Green Processing and Synthesis*, 12, 20230001.

Meijers, B. K., Poesen, R. & Evenepoel, P. 2014. Heparin-coated dialyzer membranes: Is non-inferiority good enough? *Kidney International*, 86, 1084–1086.

Metselaar, J. M., Wauben, M. H., Wagenaar-Hilbers, J. P., Boerman, O. C. & Storm, G. 2003. Complete remission of experimental arthritis by joint targeting of glucocorticoids with long-circulating liposomes. *Arthritis & Rheumatism: Official Journal of the American College of Rheumatology*, 48, 2059–2066.

Mishra, M., Sharma, M., Dubey, R., Kumari, P., Ranjan, V. & Pandey, J. 2021. Green synthesis interventions of pharmaceutical industries for sustainable development. *Current Research in Green and Sustainable Chemistry*, 4, 100174.

Mohammadi Ziarani, G., Kheilkordi, Z. & Gholamzadeh, P. 2020. Ultrasound-assisted synthesis of heterocyclic compounds. *Molecular Diversity*, 24, 771–820.

Mozaffari, S., Yasuj, S. R., Motaghinejad, M., Motevalian, M. & Kheiri, R. 2019. Crocin acting as a neuroprotective agent against methamphetamine-induced neurodegeneration via CREB-BDNF signaling pathway. *Iranian Journal of Pharmaceutical Research: IJPR*, 18, 745.

Mukherjee, A., Sarkar, D. & Sasmal, S. 2021. A review of green synthesis of metal nanoparticles using algae. *Frontiers in Microbiology*, 12, 693899.

Mureşan, A. C. 2018. Ibuprofen: Original versus green synthesis. *The Annals of "Dunarea de Jos" University of Galati. Fascicle IX, Metallurgy and Materials Science*, 41, 30–34.

Nagore, P., Ghotekar, S., Mane, K., Ghoti, A., Bilal, M. & Roy, A. 2021. Structural properties and antimicrobial activities of *Polyalthia longifolia* leaf extract-mediated CuO nanoparticles. *BioNanoScience*, 11, 579–589.

Nath, D. & Banerjee, P. 2013. Green nanotechnology–a new hope for medical biology. *Environmental Toxicology and Pharmacology*, 36, 997–1014.

Nerkar, A. G., Pawale, D., Ghante, M. R. & Sawant, S. D. 2013. Microwave assisted organic synthesis of some traditional and named reactions: A practical approach of green chemistry. *International Journal of Pharmaceutical Sciences*, 5, 564–566.

Nethi, S. K., Das, S., Patra, C. R. & Mukherjee, S. 2019. Recent advances in inorganic nanomaterials for wound-healing applications. *Biomaterials Science*, 7, 2652–2674.

Ovais, M., Ahmad, I., Khalil, A. T., Mukherjee, S., Javed, R., Ayaz, M., Raza, A. & Shinwari, Z. K. 2018. Wound healing applications of biogenic colloidal silver and gold nanoparticles: Recent trends and future prospects. *Applied Microbiology and Biotechnology*, 102, 4305–4318.

Pagar, T., Ghotekar, S., Pagar, K., Pansambal, S. & Oza, R. 2019. A review on bio-synthesized Co3O4 nanoparticles using plant extracts and their diverse applications. *Journal of Chemical Reviews*, 1, 260–270.

Panara, K., Joshi, K. & Nishteswar, K. 2012. A review on phytochemical and pharmacological properties of *Citrus medica* Linn. *International Journal of Pharmaceutical & Biological Archive*, 3, 1292–1297.

Pansambal, S., Oza, R., Borgave, S., Chauhan, A., Bardapurkar, P., Vyas, S. & Ghotekar, S. 2023. Bioengineered cerium oxide (CeO2) nanoparticles and their diverse applications: A review. *Applied Nanoscience*, 13, 6067–6092.

Patel, R. N. 2016. Green processes for the synthesis of chiral intermediates for the development of drugs. *Green Biocatalysis*. Wiley; 71–114.

Patneedi, C. B. & Prasadu, K. D. 2015. Impact of pharmaceutical wastes on human life and environment. *Rasayan Journal of Chemistry*, 8, 67–70.

Patra, J. K., Das, G., Fraceto, L. F., Campos, E. V. R., Rodriguez-Torres, M. D. P., Acosta-Torres, L. S., Diaz-Torres, L. A., Grillo, R., Swamy, M. K. & Sharma, S. 2018. Nano based drug delivery systems: Recent developments and future prospects. *Journal of Nanobiotechnology*, 16, 1–33.

Pedro, J., Cardoso, C., Afonso, F. & Bandarra, N. 2022. Season affects three insufficiently studied seaweed species (*Bifurcaria bifurcata, Codium sp., Ericaria selaginoides*): Bioactivity alterations. *Applied Phycology*, 3, 98–108.

Peirouvi, T. & Razi, M. 2022. Molecular mechanism behind methamphetamine-induced damages in testicular tissue: Evidences for oxidative stress, autophagy, and apoptosis. *Andrologia*, 54, e14534.

Pourmadadi, M., Rahmani, E., Eshaghi, M. M., Shamsabadipour, A., Ghotekar, S., Rahdar, A. & Ferreira, L. F. R. 2023. Graphitic carbon nitride (g-C3N4) synthesis methods, surface functionalization, and drug delivery applications: A review. *Journal of Drug Delivery Science and Technology*, 79, 104001.

Rafique, M., Sadaf, I., Rafique, M. S. & Tahir, M. B. 2017. A review on green synthesis of silver nanoparticles and their applications. *Artificial Cells, Nanomedicine, and Biotechnology*, 45, 1272–1291.

Ravindran, A., Chandran, P. & Khan, S. S. 2013. Biofunctionalized silver nanoparticles: Advances and prospects. *Colloids and Surfaces B: Biointerfaces*, 105, 342–352.

Richards, P. J., Williams, A., Goodfellow, R. M. & Williams, B. D. 1999. Liposomal clodronate eliminates synovial macrophages, reduces inflammation and ameliorates joint destruction in antigen-induced arthritis. *Rheumatology*, 38, 818–825.

Saghafi, T., Taheri, R. A., Parkkila, S. & Zolfaghari Emameh, R. 2019. Phytochemicals as modulators of long non-coding RNAs and inhibitors of cancer-related carbonic anhydrases. *International Journal of Molecular Sciences*, 20, 2939.

Sargazi, S., Laraib, U., Er, S., Rahdar, A., Hassanisaadi, M., Zafar, M. N., Diez-Pascual, A. M. & Bilal, M. 2022. Application of green gold nanoparticles in cancer therapy and diagnosis. *Nanomaterials*, 12, 1102.

Sathyanarayanan, S., Paul, B. M. & Thangaraj, P. 2022. Green synthesis of nanoparticles from herbals for biomedical applications: A systematic review. *Nanophytomedicine*, 89–96.

Sepehr, A., Taheri, F., Heidarian, S., Motaghinejad, M. & Safari, S. 2020. Neuroprotective and neuro-survival properties of safinamide against methamphetamine-induced neurodegeneration: Hypothetic possible role of BDNF/TrkB/PGC-1α signaling pathway and mitochondrial uncoupling protein– 2 (UCP-2). *Medical Hypotheses*, 143, 110094.

Shailaja, P. & Priyanka, P. 2021. Review on regulatory insights of green chemistry and sustainability. *International Journal of Biology, Pharmacy and Allied Sciences*, 10(12), 98–109.

Shalaby, E. A., Shanab, S. M., El-Raheem, W. M. A. & Hanafy, E. A. 2022. Biological activities and antioxidant potential of different biosynthesized nanoparticles of *Moringa oleifera*. *Scientific Reports*, 12, 18400.

Shan, Y., Lee, M. & Chang, E. B. 2022. The gut microbiome and inflammatory bowel diseases. *Annual Review of Medicine*, 73, 455–468.

Sharma, A., Priya, A., Kaur, M., Singh, A., Kaur, G. & Banerjee, B. 2021a. Ultrasound-assisted synthesis of bioactive S-heterocycles. *Synthetic Communications*, 51, 3209–3236.

Sharma, D., Sharma, V., Sharma, A., Goyal, R., Tonk, R. K., Thakur, V. K. & Sharma, P. C. 2021b. Green chemistry approaches for thiazole containing compounds as a potential scaffold for cancer therapy. *Sustainable Chemistry and Pharmacy*, 23, 100496.

Sharma, P. C., Bansal, K. K., Sharma, A., Sharma, D. & Deep, A. 2020. Thiazole-containing compounds as therapeutic targets for cancer therapy. *European Journal of Medicinal Chemistry*, 188, 112016.

Sheldon, R. A. 2016. Green chemistry and resource efficiency: Towards a green economy. *Green Chemistry*, 18, 3180–3183.

Sheldon, R. A., Bode, M. L. & Akakios, S. G. 2022. Metrics of green chemistry: Waste minimization. *Current Opinion in Green and Sustainable Chemistry*, 33, 100569.

Singh, H., Kumar, A., Thakur, A., Kumar, P., Nguyen, V.-H., Vo, D.-V. N., Sharma, A. & Kumar, D. 2020. One-pot synthesis of magnetite-ZnO nanocomposite and its photocatalytic activity. *Topics in Catalysis*, 63, 1097–1108.

Singh, K., Srichairatanakool, S., Chewonarin, T., Prommaban, A., Samakradhamrongthai, R. S., Brennan, M. A., Brennan, C. S. & Utama-ang, N. 2022. Impact of green extraction on curcuminoid content, antioxidant activities and anti-cancer efficiency (in vitro) from turmeric rhizomes (*Curcuma longa L.*). *Foods*, 11, 3633.

Souza, J., Santos, G. S., Barbosa, T. C., Santos, A. L., Menezes, L. G., Andrade, L. N., Bani, C., Sarmento, V. H., Nunes, R. R. & Zielinska, A. 2021. Citrus sinensis essential oil-based microemulsions: Green synthesis, characterization, and antibacterial and larvicide activities. *ACS Food Science & Technology*, 1, 462–469.

Stevenson, R., Hueber, A. J., Hutton, A., Mcinnes, I. B. & Graham, D. 2011. Nanoparticles and inflammation. *The Scientific World Journal*, 11, 1300–1312.

Suffredini, E., Mioni, R., Mazzette, R., Bordin, P., Serratore, P., Fois, F., Piano, A., Cozzi, L. & Croci, L. 2014. Detection and quantification of *Vibrio parahaemolyticus* in shellfish from Italian production areas. *International Journal of Food Microbiology*, 184, 14–20.

Sukweenadhi, J., Setiawan, K. I., Avanti, C., Kartini, K., Rupa, E. J., & Yang, D. C. 2021. Scale-up of green synthesis and characterization of silver nanoparticles using ethanol extract of *Plantago major* L. leaf and its antibacterial potential. *South African Journal of Chemical Engineering*, 38(1), 1–8.

Sulzer, D., Sonders, M. S., Poulsen, N. W. & Galli, A. 2005. Mechanisms of neurotransmitter release by amphetamines: A review. *Progress in Neurobiology*, 75, 406–433.

Tata, D. A. & Yamamoto, B. K. 2007. Interactions between methamphetamine and environmental stress: Role of oxidative stress, glutamate and mitochondrial dysfunction. *Addiction*, 102, 49–60.

Taylor, J. P., Brown JR, R. H. & Cleveland, D. W. 2016. Decoding ALS: From genes to mechanism. *Nature*, 539, 197–206.

Thongtra, P. & Sapronova, A. June 26–29, 2017. Time-Series Data Analytics using Spark and Machine Learning. Foundations of Intelligent Systems: 23rd International Symposium, ISMIS 2017, Warsaw, Poland, Proceedings 23. 509–515.

Varga, Z., Rafay Ali Sabzwari, S. & Vargova, V. 2017. Cardiovascular risk of nonsteroidal anti-inflammatory drugs: An under-recognized public health issue. *Cureus*, 9(4), e1144.

Venkatas, J., Daniels, A. & Singh, M. 2022. The potential of curcumin-capped nanoparticle synthesis in cancer therapy: A green synthesis approach. *Nanomaterials*, 12, 3201.

Virk, P., Awad, M. A., Alsaif, S. S. A.-L., Hendi, A. A., Elobeid, M., Ortashi, K., Qindeel, R., El-Khadragy, M. F., Yehia, H. M. & El-Din, M. F. S. 2023. Green synthesis of *Moringa oleifera* leaf nanoparticles and an assessment of their therapeutic potential. *Journal of King Saud University-Science*, 35, 102576.

Vizetto-Duarte, C., Custódio, L., Barreira, L., Moreira Da Silva, M., P. Rauter, A., Albericio, F. & Varela, J. 2016. Proximate biochemical composition and mineral content of edible species from the genus *Cystoseira* in Portugal. *Botanica Marina*, 59, 251–257.

Wang, F., Li, Q., Xu, T., Li, Z., Jiang, Y., Ma, Y., Li, X., Wang, W. & Qian, H. 2022. An orally administered gold nanocluster with ROS scavenging for inflammatory bowel disease treatment. *Fundamental Research*.

Wang, R., Li, Z., Liu, S. & Zhang, D. 2023. Global, regional and national burden of inflammatory bowel disease in 204 countries and territories from 1990 to 2019: A systematic analysis based on the Global Burden of Disease Study 2019. *BMJ Open*, 13, e065186.

Wang, Z., Zhang, J., Luo, H., Ye, Y., Yan, J. & Hou, Y. 2013. Screening and confirmation of microRNA markers for forensic body fluid identification. *Forensic Science International: Genetics*, 7, 116–123.

Yang, Y., Liang, M., Ouyang, D., Tong, H., Wu, M. & Su, L. 2022. Research progress on the protective effect of brown algae-derived polysaccharides on metabolic diseases and intestinal barrier injury. *International Journal of Molecular Sciences*, 23, 10784.

Yen, T. H., Van Thuan, D., Hanh, N. T., Vy, N. H. T., Hang, T. T. M., Van Ha, H., Pham, T.-D., Sharma, A. K., Nguyen, M.-V. & Dang, N.-M. 2020. Synthesis of N and S Co-doped TiO 2 nanotubes for advanced photocatalytic degradation of volatile organic compounds (VOCs) in gas phase. *Topics in Catalysis*, 63, 1077–1085.

Zeng, X.-S., Geng, W.-S., Jia, J.-J., Chen, L. & Zhang, P.-P. 2018. Cellular and molecular basis of neurodegeneration in Parkinson disease. *Frontiers in Aging Neuroscience*, 10, 109.

Zenych, A., Fournier, L. & Chauvierre, C. 2020. Nanomedicine progress in thrombolytic therapy. *Biomaterials*, 258, 120297.

Zhang, M. & Merlin, D. 2018. Nanoparticle-based oral drug delivery systems targeting the colon for treatment of ulcerative colitis. *Inflammatory Bowel Diseases*, 24, 1401–1415.

Zhang, R.-L., Lei, B.-X., Wu, G.-Y., Wang, Y.-Y. & Huang, Q.-H. 2023. Protective effects of berberine against β-amyloid-induced neurotoxicity in HT22 cells via the Nrf2/HO-1 pathway. *Bioorganic Chemistry*, 133, 106210.

Zhang, X., Servos, M. R. & Liu, J. 2012. Ultrahigh nanoparticle stability against salt, pH, and solvent with retained surface accessibility via depletion stabilization. *Journal of the American Chemical Society*, 134, 9910–9913.

Zhao, W., Liu, Q., Zhang, X., Su, B. & Zhao, C. 2018. Rationally designed magnetic nanoparticles as anticoagulants for blood purification. *Colloids and Surfaces B: Biointerfaces*, 164, 316–323.

Zhu, S., Jiang, X., Boudreau, M. D., Feng, G., Miao, Y., Dong, S., Wu, H., Zeng, M. & Yin, J.-J. 2018. Orally administered gold nanoparticles protect against colitis by attenuating Toll-like receptor 4-and reactive oxygen/nitrogen species-mediated inflammatory responses but could induce gut dysbiosis in mice. *Journal of Nanobiotechnology*, 16, 1–18.

Zimmerman, J. B., Anastas, P. T., Erythropel, H. C. & Leitner, W. 2020. Designing for a green chemistry future. *Science*, 367, 397–400.

8 Phytomedicines and Management of Disease

Malik Saadullah, Saher Madni, Areej Asghar, Sameen Qamar, Shaḥid Shah, and Zia Ur Rehman*

8.1 INTRODUCTION

Phytomedicines, or herbal medicines, are concoctions of secondary metabolites present in plants. These metabolites have substances that show pharmacological activeness as well as therapeutic features. These medicines have a wide range of biological effects. They have been used extensively throughout history. Plant parts like roots, stems, leaves, barks, tubers, herbs, or whole plant extracts are used to make phytomedicines. The active substances are terpenes, glycosides, phenolic compounds, alkaloids, basic metabolites, and secondary metabolites secreted by these plant parts. Plant-based medicinal preparations include decoctions, emulsions, apozems, liniments, electroactive, and powders. Developing new and cutting-edge therapeutic goods and strategies is likely to be one of the most exciting aspects of the future of plant-derived medicines. The significant secondary metabolites have unique pharmacological characteristics that are beneficial to the human body [1].

In ancient times, the isolation, extraction, and purification of medications from medicinal plants led to the creation of miraculous chemicals that treat human illness directly or indirectly today. Biomolecules are continually being analysed and characterised in phytomedicine in order to find new uses for them. Natural treatments have given rise to new hopes in the context of rising superinfections, new superbugs, and strains that are resistant to all infectious species. If phytomedicine is merged with other medical science disciplines while maintaining a more comprehensive multidisciplinary approach, it could show to be a game-changing move in the current healthcare landscape [2].

Disease management refers to a collection of coordinated actions intended to prevent or control one or more chronic disorders through the use of a methodical, interdisciplinary approach and maybe a number of different treatment modalities. Disease management's primary goal is to provide effective and efficient care for people with one or more chronic conditions, regardless of the location of the patient's care or the usual payment methods. Disease management uses a multidisciplinary strategy for certain patients to enhance care quality and lower costs. It is a strategy to managing or preventing chronic disorders that emphasizes patient self-management, risk stratification, and the best clinical results, efficacy, and efficiency feasible [3].

After the rise in ayurvedic and traditional Chinese medicine (TCM) acceptance, data shows that about 70 to 80% of the population relies on plant remedies for their healthcare as a primary tool. Africa has a very high utilisation of traditional medicine. Furthermore, even those living in industrialised nations rely on plants in one way or another for their health. Twenty-five per cent of the prescriptions filled in neighbourhood pharmacies in the US contained plant extracts or active substances taken from plants. Since ancient times, medicinal plants have been used as trusted natural sources to cure a variety of illnesses. The World Health Organization (WHO) has reported that 20,000 different species of plants are used for therapeutic purposes. In addition, due to their unrivalled abundance variety of chemical, and natural products. They may be raw chemicals or standardised extracts of plants which offer limitless prospects for new drug leads [4].

*Corresponding author

DOI: 10.1201/9781032707273-8

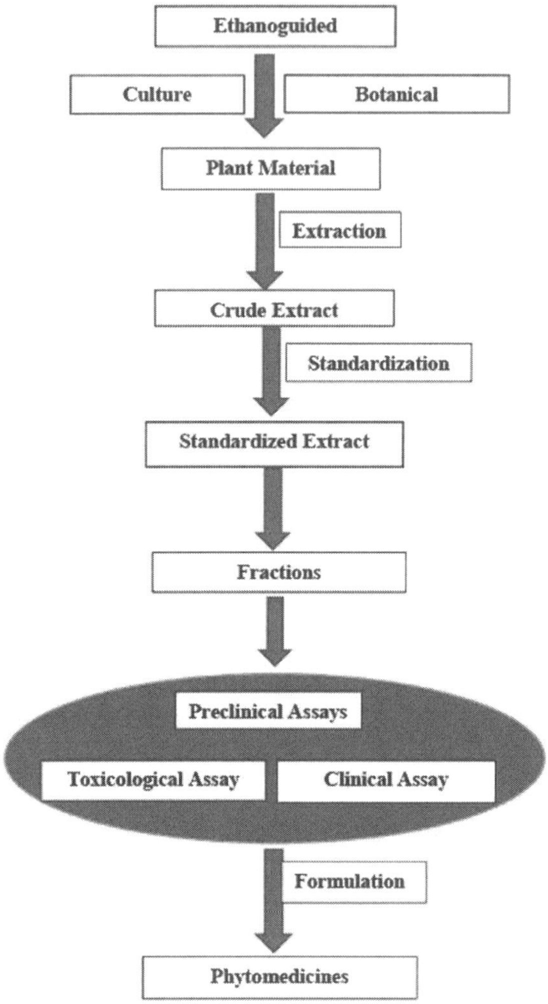

Figure 8.1 Flow chart for the development of phytomedicines.

8.2 STEPS INVOLVED IN DEVELOPING PHYTOMEDICINES

The steps involved in developing phytomedicines are given in Figure 8.1.

8.3 PHYTOMEDICINE USED IN JOINT DISORDERS

An increase in oxidative stress as well as inflammation are responsible for such chronic inflammatory joint diseases like osteoarthritis (OA) and rheumatoid arthritis (RA).

OA affects about 15% world population. This is represented by irreparable damage of bone erosion and articular cartilage. The factors responsible for this disease are interleukin-1 (IL-1), interleukin-6 (IL-6), and tumour necrosis factor (TNF) which are pro-inflammatory cytokines. These mediators not only boost collagenase or matrix metalloproteinase (MMP) production but also collagen type II breakdown by decreasing inhibitor of collagenase, proteoglycan synthesis and collagen. Most of the factors that trigger OA are biomechanical stress (e.g. pressure, load-holding capacity), sex, age, obesity or higher value of body mass index (BMI), genetics, food habits, trauma, race and definite bodily or work-related activities across the joints.

One per cent of the population is affected by RA that is a chronic and progressive autoimmune disease of a system that not only increases the risk of cancer but also causes heart diseases and ultimately death. It is frequently linked to oxidative stress as well as inflammatory mediators. Nonsteroidal anti-inflammatory drugs (NSAIDs, also known as pain removers), along with potent biological therapies that target specific pathways of immunological and inflammatory systems, for example, α-TNF, TNF inhibitors and IL-1 receptor antagonists, are used to treat RA nowadays. Acetaminophen is the most commonly used NSAID in extremely high doses (4000 mg/day). Tramadol, as well as other opioids (such as morphine), are widely recommended as pain relievers. TNF inhibitors used in the treatment of severe RA include etanercept, infliximab, and rituximab. Anakinra (an IL-1 receptor antagonist) and methotrexate are found to be two more RA treatments. Biologic therapy has been shown to be highly successful and effective in the vast majority of RA cases, including the most severe.

Unluckily, serious side effects that occur after the use of recommended drugs in these arthropathic diseases include: ulcerations of the gastrointestinal tract, haemorrhagic events, NSAID-related nephrotoxicity, hypersensitivity reactions as well auto-immune responses (e.g. lupus-like syndrome) which is triggered by inhibitors of TN. These biological drugs (anakinra) are responsible for higher risk of severe infection, greatly targeting the tract of respiratory system. As a result, there is increasing interest in plant treatments, which have no serious side effects and have been used for millennia. These therapies may help not just with symptoms but also with the progression of the condition (Table 8.1).

Table 8.1: Phytochemicals Responsible for Management of Osteoarthritis and Rheumatoid Arthritis [4])

Plant Name	Family	Part Used	Phytochemicals
Arnica montana	Asteraceae	Fresh plant	Flavonoid as well as phenolic compounds
Boswellia spp.	Burseraceae	Resins (gums)	Boswellic acids (pentacycyclic triterpenes)
Curcuma spp.	Zingiberaceae	Rhizome	Diferuloylmethane (curcumin)
Equisetum arvense (horsetail)	Equisetaceae	Plant	Kynurenic acid
Harpagophytum procumbens (devil's claw)	Pedaliaceae	Tubers and roots	Glycosides (harpagoside, harpagide, and procumbide)
Panax notoginseng	Araliaceae	Plant	Saponins
Salix spp.	Salicaceae	Willow bark	Salicin
Sesamum indicum	Pedaliaceae	Plant	Lignans (sesamin, sesamolin)
Symphytum officinalis (comfrey)	Boraginaceae	Roots	Phenolic acids (rosmarinic acid), glycopeptides
Zingiber officinalis (ginger)	Zingiberaceae	Rhizome	1-Dehydro-gingerdione
Whitania somnifera (ashwagandha)	Solanaceae	Roots	Withaferin A

8.4 PHYTOMEDICINES USED FOR MANAGEMENT OF DIABETES MELLITUS

Persistent hyperglycaemia caused by abnormalities in secretions of insulin by cells of pancreas, decreased insulin sensitivity of cell surface receptors, or both characterizes diabetes mellitus, a metabolic and/or hormonal disorder. Diabetes mellitus comes in four main forms: type 1, type 2, gestational diabetes mellitus, and "other specific types of diabetes mellitus." Signs and symptoms are the acute effects of incorrect management or uncontrolled hyperglycaemia. If ignored or overlooked, these warning signs and symptoms may lead to the development of chronic illnesses like hypertension, stroke, blindness, erectile dysfunction, and renal failure. Everyone is impacted by this degenerative metabolic condition, which is on the rise globally; 592 million individuals are expected to develop diabetes mellitus by the end of the year. The immediate effects of improper treatment or uncontrolled hyperglycaemia are signs and symptoms. These warning signs and symptoms, if disregarded or disregarded, may result in the onset of chronic illnesses like hypertension, stroke, blindness, erectile dysfunction, and renal failure. This degenerative metabolic disorder, which is spreading over the world, affects everyone. By year's end, 592 million people are anticipated to have diabetes mellitus.

Medication, exercise, and dietary modifications are used to address this chronic condition. The type of diabetes mellitus, the patient's age, their response, and other factors all affect how anti-diabetic pharmacotherapeutic medicines are administered. Pharmacotherapy typically entails (i) drugs that facilitate or facilitate the release of insulin from pancreatic islet cells, (ii) drugs that increase the sensitivity of insulin receptors or decrease insulin resistance, (iii) drugs that slow the absorption of glucose, and (iv) drugs that prevent protein glycation. Currently, orthodox treatments for diabetes mellitus include insulin, biguanides, sulfonylureas, inhibitors of -glucosidase and -amylase, aldose reductase inhibitors, thiazolidinediones, dipeptidyl peptidase-4 (DPP-4) inhibitors, carbamoylmethyl benzoic acid, and insulin-like growth factors.

The majority of commonly prescribed medications have a long list of side effects, including hypoglycaemia, diarrhoea, weakness, exhaustion, shortness of breath nausea, weight gain, dizziness, hepatotoxicity, lactic acidosis, lactic acid intoxication, an increase in LDL cholesterol levels and renal toxicity. These medications are frequently quite pricey as well. Due to the high cost of managing diabetes mellitus, many natural remedies that are less expensive must be used. Diabetes mellitus can be treated using herbal remedies in both rural as well as urban areas. According to the recommendations of WHO, a variety of factors, including economic hardships, the population's geographic difficulty in accessing conventional antidiabetics, insufficient healthcare systems, the ease of access to herbal medicine, and local knowledge of the people, as well as other factors, contribute to the majority of Africans having diabetes (Table 8.2).

8.5 PHYTOMEDICINE IN THE MANAGEMENT OF NEUROLOGICAL DISORDERS

Nervous system is a complex and intricate system that regulates and coordinates the basic functions of the body. It governs our entire body. If there is a problem with our nervous system, all of our body's functions are disrupted. As a result, neurological problems occur, and treatment is critical. Nowadays, most individuals rely on herbal medications for primary healthcare, despite the fact that these conditions are no longer considered common. Various individuals with mental disabilities, such as epilepsy, are severely harmed by health-related stigma and intolerance regulations. Thus, epilepsy is the most dangerous

Table 8.2: Phytochemicals Used for Diabetes Mellitus Management [5]

Plant Name	Family	Part Used	Phytochemicals
Adinia lobata	Passifloraceae	Bark and stem	Palmitic acid
Allium cepa	Amyryllidaceae	Bulb	Allyl propyl disulphide
Allium sativum	Liliaceae	Bulbs	Allyl propyl disulphide, allicin
Aloe barbadensis	Asphodelaceaceae	leaf	Lophenol, 24-dimethyl lophenol, 24-methylene cyclartenol, cyclartenol, 24-ethyl lophenol
Azadirichta indica	Maliaceae	Leaf, flower, seed	Nimbidin, β-sitosterol
Cassia auriculata	Fabaceae	Flower	β-sitosterol
Cinamomum zeylancium	Lauraceae	Bark	Cinnamaldehyde, eugenol
Costus afer Ker Gawl	Costaceae	Whole plant	Diosgenin
Mangifera indica	Anacardiaceae	Leaf, stem, bark, fruit	Mangiferin
Momordica charantia	Cucurbetaceae	Leaves, whole plant, fruit	Charantin, momordicin, Oleanolic acid, vicine
Scoparia dulcis	Scrophulariaceae	Whole plant	Apigeninen, luteolin, scoparic acid dcoxicol, vicine
Zingiber officinale	Zingiberaceae	Bulb	Gingerol

common chronic neurological condition, affecting around 10 million people worldwide. Other illnesses like Parkinson's and Alzheimer's are also becoming more prevalent. Breathing, mood, speaking, movement, and memory are all affected by neurological problems. Our bodies' brain and spinal cord are affected in neurological illnesses. As a result, neurological disorders are more expensive to treat and have more adverse effects. Around 2900 herbal medicines are used to treat neurological problems by reducing adverse effects and curing the ailment. In comparison to conventional drugs, herbal medicines have a wide range of safety.

8.5.1 Migraine

Migraine is a form of headache characterised by recurrent moderate to severe headaches. Migraine is usually one-sided and is accompanied by photophobia and nausea, all of which have a detrimental impact on the patient's life. It can begin at any age, but the first attack usually happens during puberty or in old age. Migraine headache is a chronic headache characterised by pulsing pain on one side of the head. However, it usually spreads all over the head. Migraine is also linked to cervical artery disease, structural changes in the brain, and stroke. Not only genetic but there are also some environmental factors that can contribute to the development and severity of headache. Sometimes optimal migraine intervention can enhance patients' health and prevent other migraine-related disorders. Numerous treatments are tried to prevent and reduce the severity of pain. Despite this, antidepressants, beta-blockers, and anticonvulsant medications are widely utilised (Table 8.3). Patients must deal with issues such as depression, tremor, constipation, and weight gain, which are unavoidable during migraine treatment.

Table 8.3: Phytochemicals Responsible for Management of Migraine [7]

Plant Name	Local Name	Family	Part Used
Matricaria recutita	Chamomile	Compositae	Patel
Paliurus spina	Sipa	Rhamnaceae	Leaf, stem
Violo tricolor	Gole banoushe	Umbelliferae	Flower, branches
Curcuma longa	Turmeric	Zingiberaceae	Rhizome
Zingiber officinale	Ginger	Zingiberaceae	Rhizome
Syzygium aromaticum	Clove	Myrtaceae	Fruit
Allium sativum	Garlic	Amaryllidaceae	Bulb
Coriandrum sativum	Coriander	Apiaceae	Seed
Angelica sinensis	Dang Gui	Umbellifereae	Dried root
Harpagophytum procumbens	Grapple	Pedaliaceae	Plant roots, tubers
Boswellia serrate	Guggulu	Burseraceae	Resin
Tanacetum parthenium	Feverfew/ Featherfew	Asteraceae	Dried leaves
Cannabis sativa	Bhang	Cannabinaceae	Fruit
Hypericum perforatum	Choli phulya	Hypericaceae	Plant extract
Corydalis yanhusuo	Asian corydalis	Papaveraceae	Roots
Bryonia alba	White bryony	Cucurbitaceae	Roots
Valeriana officianalis	Tager	Caprifoliaceae	Roots
Eschscholtzia californica	California poppy	Papaveraceae	Areal part
Salix alba	Bhushan	Salicaceae	Bark
Populus tremuloides	White poplar	Salicaceae	Bark
Capsicum frutescens	Cayenne pepper	Solanaceae	Fruit
Symphytum officinale	Comfrey	Boraginaceae	Roots and leaves
Ginkgo biloba	Ginkgo	Ginkgoaceae	Leaf
Centella asiatica	Mandookparni	Umbellifers	Whole plant
Viburnum opulus	Cramp bark	Adoxaceae	Bark and root
Viburnum prunifolium	Black haw	Adoxaceae	Bark
Scutellaria laterifolia	Skullcap	Lamiaceae	Arial parts
Scutellaria baicalensis	Huang qin	Lamiaceae	Arial parts
Rosa canina	Dog rose	Rosaceae	Fruits
Solidago chilensis	Brazilian arnica	Asteraceae	Arial parts
Piper Longum	Piperaceae	Piperaceae	Fruit
Eclipta alba	Bhringaraj	Asteraceae	Whole plant
Cinnmomum zeylanica	Dalchini	Lauraceae	Bark
Bambusa arundinaceae	Bamboo	Poaceae	Vansh lochan
Ellettaria cardamomum	Elaichi	Zingiberaceae	Fruit
Cinnamomum Cassia	Chinese cassia	Lauraceae	Bark
Ferula northax	Devil's dung	Apiaceae	Resin
Citrus acida	Lamon	Rutaceae	Fruit
Petasites Hybridus	Butterbur	Asteraceae	Leaves
Sapindus trifoliatus	Reetha	Sapindaceae	Fruit
Citrus limon	Nimboo	Rutaceae	Fruit

8.5.2 Epilepsy

It is the most prevalent chronic neurological illness and is distinguished by recurring epileptic seizures without cause. These brief signs or symptoms, known as seizures, are brought on by abnormally high or coordinated neuronal activity found in the brain. Around 50 million people worldwide are impacted by it. It is described as evidence of irregular and paroxysmal brain neuronal discharges. On the basis of their clinical phenomenon, the various seizure types can be identified. Although not all epilepsy patients respond well to conventional medical treatments, anti-epileptic medications are widely accessible to treat epileptic symptoms by reducing adverse effects on the body. The best and healthiest medications work well to treat anti-epileptic seizure problems. Therefore, ayurvedic herbal medicines have traditionally been used to treat and inhibit patients' seizure conditions. It is a neurological illness characterised by abrupt, recurrent bouts of sensory disturbance, aberrant, chaotic nerve cell discharge in the brain, and a temporary impairment of motor, sensory, or mental function (Table 8.4).

Table 8.4: Phytochemicals Responsible for Management of Epilepsy [7]

Plant Name	Local Name	Family	Part Used
Bocopa monneria	Brahmi	Scrophulariaceae	Whole plant
Cissus sicyoides	Princess vine	Vitaceae	Areal part
Rosa domescana	Rose	Rosaceae	Flower
Glycerrhiza glabra	Yasthimadhu	Fabaceae	Root
Oscimum sanctum	Tulsi	Lamiaceae	Leaves
Pongamia pinnata	Karanj	Papilionaceae	Leaves
Berberis vulgeris	Daruhaldi	Berberidaceae	Roots
Boerhaavia diffusa	Punarnava	Nyctaginaceae	Roots
Butea monosperma	Plash	Fabaceae	Flower
Valeriana officinalis	Tagar	Valerianaceae	Roots
Saussurea lappa	Kushtha	Asteraceae	Roots
Cymbopogon Winterianus	Lemon grass	Poaceae	Leaves
Taxus wallichiana	English yew	Taxaceae	Extract
Nardostachys jatamansi	Jatamansi	Valerianaceae	Root
Dorstenia arifolia	Dorstania	Moraceae	Rhizome
Scutellaria lateriflora	Blue skullcap	Lamiaceae	Arial part
Sutherlandia frutescens	Aavartaki/tarwar	Fabaceae	Shoot
Echium amoenum	Red feathers	Boraginaceae	Flowers
Clerodendrum Infortunatum	Bhantaka	Verbenaceae	Leaves
Argyreia Speciosa	Vriddaru	Convolvulaceae	Leaves
Drosera Burmannii	Mukhjali	Droseraceae	Whole plant
Passion flower	Vidari	Passifloraceae	Leaves and flower
Acanthus montanus	Amarkosh	Acanthaceae	Leaves
Matricaria recutita	Chamomile	Asteraceae	Flowers
Piper methysticum	Kava	Piperaceae	Root
Foeniculum vulgare	Mishreya	Umbelliferea	Fruit

(Continued)

Table 8.4 *(Continued)*: Phytochemicals Responsible for Management of Epilepsy [7]

Plant Name	Local Name	Family	Part Used
Hyssopus officinalis	Hyssop	Lamiaceae	Leaves
Eucalyptus globules	Southern blue gum	Myrtaceae	Leaves
Salvia officinalis,	Sage	Lamiaceae	Leaves
Juniperus Sabina	Savin	Juniperus Sabina	Leaves
Withenia somnifera	Ashwagandha	Solanaceae	Bark
Ficus religiosa	Ashwatha	Moraceae	Bark
Cannabis sativa	Cannabis	Cannabinaceae	Leaves, fruit
Alium sativum	Lehsun	Amaryllidaceae	Bulb
Alium cepa	Pyaaz	Amaryllidaceae	Bulb
Zingiber officinale	Adarak	Zingiberaceae	Rhizome

8.5.3 Parkinson's Disease

One of neurodegenerative ailment known as Parkinson's disease is brought on by the degeneration of brain's neurons. Dopamine levels in the brain start falling as dopaminergic neurons are destroyed by this illness. Parkinson's disease develops because the brain doesn't produce enough dopamine. Dopamine aids in the smooth contraction of human body muscles. The signs and symptoms of this illness appear gradually. Tremor, rigidity, akinesia (loss of movement), and postural instability are just a few of the symptoms that can arise when a person's brain is depleted of dopamine. Various illness phases can cause the symptoms to vary from person to person. Herbal medications are typically used to treat this diseases (Table 8.5).

8.5.4 Alzheimer's Disease

The most frequent mental illness that affects older population is Alzheimer's disease. It is a neurogenerative condition that develops because of brain neuronal degeneration; brain cells suffer damage in this condition. Alzheimer's disease is also considered to be the primary contributor to dementia in elderly people. The two main signs of Alzheimer's disease are emotional instability and difficulty adjusting to family routines. Alzheimer's disease destroys thinking abilities and memory, making it impossible for a person to perform even simple tasks. Because of ageing, this condition may get worse as it is irreversible. Many herbal medications are used to treat Alzheimer's disease because they are less likely to cause side effects than the modern medications and are also much safer (Table 8.6).

8.5.5 Phytomedicines Used in Cardiovascular Diseases

Cardiovascular diseases (CVDs) are the leading cause of death worldwide. In 2013, approximately 17.3 million individuals globally died from CVDs, and by 2030, this number is projected to increase to 23.3 million, according to a WHO bulletin. The underlying pathology of CVD, atheromatous vascular disease, causes a variety of other illnesses, such as coronary artery disease (CAD), cerebrovascular disease, peripheral vascular disease, and the later onset of heart failure and cardiac arrhythmias. The most common form of treatment for CVDs continues to be standard pharmacological therapy, which includes diuretics, vasodilators, anticoagulants, antiplatelet medications, and beta-blockers.

Since the dawn of time, phytomedicines, also referred to as herbal medicine, have helped relieve human suffering, and their use is still expanding across

Table 8.5: Phytochemicals Responsible for Management of Parkinson's Disease [7]

Botanical Name	Local Name	Family	Part Used
Acanthopanax senticosus	Siberian Ginseng	Araliaceae	Root, stem
Alpinia oxyphylla	Alpinia	Zingiberaceae	Fruit
Anemopaegma mirandum	Catuaba	Bignoniaceae	Bark
Mucuna pruriens	Kapikacchu	Fabaceae	Fruits hairs
Withania somnifera	Ashwagandha	Solanaceae	Bark
Vitis Vinifera	Draksha	Vitaceae	Fruit
Centella asiatica	Mandookparni	Apiaceae	Whole plant
Valeriana officinalis	Tagar	Valerianaceae	Root
Bacopa monnieri	Brahmi	Sacrophulaliaceae	Whole plant
Astragalus membranaceus	Astragalus	Leguminosae	Root
Ocimum sanctum	Tulsi	Lamiaceae	Leaves
Cassia obtusifolia	Senna	Leguminosae	Leaves
Camellia sinensis	Caha	Theaceae	Leaves
Ginkgo biloba	Rason	Ginkgoaceae	Bulb
Hypericum perforatum	Choli phuliya	Guttiferae	Arial roots
Ligusticum chuanxiong Hort	Szechuan lovage	Umbelliferae	Root
Sida cordifolia	Bala	Malvaceae	Seed, roots
Plumbago scandens	Chitraka	Plumbaginacea	Leaves, roots
Morus alba	Shahtoot	Moraceae	Fruit
Uncaria rhynchophylla	Khadir	Rubiaceae	Root, bark
Toxicodendron vernicifluum	Rungla	Anacardiaceae	Seed, leaves
Trifolium pretense	Red clover/wild clover	Fabaceae	Flower tops
Scutellaria baicalensis Georgi	Skullcap	Labiatae	Dried root
Psoralea corylifolia	Bakuchi	Leguminosae	Seeds
Chrysanthemum indicum	Chrysanthemum	Asteraceae	Flower
Cistanche deserticola	Rou cong rong	Orobanchaceae	Stem
Citrus tangerine	Mandarin	Rutaceae	Fruits
Clausena lansium	Wampee	Rutaceae	Fruits
Cynodon dactylon	Bermuda grass	Poaceae	Whole plant, root, rhizome
Evolvulus alsinoides	Vishnukrantha	Convolvulaceae	Whole plant
Gynostemma pentaphyllum	Sweet tea vine	Cucurbitaceae	Roots, areal parts
Murraya koenigii	Kadi patta, sweet neem	Rutaceae	Leaves, bark, roots
Nardostachys jatamansi	Jatamansi	Valirenaceae	Rhizome
Paeonia lactiflora	Garden peony	Ranunculaceae	Roots, flowers
Vaccinium cyanococcus	Blueberries	Ericaceae	Leaves, fruits
Selaginella delicatula	Expand	Selaginellaceae	Whole plant
Polygala sibirica	Sanjivani	Polygalaceae	Leaves, roots
Polygonum cuspidatum	Sarpakshee	Polygonaceae	Roots
Pueraria lobata	Bhukushmandi	Leguminosae	Root, leaves, flower

Table 8.6: Phytochemicals Responsible for Management of Alzheimer's Disease [7]

Botanical Name	Local Name	Family	Part Used
Salvia officinalis	Sage	Lamiaceae	Leaves
Melissa officinalis	Lemon balm	Lamiacea	Leaves
Ginkgo biloba	Ginkgo	Ginkgoaceae	Leaf
Polygala tenuifolia	Senega	Polygalaceae	Root
Uncaria rhynchophylla	Fish hook vine	Rubiaceae	Stem, thorn
Huperzia serrate	Toothed clubmoss	Lycopodiaceae	Leaf
Galanthus woronowii	Green snowdrop	Amaryllidaceae	Bulbs
Curcuma longa	Haldi	Zingiberaceae	Rhizome
Vitis Vinifera	DrakshaVitis	Vitaceae	Fruit
lily Lycoris radiate	Red spider-	Amaryllidaceae	Bulb
Lycoris squamigeric	Surprise lily	Amaryllidaceae	Roots
Physostigma venensosum	Calabar bean	Leguminosae	Seeds
Bacopa monniera	Brahmi	Scrophulariaceae	Whole plant
Withania somnifera	Ashwagandha	Solanaceae	Bark
Centella asiatica	Mandookparni	Umbelliferae	Whole plant
Panaxi ginseng	Ginseng	Araliaceae	Root

the globe. This increase in interest is due to the fact that many contemporary ailments are chronic and modern medicine has struggled to effectively treat them. Since half of all of medications approved by FDA are herbal (natural) products and their derived compounds, the scientific community must now acknowledge the exponential global rise in public enthusiasm for using herbal remedies to treat disease. Numerous studies have shown that people who suffer from chronic conditions like lack of sleep (insomnia), chronic fatigue, and CVD frequently turn to natural remedies to treat underlying signs and symptoms.

In order to generate commercial drug formulations, many drug leads have been or are being derived from the variety of biologically active natural compounds found in phytomedicines. Few examples include ephedrine from the plant *Ephedra sinica* (ma-huang), digitoxin from the plant *Digitalis purpurea* (foxglove), salicin (the active component in aspirin) from the plant *Salix alba* (willow bark), and reserpine from the plant *Rauwolfia serpentina* (snakeroot). Similarly, a variety of herbal products have been used to manage CVD on a global scale, and this tradition has been carried down to the current generation (Table 8.7).

8.5.6 Phytomedicines Used Against Covid-19

The zoonotic virus that causes COVID-19, a pandemic illness of the twenty-first century that was recognised by the WHO and spread over the world in late 2019, is the severe acute respiratory syndrome coronavirus 2 (SARS-CoV-2). The COVID-19 symptoms, which are brought on by SARS-CoV-2, have been divided into three groups as per the WHO: the most common symptoms (fever, dry cough, fatigue), the less common symptoms (aches and pains, sore throat, diarrhoea, headache, loss of taste or smell, rash on skin), and the serious symptoms (breathing issues, chest pain or pressure, loss of speech, or movement).

It is fascinating that traditional medicine research is being funded by the WHO for the cure of COVID-19. In addition to its direct application for COVID-19, phytomedicines, or medicinal plants, are used in supportive therapy to increase

Table 8.7: Phytochemicals Responsible for Cardiovascular Diseases Management [8]

Botanical Name	Local Name	Family	Part Used	Phytochemicals
Aconitum spp.	Monk shood	Ranunculaceae	Roots	Aconitine (diterpenoid ester alkaloids)
Allium sativum	Garlic	Alliaceae	Bulb	Allicin, ajoene, and alliin
Citrus paradisi Macfad.	Grapefruit	Rutaceae	Fruit	Bergamottin naringenin
Crataegus spp.	Hawthorn	Rosacea	Whole plant extract	Phenolic acids, flavones
Ephedra sinica	Ma huang or Chinese ephedra	Ephedraceae	Whole plant extract	Alkaloids
Ginkgo biloba	Ginkgo	Ginkgoaceae	Roots and kernels	Quercetin, kaempferol, isorhamnetin and terpene trilactones
Panax quinquefolius and Panax ginseng	American ginseng and Asian ginseng	Araliaceae	Whole plant extract	Terpene glycosides (ginsenosides)
Hypericum perforatum	St John's Wort	Hyperiacea	Flowers and leaves	Naphthodianthrones (hypericin and pseudohypericin, phloroglucinols), flavonoids (quercitin)
Leonurus cardiaca	Motherwort	Lamiaceae-	Aerial parts	Phenylpropanoid, lavandulifolioside
Stephania tetrandra Stephania moore	Agrimony	Menispermaceae	Roots	Alkaloid tetrandrine

immunity against viral infections. About one-third of the 301 Egyptian adults who took part in the study mentioned taking vitamins, zinc, and immune-booster drinks made from plants that have long been used as medicines to lessen or prevent the severity of COVID-19. In China, the United States, and South Asia, demand for medicinal herbs has increased incredibly due to the SARS-CoV-2 pandemic. Traditional Chinese medicine may have greatly improved symptoms and reduced mortality and reinfection rates in the 91.50% of COVID-19 cases in China that used it. The study of foods, phytomedicines, and COVID-19 has all exploded at the same time (Table 8.8). Plant-based diets may have immunomodulatory and antiviral effects for a potential COVID-19 preventive strategy. According to research by Yang et al., Sambucol Black Elderberry in liquid form is being studied by researchers at East Kent Hospitals in England as a COVID-19 treatment. Plant-based functional foods include those made from *Allium cepa, Citronella cassia, Piper nigrum L., J. Presl Allium sativum, Curcuma longa, Camellia sinensis Kuntze, Glycine max Merr., Mangifera indica, and Abelmoschus esculentus Moench.*

8.5.7 Phytomedicines Used Against Obesity

The "New World Syndrome," or obesity, has been labelled a global epidemic by the WHO and is now acknowledged as a serious health threat, particularly in developing nations. According to a WHO report, there are 1.6 billion overweight adults and over 400 million people who are clinically obese. It is rising more

Table 8.8: Phytomedicines Responsible for Management of COVID-19 [9]

Botanical Name	Family	Phytochemicals
Rheum palmatum	Polygonaceae	Emodin
Rhodiola kirilowii	Crassulaceae	Luteolin
Allium cepa	Amaryllidaceae	Quercetin
Galla chinensis	anacardiaceae	Tetra-O-galloyl-b-D-glucose (TGG)
Broussonetia papyrifera	Moraceae	3′-(3-Methylbut-2-enyl)-30,4,7-trihydroxyfavane, broussochalcone and broussoflavan, Kazino
Angelica keiskei	Umbellifere	4-Hydroxyderricin, isobavachalcone, xanthoangelol
Breynia fruticosa	Phyllanthaceae	Betulinic acid
Salvia miltiorrhiza Bunge	Labiate	Dihydrotanshinone I, methyl tanshinonate, rosmariquinone, tanshinone
Isatis indigotica	Cruciferae	Hesperetin
Alnus japonica	Betulaceae	Hirsutenone
Glycyrrhiza Glabra	Fabaceae	Isoliquiritigenin
Machilus Zuihoensis Hayata	Lauraceae	Quercetin-3-b-galactoside
Chamaecyparis obtusa	Cupressaceae	Savinin
Paulownia tomentosa Steud.	Paulowniaceae	3′0-O-Methyldiplacol, 3′0-O-methyldiplacone, 4′0-O-methyldiplacol, 4′0-O-methyldiplacone, 6-geranyl-40,5,7-trihydroxy-30,50-dimethoxyfavanone, diplacone, mimulone, tomentin
Psoralea corylifolia	Fabaceae	4′0-O-Methylbavachalcone, isobavachalcone, neobavaisofavone, psoralidin
Broussonetia papyrifera	Moraceae	Broussochalcone, biphenyl propanoids, papyrifavonol
Salvia miltiorrhiza Bunge	Labiateae	Cryptotanshinone, dihydrotanshinone, methyl tanshinonate, tanshinone
Curcuma longa	Zingeberaceae	Curcumin
Alnus japonica	Betulaceae	Hirsutanonol, hirsutenone, rubranol, rubranoside
Glycyrrhiza glabra	Fabaceae	Isoliquiritigenin
Zingiber officinale Roscoe	Zingiberaceae	Kaempferol
Angelica keiskei	Umbelliferae	Xanthoangelol
Tribulus terrestris	Zygophyllaceae	Terrestrimine
Camellia sinensis	Theaceae	Myricetin
Scutellaria baicalensis Georg.	Lamiaceae	Scutellarein

quickly in industrialised nations like the USA, Australia, and Canada. Obesity has been implicated as a significant cause of fatalities in the United States. Obesity affects one in three Americans. Around 20% of school-age children and 50% of adults in the European Union are reportedly overweight. In 2010, it was estimated that 26% of adults in England were obese. Deaths from obesity are increasing year over year. serious health. Recently, it was discovered that it also encourages prostate cancer. It is worthy to note that only five to ten percent obese persons can reduce their chance of developing diabetes of type 2 as well as heart or cardiovascular disease. Additionally, there are medications as well as therapeutic substances that can stop or inhibit the digestion of fat. When taken in accordance with appropriate dietary guidance, orlistat, a saturated derivative of lipstatin, reduces body weight via enhancing oral glucose tolerance. A clinical solution for the treatment of obesity can be created using natural substances from plants and bacteria. One method of treating obesity and associated disorders is to slow down nutrition digestion and absorption. For instance, acarbose, chestnut astringent skin extract, white bean amylase inhibitor, and valoneic acid dilactone from banaba all demonstrated clinical efficacy against diabetes and cardiometabolic illnesses. Anti-lipase activity in marine algae, fungi, and certain plants has been investigated. The key factors contributing to the prevalence of obesity in westernised civilizations are high dietary fat consumption, overeating, and decreased physical activity. Fat percentages for men should be 18–23% and for women, it should be 25–30%. Having more than 30% body fat in women and more than 25% in males comes under definition of Obesity. To calculate obesity, a term 'body mass index' (BMI) is used which is actually a derivation of weight and height. This BMI is used to identify individuals who are overweight or obese. If an adult's BMI is between 25 and 29, that person is deemed overweight.

Through various mechanisms, phytomedicines or phytochemicals might be the key component of an anti-obesity agent. They may lessen dietary lipid absorption, lower caloric intake, increase energy expenditure by breaking down fat reserves. It may also decrease differentiation and proliferation of pre-adipocyte, and raises the lipolysis of body fat stores. Since dietary fat is the primary cause of adiposity, pancreatic lipase may be the target of drug development in order to create inhibitors of lipid digestion and absorption. There are numerous identified phytochemicals and compounds with pancreatic lipase inhibitory effects. Reduce calorie consumption by decreasing appetite as an additional method of managing obesity. The neurotransmitters dopamine, noradrenaline, serotonin, and hydroxyl tryptamine are controlled by the appetite suppressor sibutramine. Natural appetite suppressants including *Hoodia gordonii*, *Garcinia cambosia*, and *Cissus quadrangularis* have also been discovered. Some herbal remedies have also been connected to increased energy expenditure and weight loss, including EGCG from green tea, *Pinellia turnate*, and *Panax ginseng*. By preventing adipogenesis and triggering apoptosis in adipocytes, some phytochemicals, including quercetin, esculetin, genistein, resveratrol, and capsaicin, have been shown to have anti-obesity properties (Table 8.9).

Table 8.9: Phytochemicals Responsible for Management of Obesity [10]

Botanical Name	Family	Phytochemicals
Oryza sativa	Poaceae	Phenylboronic acid
Salvia officinalis	Lamiaceae	Carnosic acid (diterpene)
Platycodon grandiflorus	Campanulaceae	Platycodin D (saponin)
Glycyrrhiza uralensis	Fabaceae	Licochalcone A (polyphenol)
Scabiosa tschiliensis	Caprifoliaceae	Prosapogenin (saponin)
Acanthopanax sessiliflorus	Araliaceae	Sessiloside and chiisanoside (saponins)
Panax japonicas	Araliaceae	Chikusetsusaponins (saponin)
Dioscorea nipponica	Dioscoreaceae	Dioscin (saponin)
Aesculus turbinate	Sapindaceae	Escins (saponin)
Cyclocarya paliurus	Juglandaceae	Cyclocariosides (saponin)
Gardenia jasminoids	Rubiaceae	Crocin (terpene)

8.5.8 Phytomedicines Used in Management Sickle Cell Disease

Sickle cell disease (SCD), a genetic haemoglobin illness, is brought on by the non-conservative replacement of a polar glutamate (Glu) with a nonpolar valine (Val) in an invariant region of the haemoglobin chain-subunit. The natural folding of haemoglobin is altered by this change, leading to issues including a sticky region on the surface of the β-chains.

Valine which is a single non-polar amino acid, and its replacement causes a metabolic change that results in the surface of the β-chains developing a sticky patch. Both oxygenated ("R") and deoxygenated ("T") forms of HbS exhibit the sticky patch. Due to the changed characteristics, the Hb molecule's folding and binding pattern are distorted. Nucleation, which results from the HbS molecules aggregation, is another recognised method of polymerization of Hb in sickle cell disease. When there is low oxygen (hypoxia), deoxyhaemoglobin S polymerizes inside red blood cells (RBC). As a result, a network of polymers forms, stiffening and distorting cells with rigid, asymmetric erythrocytes. The formation of critical Hb polymer aggregate as a result of oxygen deprivation in the RBC disrupts the membrane of cell and encourages cellular proteins aggregation which prevents blood flow through the small capillaries that ultimately results in localised oxygen deprivation (anoxia). Due to the development of the deoxy-HbS polymer, which also changes the biconcave structure of red blood cells and ultimately results in cell lysis, the Hb becomes insoluble.

The root extract of *Fagara xanthoxyloides* was used decades ago to illustrate the potency of medicinal plants as anti-sickling drugs. According to clinical research in SCD patients, the drug significantly reduces the frequency of clinical episodes. In Southwest Nigeria, a herbal hematinic called Jawaron/Jobelyn is used to treat SCD. White blood cell count, packed cell volume, and haemoglobin have all been demonstrated to improve using *Cassus populnea* root extract—all of which are frequently lower in SCD patients. Additional anti-illness drugs made from medicinal plants include Ciklavit, Diascovite, and Solamin, which are created by Neimeth Pharmaceuticals in Lagos. Although synthetic drug

Table 8.10: Phytochemicals Responsible for Sickle Cell Disease Management [11]

Scientific Name	Family	Part Used	Phytochemicals
Carica papaya	Caricaceae	Dried leaves	5,7-Dimethoxycoumarin, kempferol, and quercetin
Piper guineense	Piperaceae	Plant extrct	B-caryophyllene (cannabinoids)
Fagara zanthoxyloides	Rutaceae	Roots	Three isomeric divanilloylquinic acids (burkinabin A, burkinabin B, and burkinabin C) coumarins, vanillic acid, parahydroxybenzoic acid
Cajanus cajan	Fabaceae	Seed	Phenylalanine
Khaya senegalensis	Maliaceae	Stem bark/leaf	Limonoids
Hymenocardia acida	Phyllanthaceae	Leaf	Anthocyanin
Trema orientalis	Cannabenaceae	Leaf	Anthocyanin
Allium sativum (garlic)	Amyryllidaceae	Bulb	Allicin
Eugenia caryophyllata	Caryophyllaceae	Plant extract	Eugenol, capsaicin, and piperine

development is now the norm, the roots of contemporary pharmacology can be found in the world of healing plants (Table 8.10).

8.5.9 Phytomedicines Used in Management of Renal Diuresis [12]

Diuretics are the chemicals that enhance flow of urine and, consequently, water excretion. One of the most significant therapeutic agents, diuretics are successfully and frequently used to treat a wide range of medical disorders worldwide. The majority of diuretics function by reducing the reabsorption of salt chloride at various locations in the nephron, which raises urine sodium and, as a result, water loss. Among the most often prescribed drugs are diuretics. Diuretics have been suggested as the first-line therapy for hypertension in addition to being helpful in the treatment of hypovolemia, electrolyte imbalances, chronic and acute renal failure, liver cirrhosis, pulmonary congestion, and peripheral oedema. One of their important and well-studied applications is the use of plant products as diuretic agents. Numerous herbal remedies, many of which had pronounced diuretic properties, were frequently used in conventional treatment for some renal diseases. Conventional diuretics like loop diuretics and thiazide diuretics specifically inhibit the Na+/Cl cotransporters in the distal convoluted tubules and the Na+/K+/2Cl cotransporters in the thick ascending limb, respectively. The ability to concentrate pee is subsequently decreased by preventing NaCl reabsorption, and a significant volume of urine that is extremely close to isosmotic conditions is evacuated. These diuretics may disturb electrolytes when taken for an extended period of time, particularly if blood potassium levels are low. This could have a serious effect on how the body functions physiologically and raise the likelihood of outcomes like hyperglycemia and hyperlipidemia that could be lethal. The usage of synthetic substances has a number of downsides, including rising costs and potential health risks. As a good substitute for synthetic medications, medicinal plants can be used because of their natural effectiveness and non-toxic ingredients.

The costs of medicinal plants are low, and their adverse effects are negligible. Herbal medicine usage offers an alternate method of treating a variety of health issues. A recently undiscovered chemical compound with potential therapeutic properties may be found in medicinal plants. The toxicity and negative side effects of allopathic medications have made the use of herbal medicine more widespread. As diuretics, mono- and poly-herbal formulations have been employed. This causes a dramatic rise in the number of companies making natural drugs. The most widely given drugs for diuretics have historically been herbal ones. These plants' medicinal potential comes from a number of chemically active compounds that have a clear physiological effect on the human body. Saponin, glycosides, alkaloids, and tannins are the four chemically bioactive components of plants that are most significant (Table 8.11).

8.5.10 Phytomedicines Used in Management of Malaria and Liver Diseases

A deadly parasite disease of the tropics, malaria is brought on by Plasmodium species and spread by Anopheles mosquitoes. There are roughly 250 million cases of malaria each year, and is found more common in most of underdeveloped nations such as Africa, South America and Asia. According to the World Malaria Report, malaria claims a million lives annually. The introduction and quick dissemination of Plasmodium falciparum strains that are multi-drug resistant pose a serious challenge for the treatment of the disease, making it more challenging and reducing the available pharmacological options. Eleven anti-malarial medications are either natural compounds,

Table 8.11: Phytochemicals Responsible for Management of Renal Diuresis [12]

Botanical Name	Family	Part Used	Phytomedicines
Alisma orientale	Alismataceae	Rhizome	Allisol, allisol A, allisol B, allisolc, allisol acetate (triterpenes)
Cissampelos pareira	Menispermaceae	Leaves and roots	Berberine (alkaloids)
Cyclea peltate	Menispermaceae	Leaves	Cycleanine, bebeerine, hayatinin, hayatidin and hayatin (alkaloids)
Acorus calamus	Araceae	Rhizome	Quinone (alkaloid), steroids, tannins, phenolic compound, terpenoids and flavonoids
Moringa stenopetala	Moringaceae	Whole plant extract	Tannin, flavonoid alkaloid
Euphorbia granulate	Euphorbiaceae	Latex	Quercetin, ferulic acid, and gallic acid (flavonoids), tannins and saponins
Crocus sativus (Saffron)	Iridaceae	Dried Stigmas	Picrocrocin (monoterpene glycoside), safranal and crocins (caratenoid)
Hibiscus sabdariffa	Malvaceae	Whole plant extract	Quercetin (flavonoid)
Smilax canariensis	Smilacaceae	Plant	Flavonoids, saponins
Ampelozizyphus amazonicus	Rhamnaceae	dried roots	Jujubogenins, s lupeol, betulinic, and melaleucic acid
Ajuga remota	Lamiaceae	Rhizome	Beta-sitosterol, gamma-sitosterol, cetyl alcohol, anthocyanidin-glucosides and iridoid glycosides, quinols
Withania aristata	Solanaceae	Aerial portion	Phytosterols, oleoresins, withaminol and withanolides
Tribulus terrestris	Zygophyllaceae	Fruits and seeds	Nitrates, flavonoids, saponins, alkaloids
Plantago major	Plantaginaceae	Leaves	Aucubin and catalpol, irioid glycosides

their equivalents, or were developed based on the pharmacophores of natural products, according to the WHO therapeutic schemes for treating malaria. The two most potent anti-malarial drugs currently in use, quinine and artemisinin, which came from ancestral therapies described in traditional medicines from Peru and China, respectively, serve as a stark illustration of the significance of plant-derived drugs for the treatment of malaria. The necessity for the discovery of novel drugs in this competition is urgent given the serious epidemiological situation of this disease and the rising prevalence of treatment resistance. Malaria and liver diseases continue to be important public health concerns. As a result, many medicinal plants are widely used in combination in local traditional medicine to treat liver ailments including hepatic syndromes and malaria as well as liver illnesses like hepatitis, jaundice, and hepatic syndromes. It is permitted to establish the effectiveness and purity of many of these medicinal plants in treating malaria and protecting the liver through clinical and preclinical research (Table 8.12).

Table 8.12: Phytochemicals Responsible for Malaria and Liver Diseases Management [13]

Botanical Name	Family	Part used	Phytochemicals
Argemone mexicana (Mexican poppy)	Papaveraceae	Leaf	Soquinoline and benzylisoquinoline alkaloids, berberine, tetrahydroberberine, protopine, and benzophenanthridines
Cochlospermum tinctorium	Cochlospermaceae		Gallic and ellagic acids, cochloxanthin, dihydrocochloxanthin, arjunolic acid, oleanane type triterpene, 3-O-E-p-coumaroylalphitolic acid, alphitolic acid, 1-hydroxytetradecan-3-one
Combretum micranthum	Combretaceae		Vitexin, choline, stachydrine, gallic acid, rutin trihydrate, (+)-catechin and benzoic acid
Entada africana	Fabaceae		Triterpene saponins
Erythrina senegalensis	Fabaceae		2,3-Dihydro-2′-hydroxyosajin, osajin, 6,8-diprenylgenistein
Mitragyna inermis	Rubiaceae		Indole alkaloids
Nauclea latifolia	Rubiaceae		Alkaloids, glycoalkaloids, saponins, nauclefine and naucletine, indole-quinolizidine, indole alkaloids: nauclefolinine, triterpenic compounds
Securidaca longepedunculata	Polygalaceae		Polymethoxylated xanthones, benzyl 3-hydroxy-2-methoxybenzoate, diphenylmethyl-substituted xanthones (muchimangines A-L), bidesmosidic saponins (securidacasides A and B), triterpene saponins, medicagenic acid
Trichilia emetica	Maliaceae		Aldehyde, limonoids, polysaccharides
Vernonia colorata	Asteraceae		Vernolide, vernodalin

REFERENCES

1. Rashid S, Majeed LR, Nisar B, Nisar H, Bhat AA, Ganai BA. Phytomedicines: Diversity, extraction, and conservation strategies. In *Phytomedicine* 2021 Jan 1 (pp. 1–33).

2. Kurhekar JV. Ancient and modern practices in phytomedicine. In *Preparation of Phytopharmaceuticals for the Management of Disorders* 2021 Jan 1 (pp. 55–75).

3. Schrijvers G. Disease management: A proposal for a new definition. *International Journal of Integrated Care*. 2009 Jan;9(06).

4. Kihagi RW, Amuka O, Machocho AK, Wafula AW. Phytomedicine and allied compounds in both human and animal healthcare. *Biomedical Journal of Scientific & Technical Research.* 2019;14(2).

5. Dragos, D., Gilca, M., Gaman, L., Vlad, A., Iosif, L., Stoian, I., Lupescu, O. Phytomedicine in joint disorders. *Nutrients.* 2017;9(1):70.

6. Adinortey, M. B., Agbeko, R., Boison, D., Ekloh, W., Kuatsienu, L. E., Biney, E. E., … & Nyarko, A. K. Phytomedicines used for diabetes mellitus in Ghana: a systematic search and review of preclinical and clinical evidence. *Evidence-Based Complementary and Alternative Medicine*, 2019;2019(1):6021209.

7. Gurria KI, Sharma S, Bhardwaj K. Wonders of phytomedicine in the management of neurological disorders. *European Journal of Molecular & Clinical Medicine.* 2020;7:2899–2914.

8. Suroowan S, Mahomoodally F. Common phyto-remedies used against cardiovascular diseases and their potential to induce adverse events in cardiovascular patients. *Clinical Phytoscience.* 2015 Dec;1(1):1–3.

9. Nasir Ahmed M, Hughes K. Role of ethno-phytomedicine knowledge in healthcare of COVID-19: Advances in traditional phytomedicine perspective. *Beni-Suef University Journal of Basic and Applied Sciences.* 2022 Aug 4;11(1):96.

10. Sukhdev S, Singh KS. Therapeutic role of phytomedicines on obesity: Importance of herbal pancreatic lipase inhibitors. *International Research Journal of Medical Sciences.* 2013;1(9):15–26.

11. Alli LA, Okoh MP. Phyto-medicine in gene (s) targeting future direction for sickle cell disease management. *Hereditary Genetics.* 2016;5(169):2161–1041.

12. Aswini EV, Vivek D, Swathilakshmi S. Role of phytochemicals in diuresis management. *World Journal of Pharmaceutical Research.* 2020;9(13):551–562.

13. Haidara M, Bourdy G, De Tommasi N, Braca A, Traore K, Giani S, Sanogo R. Medicinal plants used in Mali for the treatment of malaria and liver diseases. *Natural Product Communications.* 2016 Mar;11(3):339–352.

9 Discovery and Development of Drugs by Computational Methods

Nasir Ali, Mohsin Ali, Abid Mahmood, Rizwan Rasheed Bazmi, Muhammad Yasir Ali, and Samina Afzal*

9.1 INTRODUCTION

Clinical trials of 90% drugs which reach consumer market flop sufficiently in getting approval from Food and Drug Administration (FDA). This is the reason of possibility of a failure in the drug innovation and progress channel. Failures along the drug finding and design channels take about 75% of the cost. By using robotic automation to screen thousands of molecules, rapid high-throughput screening (HTS) experimentations we can eliminate the need for human efforts associated with compound screening.

Although, HTS is costly and needs a lot of means of ligands along with targets. It is often the case that these resources are not accessible within educational institutions. To avoid wasting time and resources on compounds that are unlikely to have successful outcomes, pharmaceutical companies are seeking more efficient methods. Computer-aided drug discovery (CADD) is gaining traction among pharmaceutical businesses and university researchers [1].

9.2 CADD TECHNOLOGIES

Computer-aided drug design (CADD) implicates the utilization of computational approaches to uncover, create, and evaluate molecules with similar chemical properties intended for drugs or other active agents. Homology modeling, virtual screening (VS) or virtual HTS (vHTS), molecular docking (MD), quantitative structure–activity relationship (QSAR), and three-dimensional (3D) pharmacophore plotting are the central elements of CADD. Out of all these procedures, VS has all the earmarks of being an essential supporter of CADD and has turned into a perceived and acknowledged computational technique equaling trial high-throughput separating hit, distinguishing proof as well as streamlining [2].

Powerful tools, e.g., CADD, can help reduce the amount of ligands which require to be tested during experiments. A demanded substitute to HTS is vHTS, also known as *in silico* HTS. CADD not only decreases the expenses of drug discovery by confirming the most optimal lead composite that moves on to testing with animals, but it can also lessen the time it requires for a drug to reach consumers. It serves as a sort of "virual shortcut" in the drug development channel. CADD methods are used to recognize appropriate lead molecules for testing, foresee their efficiency and potential adverse effects, and even boost their bioavailability when necessary.

For example, in a latest research of CADD, it was discovered that by base molecule pyridazinone with the insertion of triphenylphosphine group, inhibitors for proteasome are obtained [3]. Analogs of the initial structure exhibited high potency. Many studies have demonstrated the impact CADD can have on the production of new medicines.

* Corresponding author

DOI: 10.1201/9781032707273-9

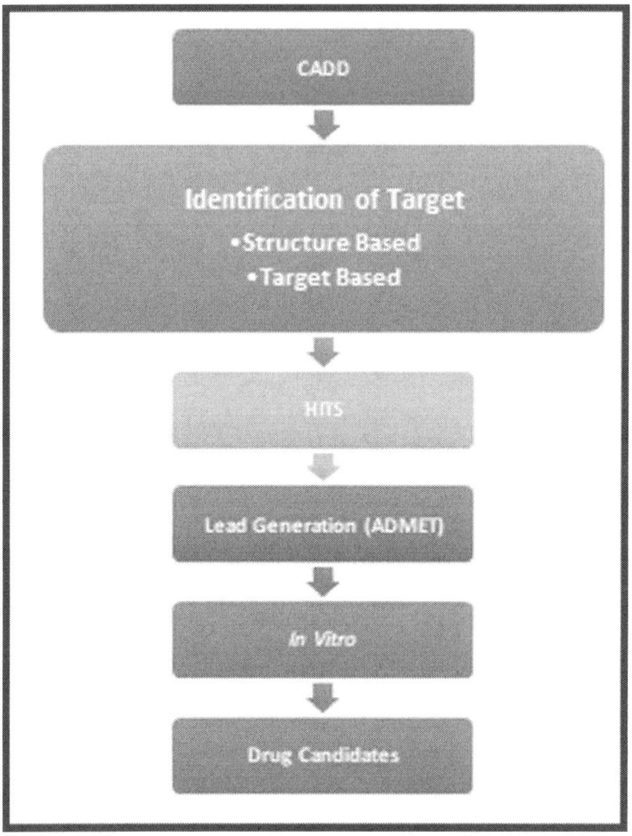

Figure 9.1 Schematic diagram for computational drug discovery.

9.2.1 Classification

There are two distinct classifications of CADD methods: structure-based (SB) and ligand-based (LB) methods (Figure 9.1). If target structure information is available, then SBDD tools can be employed. Structural data is usually acquired through nuclear magnetic resonance (NMR) or X-ray crystallography, but homology modeling, which can predict 3D structure of target, can be utilized if neither option is viable. With a known target structure, vHTS and direct docking procedures, to assess the binding capabilities of drug molecules, become available. The affinity of molecules for a given target can then be evaluated by estimating binding free energies. Potential therapeutic molecules are then optimized and filtered before the final lead compounds are tested for activity in vitro. When the desired structure cannot be predicted computationally or determined empirically, LB methods are the standard approach (Figure 9.1). These strategies rely upon the accessibility of dynamic folios connected to the objective, as well as their insight [4].

9.2.1.1 Structure-based Drug Discovery

SB-CADD involves detecting and probing the biological molecule's 3D structure. The central idea of this method is that a molecule's capability to bind to a particular protein and produce the wanted biological outcome is based on how well it connects with a certain binding spot on the protein. Compounds that interact in this promising way will usually have same biological effects. Thus, new substances can be revealed through careful assessment of the protein's binding position. Information about the structure of target is necessary for any SB-CADD task. Scientists began using target protein structure to help in drug evaluation in the early 1980s (NIH-structure-based). Meanwhile, SB-CADD is a popular approach for drug discovery due to expansions in genomics and proteomics which have resulted in the identification of numerous potential drug targets [1].

The most popular SB-CADD approaches rely on structure when information about the 3D structure of a pharmacological target is present. Our comprehension of the target's structure is used to develop the therapies in SBDD. Two distinctive techniques are molecular docking (antagonists, agonists, inhibitors, etc.) and de novo ligand design. SBDD uses molecular dynamics (MD) simulations to gain a better understanding of how ligands interact with proteins, as well as for accounting for target flexibility and pathways of interaction—this is especially critical when dealing with membrane proteins where permeability is necessary for effective drugs [5, 6].

9.2.1.2 Target Structure Grounding

Target structures which are determined experimentally from X-ray crystallography or NMR techniques that are added in the Protein Data Base (PDB) are great for initiating with docking. Structural genomics has been very helpful in speeding up the rate at which these target structures are being determined. For VS to be successful, it is essential to have a good sum and value of structural information about the target and small molecules being docked. The first step is to determine whether the target has a good binding pocket [7].

9.2.2 Comparative Modeling

Because protein arrangement is more conserved than the sequence and proteins with similar sequences typically have same structures, comparative modeling is used to estimate target structure by developing a comparable sequence template. Homology modeling is a kind of comparative modeling where proteins to be targeted and template are from the same evolutionary source. To carry out comparative modeling, five steps are followed:

1. Tracing similar proteins that can be used as templates,
2. Aligning sequences of target and template proteins,
3. Replicating coordinates for sections that align well,
4. Building atom coordinates for any missing parts of the target structure, and
5. Refining and evaluating the model.

9.3 CHARACTERIZATION AND DETECTION OF BINDING SITE

Computational approaches for recognizing as well as characterizing the binding locations can be classified into three broad categories:

1. Geometric algorithms to locate shape concave invaginations in target
2. Methods relying on energetic consideration
3. Methods based on dynamics of protein assemblies.

Computational methods, for example, POCKET, SURFNET, Q-SITEFINDER, etc. [8] are also used for knowing the binding sites.

9.3.1 Characterizing Target Protein and Minute Molecules for Docking Models

Three simple methods to characterize target and ligand structures are in silico atomic, surface, and grid representations [9].

When ranking and scoring rely on potential energy functions, the surface of target is typically represented as an atomic structure. A fragment of the atom's van der Waals surface was mapped to produce all of the smooth convex, concave, and saddle-shaped faces on this surface. To guide the process, docking takes advantage of complementary arrangement of ligand and surfaces of binding site. It starts with systematically pairing a ligand sphere a1 with one receptor sphere b1, tailed by another second pair, a2 and b2. If the modification in atomic voids is below an experimentally defined cut-off value, the relocation is favored. The cut-off value stipulates the full allowable deviation among ligand and receptor's internal distance. With the same internal distance checks as before, pairing procedure is performed for the third atomic pair. If at least four assignable pairings can be found to determine orientation, the match is kept; otherwise, it is rejected.

9.3.2 Protein–Ligand Docking Sampling Algorithms

Considering the ligand and protein flexibility during process of docking, docking methods can be categorized as rigid body or flexible applications [10].

Rigid body docking approaches only reflect the static geometric or physiochemical aspects between target and ligand, disregarding flexibility and induced-fit binding models. Rigid docking is usually chosen when speed is important, e.g., during vHTS. However, flexible docking methods may be needed to refine and optimize poses gained from the primary rigid docking process. With increasing computational capabilities, flexible docking is becoming increasingly commonplace.

9.3.3 Scoring Functions for Assessing Protein–Ligand Complexes

To further improve hit-to-lead and lead-to-drug optimization, more advanced scoring roles have been established that predict target–ligand binding affinities. These scoring functions are categorized into four categories:

1. Scoring functions based on force field and molecular mechanics
2. Empirical scoring functions
3. Scoring functions based on knowledge
4. Consensus scoring functions.

9.3.3.1 Scoring Functions Based on Force Field or Molecular Mechanics

The energy approximations in force field scoring functions are constructed on conventional molecular system. The parameters that were analyzed using ab initio quantum mechanics and experimental data were used by these functions.

9.3.3.2 Empirical Scoring Functions

Empirical scoring functions employ parameters taken from experiments or ab initio quantum mechanics computations, such as the energy required for binding, determined by the weighted sum of hydrogen bond interactions, desolvation effects, hydrophobic contacts and entropy. Such expressions are easy to calculate, but the weights are usually obtained from regression analysis with molecular data.

9.3.3.3 Scoring Functions Based on Knowledge

Knowledge-based scoring functions meanwhile use report from experimentally resoluted complex structures. They are designed in the supposition that interatomic distances appearing more often than average distances show promising relations. While, affinity is declined by interactions that exists with lower frequencies.

9.3.3.4 Consensus Scoring Functions

Consensus methods rescore their predictions with various scoring functions which results are then merged in different techniques for ranking solutions [11]. Some approaches for combining scores contain weighted combinations of the scores, a voting strategy based on cutoff values for each scoring method, rank by number where compound average normalized scores are utilized, and rank by rank which sorts compounds centered on average ranks based on distinct scoring functions.

9.4 STRUCTURE-BASED VIRTUAL HIGH-THROUGHPUT SCREENING

Structure-based vHTS (SB-vHTS) is an *in-silico* approach for finding intensifies that are probably going to bind to an objective protein of known structure. Unlike traditional HTS, this method searches for ligands expected to bind the specific binding site, in contrast to determining general capability of a ligand to interact with the protein. To facilitate the screening process, SB-vHTS relies on restricted conformational sampling of ligand and protein with simple estimate of its binding energy. The three key steps involved in this screening process are:

1. Target protein preparation and chemical library for docking
2. Defining promising binding pose for every compound
3. Docked structures ranking.

9.5 VOLUME CALCULATION AND IDENTIFICATION OF BINDING POCKET

Once the 3D representation of a protein is established, it is crucial to locate its potential binding sites since they can explain why some small molecules can increase or decrease its activity. Experimental techniques, such as X-ray crystallography or site-directed mutagenesis, can be used; however, two broad categories of algorithms can also be employed: geometry-based and energy-based methodology. In most instances, 83% actually, the largest pockets or clefts are found to be the actual binding sites; yet, relying on geometry alone is not always enough since there exist exceptions where energy-based methods have been more successful than geometry-based ones. As a result, the pocket's size matters for function too. A geometry-based binding site recognition server is 3V [12]. Although for few cases the biggest pocket/cleft of any protein is that protein's binding pocket, it is not essentially right for all proteins which are targeted. In contrast to geometry-based techniques, approaches based on energy have been created in order to solve this problem and have shown to be effective [13]. In order to localize the protein's binding sites, Q-Site Finder employs van der Waals probe and determination of the interaction energy between the probe and the protein.

9.6 DE NOVO LIGAND DESIGN

De novo ligand designs employ either growing of a ligand or linking of ligand. By adopting the ligand-growing technique, a tiny section is inserted inside the binding site and expanded upon to form the ligand structure. This is comparable to the linking process, which docks numerous fragments to different locations on the target molecule before connecting them to create a single product. This technique can be seen as a computational form of the NMR structure activity relationship method. Different approaches have been created to develop ligands which binds to a particular target, such as ligand-growing and ligand-linking strategies. A genetic algorithm controls a variety of processes, including growth and connecting, that are used in the designing process. Another use of empirical scoring function is to gauge target-ligand complex's affinity for binding. The sequencer will examine the binding pocket after receiving the info about the target protein. On the user's demand, it can then apply either growing or connecting technique. A structural seed is inserted into a specific binding pocket during the growth method, and the program then substitutes candidate fragments for the user-defined growing locations. As a result, an innovative seed arrangement emerges, which can be utilized in subsequent development cycles.

In case of linking strategy, many fragments located at varying positions at the aimed protein are used as the seed structure. The growing process takes place concurrently on each fragment. The complexity of this method is one of its disadvantages. Due to the availability of high-resolution target structure, biochemical and organic model builder (BOMB) is used to propose ligands which bind to the target exclusive of having to utilize ligand databases [14]. De novo drug design results in compounds that must typically be synthesized. While performing VS, molecules are purchased from databases of readily available, commercially available compounds.

9.7 ADME EVALUATION, DRUG SAFETY AND LEAD OPTIMIZATION

Lead optimization is usually the next step once hits from screening small molecule databases have been discovered. To attain the necessary pharmacological profiles, such as affinity, pharmacokinetic characteristics, safety, and ADME, this entails enhancing the efficacy of prospective hits. It is possible to improve a drug's potency (efficacy) by enhancing its affinity for its target. The free energy of binding a medicine to the target is a sign of potency. We further explore this link using alchemical free energy estimates and MD simulations.

One simulation begins a target-ligand complex, while the other eliminates the ligand completely from solution. The chemical modifications of a promising drug candidate which might increase its effectiveness can then be evaluated using alchemical free energy calculations. One can use these affinity changes as a guide to increase the potency of drug candidates by progressively changing a ligand atom and calculating the corresponding change in binding affinity [15].

Drug absorption is significantly influenced by two key factors: a drug's solubility and intestinal permeability. Because of this, lead optimization includes the *in-silico* estimation of medicines' solubility with membrane permeability [16]. A medicine that is taken orally is more likely to be eliminated from the body without reaching the bloodstream if it has lesser solubility or a excessive dissolution rate, rendering it ineffective and perhaps resulting in additional negative side effects. A molecule must be manufactured, which is a time consuming and costly process, in order to test its solubility. Fortunately, computational solubility prediction is significantly quicker, and it is possible to quickly calculate the solubility data for huge libraries of compounds. This saves money and time by allowing medical chemists to analyze drug compounds without having to generate them.

Huynh et al. used an *in silico* technique for estimation of solubility of docetaxel (DTX) which is an anti-cancer molecule utilized to treat diverse types of cancer [17]. This work used MD simulations to determine the solubility parameters for DTX. This computer model matched DTX's experimental solubility. Human intestinal absorption of a potential medicine has a significant impact on bioavailability of the drug.

According to the Lipinski's "Rule of 5," poor absorption/permeation is likely when, there are above 10 Hydrogen bond acceptors, more than 5 hydrogen bond donors, Log *P* is above 5, and the molecular weight is above 500 There are extensions of the Rule of 5 in envisaging drug likeliness [18]. One of them later anticipated is the "Rule of 3" which was used in the formation of fragment libraries in order to generate lead [19]. These principles serve as broad recommendations for characterizing compounds' drug likeness and bioavailability. On the basis of these guidelines, mathematical and statistical models have been created and improved, and machine learning (ML) methods like neural networks are used to predict drug similarity and bioavailability.

QikProp, a program by Schrodinger, can be used to determine pharmaceutically related and physically significant descriptors for minute drug resembling molecules. VolSurf is an ADME package that calculates ADME properties and creates ADME models which can then be applied to unknown compounds. Additionally, it finds active ligands of interest as compounds with similar ADME characteristics. FAFDrugs2 is another ADME and toxicity categorizing tool which computes physical and chemical characteristics, toxic entities, and key functional constituents.

Despite various drug moieties passing through the experimental authentication process or animal tests, there are still many that don't reach clinical trials due to their poor pharmacokinetic profiles or toxicity. This is why filters for ADME characteristics are essential for drug detection. During past decades, computational ADME techniques have improved significantly and pharmaceutical corporations are demonstrating a great significance in this field (Figure 9.2) [4].

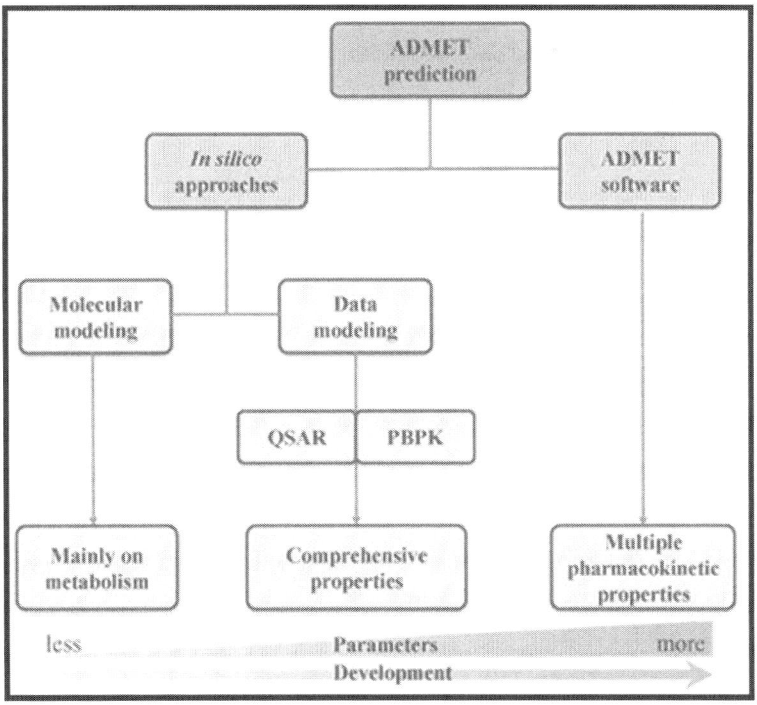

Figure 9.2 Steps for in-silico drug development.

9.7.1 2 Ligand-Based Drug Design (LBDD)

When it is impossible or undesirable to anticipate a drug target structure using homology modeling or ab initio structure prediction, the alternative to SBDD, ligand-based drug design (LBDD), is used [20]. Knowing tiny compounds which shows good bind efficiency to the target is necessary for this strategy; some well-liked techniques include pharmacophore modeling, molecular similarity algorithms, and QSAR modeling. Molecular resemblance involves using the molecular fingerprint of familiar ligands binding to the target in order to find various molecules with similar fingerprints from screening molecular libraries. Usual structural characteristics of ligands which effectively bind to a target are employed for screening in LBDD pharmacophore modeling. Lastly, QSAR is known to be a computational tool for understanding the relationship of the structural features of ligands and their corresponding biological behaviors.

9.7.2 Pharmacophore Model

Pharmacophore models of the target binding site are useful for summarizing steric and electronic properties which are necessary for a ligand to interact optimally with a target. Common features of pharmacophores include acceptors and donors of hydrogen bonds, cationic, anionic, and partially charged groups, aliphatic hydrophobes, and aromatic hydrophobic moieties. These pharmacophore descriptions have seen plenty of use in drug discovery. They are utilized to virtually screen compounds against a library for hits, in addition, inform the design of new compounds through de novo algorithms. Analyzing the structure of target-ligand complexes or target binding sites can serve as the foundation for pharmacophore techniques.

Pharmacophore model assembly procedures are recapitulated as follows:

- Finding compounds with activity that are recognized to bind to intended target and share the same interaction mechanism can be done by searching the literature or databases.
- For a 2D pharmacophore model fundamental atom forms and their connectivity are described
- For a 3D pharmacophore model the conformations are outlined using IUPAC nomenclature.
- To identify common features needed in binders, utilize Ligand Alignment or Superimposition.
- Constructing a pharmacophore model.
- Pharmacophore models are ranked, and the best models are chosen.
- Pharmacophore model validation.

9.7.2.1 Quantitative Structure–Activity Relationships

The foundation of QSAR approaches is statistical data that links target drug interaction activities to distinct molecular descriptors. The QSAR approach rely on the observation that molecules with comparable structural characteristics display similar biological activity. These models provide a mathematical explanation of how the structural characteristics of the ligand affect the activity feedback of a target that binds to it. By calculating the relationship between numerous characteristics of tiny ligand binders and empirically confirmed biological activity, QSAR is produced. QSAR relationships are also be used to calculate the activity of novel drug moiety analogues (Figure 9.3).

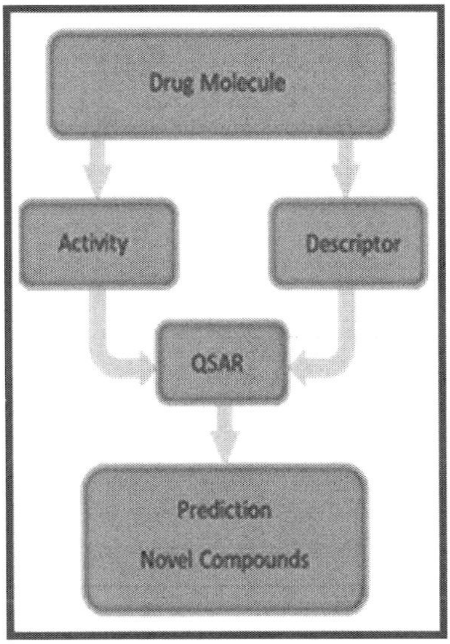

Figure 9.3 Quantitative structure–activity relationship.

Several values are used for quantifying drug molecule's activity. The two measurements that are most frequently employed includes inhibition constant (Ki) and half maximum inhibitory concentration (IC50). Just like pharmacophore models, QSAR models are also used to determine if a specific property of a drug molecule has a favorable or unfavorable impact on its activity. Thrombin, renin, and carbonic anhydrase are just a few of the therapeutic targets that have shown success when QSAR approaches have been applied. Numerous machine-acquiring techniques have been utilized to create QSAR models. Traditional or 2D QSAR approaches examine how a molecule's electrical, hydrophobic, and steric characteristics relate to biological activity. By examining the active drug molecules' quantum chemistry as well as their physical and geometric characteristics, 3D QSAR algorithms go one step further. Recently, membrane system QSAR models have also been created.

In a nutshell, QSAR involves these steps: (1) the Recognition of molecules that interact with a target drug actively and their activities via database searches, literature searches and HTS experiments; (2) the identification of structural or physicochemical molecular sorts, such as bond counts, atom counts, functional group counts, and surface area, that affect biological activity; (3) the creation of a correlation between molecular features and biological activity; (4) the authentication of the QSAR predictions; and (5) the use of the QSAR.

One dissimilarity between pharmacophore models and QSAR is that the former is built using the crucial or essential characteristics of an active ligand, whereas the latter considers both the crucial and activity-affecting characteristics. The volume of the binding site is a crucial structural element that is utilized in both the pharmacophore model and QSAR. It is widely known that the biological activity is significantly influenced by the binding pocket volume. When the binding pocket volume is determined, it is possible to eliminate molecules from the early stages of the drug development method if they are very huge to fit in the binding pocket.

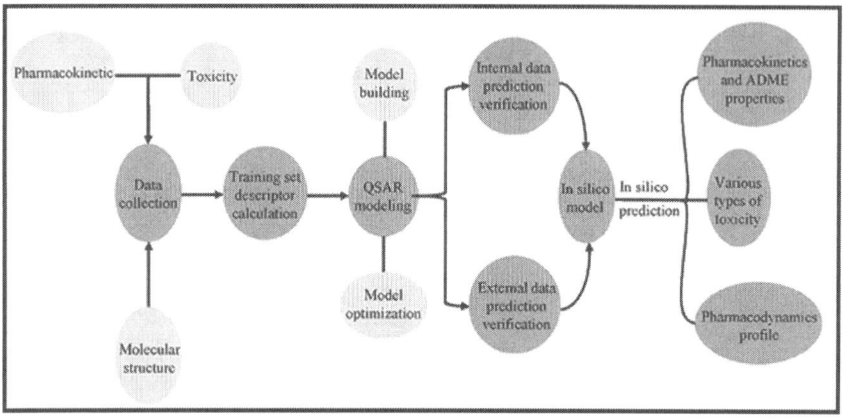

Figure 9.4 Scheme of QSAR modeling.

Using different biochemical and physical data to build models is a technique called QSAR. In QSAR investigations, compounds can be mathematically encoded as molecular descriptors. The association between molecular descriptors and defined qualities is then built using statistical methods, and a created model is then used to foresee the related attributes of new compounds. The process described here first converts molecular structure into molecular descriptors, which are then utilized to create prediction models using statistical methods or ML practices like support vector machine (SVM) and K-nearest neighbor (kNN) (Figure 9.4) [21].

9.7.2.2 Current Developments in CADD and Drugs Discovered Through Computational Techniques

CADD is use to create a number of authorize medications. In order to gather information about specific features of CADD, more than 600 papers were assessed on SciFinder® through methodical filtering and searching using the terminologies VS, software methodology, computational learnings, and publication year. Additionally, the research that later used software to do MD simulations, use density functional theory (DFT) calculations, and conduct experimental evaluates were appraised. Between the years 2015 and 2020, most updated data was available, hence most often used techniques and software packages were filtered in order to represent the most updated trends. The ZINC database was the one that was most recurrently used, with an average use of 31.2 percent. The most widely used type of VS was structure-based VS (SBVS), which reported for an average of 57.6 percent of usage. The most known VS or MD software was AutoDock, which received 41.8 percent of usage. After the process of screening, 38.5% of the studies carried out MD simulations to accompaniment the VS and GROMACS with 39.3% use was the famous MD software platform. DFT was the computing method that was used minimum, with an average usage of just 0.02 percent. In the research, reports on experimental assessments following VS were included in an average of 36.5% of the studies. At the end, above seventy authorized medications have been invented since the start and application of CADD in pre-clinical drug advancement, and this numeral is constantly rising over time. It's crucial to remember that databases or screening pools are chosen based on the objective and issue aroused. These databases contain lead and drug-like small compounds and can be used for VS; as a result, they show different degrees of scaffold variety [17]. It has been presented that some of the

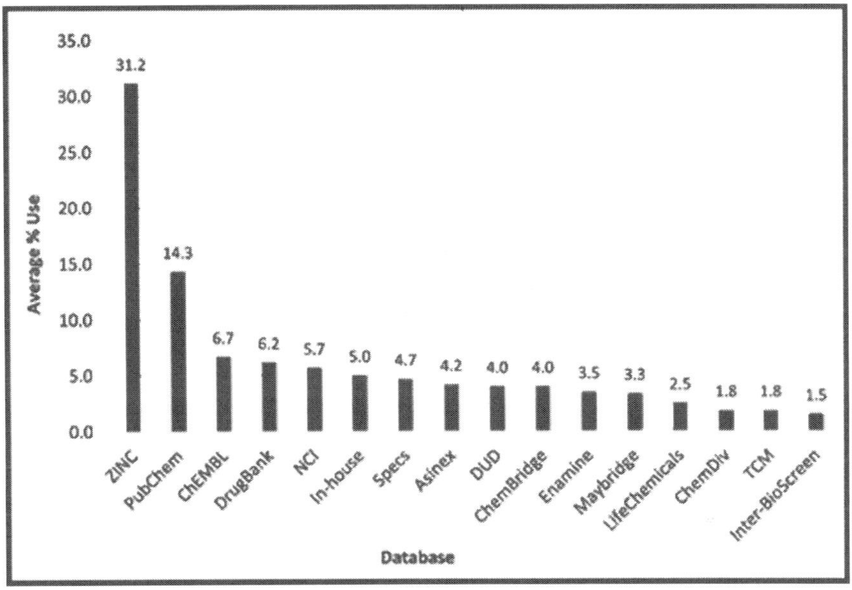

Figure 9.5 Percentage for the use of different databases.

effective databases have been found to have a balance between synthetic and natural compounds. Additionally, it has been observed that there can be chemical uniqueness and some degree of overlap within databases. Since no two databases are alike and certain compounds in them are not suited for VS, filtering is necessary. Chemical structure sources e.g, natural materials, or the specifics pertaining to the structure, are variables that may have an impact on the database selection. Characters such as availability, access and drug-likeness all impact the choice of the database, along with other factors such as chemical structure source (natural products) or features linking to the structure (approved drugs) (Figure 9.5) [17].

9.7.2.3 Software for Virtual Screening and Graphical User Interfaces in CADD

VS software programs include:

- AutoDock (AutoDock Vina, AutoDock 4, and PyRx) CDocker
- DOCK
- FRED
- GLIDE (Schrödinger® Maestro)
- GOLD
- ICM (Internal Coordinate Mechanic) LibDock
- LigandScout
- Surflex-Dock, MOE [Molecular Operating Environment] MVD (Molegro-Virtual Docker)

A computational method called VS is utilized to find new hits as bioactive compounds. Amid the VS strategies, the structure-based methodology makes use of MD as one of the highest used methods. The algorithms and scoring formulas used by docking programs to carry out the mathematical approximations necessary to calculate the computational similarities vary widely. This can specify which systems a particular program should be used on. Every algorithm and scoring function have specific biases and presumptions that are encoded into them before they execute calculations. This may affect how they are preferred to be used for VS in particular circumstances. Following docking, VS/MD simulations, PyMOL and Discovery Studio Visualizer are excellent tools for visualizing the docked conformations as well as MD snapshots [2].

9.7.2.4 *Cancer Modeling, Network Biology and Anti-Cancer Drugs Discovery through Computational Approaches*

As stated by World Health Organization (WHO), one of the most hazardous diseases in existence nowadays is cancer. It can be seen that there is a rising concern in multi-target drugs which can aim at several intracellular pathways since cancer is a complex illness. However, due to the volume and complexity of the data, studying big datasets for the investigation of anticancer complexes is challenging. For instance, the ChEMBL database assembles enormous datasets of incredibly diverse preclinical experiments. Also published is a PTML-LDA model using the ChEMBL dataset for the preclinical identification of anticancer drugs. To handle parallel issues, the concepts of perturbation theory (PT) with ML techniques are fused by the PTML model. When this model was compared to other PTML systems, Speck-Planche et al. found that it was the only one that could predict action against different forms of cancer. Using a dataset of 1933 peptide instances, they developed a multi-task (mtk) chemical information model by combining Broto Moreau auto-correlation with an artificial neural network [22]. This model is used to replicate the development and evaluation of peptides that have the potential to be both low in toxicity to various types of mammalian cells and anticancer action against various malignancies. With over 92% accuracy in both training and prediction, the model displayed excellent performance (test).

Analyzing tumor growth at many scales, from tissue to molecule length, is necessary due to the intricacy of malignancies. Microscale, mesoscale, and macroscale are the three different sizes that may be mathematically modeled to provide further information regarding the genesis of cancer. Wang et al. explored a method that applies multiscale models to identify and develop combination therapy drugs. The link between intracellular epidermal growth factor receptor (EGFR) signaling kinetics, external EGF stimulation, and multicellular development was quantified using this technique. Smarter medications that offer a thorough system-level method to assess the dynamics and effects of prevailing and future treatments in preclinical trials, model species, and specific patients are anticipated to result from combining multiscale modeling of malignancies with systemic pharmacology. These mathematical and computational research will improve our comprehension of the variables affecting medication effects, ultimately resulting in more efficient approaches to treat disease [23].

A ground-breaking advancement in this field has been the successful discovery of several novel anticancer medicines by computational drug design. Research on medications that treat cancer is currently moving swiftly, with promising discoveries coming from computational and AI techniques. SR13668, for instance, was enhanced from indole-3-carbinol (I3C) using PH4 design and has proven effective against various cancers in phase I [24]. Utilizing ML methods, a powerful inhibitor of 5-lipoxygenase has been identified that leveraged physicochemical and pharmacophore characteristics. The advent of AI has revolutionized the in-silico design of anticancer drugs, and modern learning tactics have the ability to generate chemical compounds with exceptional properties. Following this progress, a ML-based compound generator which uses variational autoencoders (VAEs) was developed and was shown to be capable of selectively generating molecules with high predicted inhibition against a certain cancer. This demonstrates that various models can be employed to generate drug candidates with vastly desired efficacy (IC50) against a certain target. This advancement is set to revolutionize the in-silico design of anticancer drugs, permitting researchers to take lead of disease bimolecular features to increase the success rate in uncovering lead compounds.

9.7.3 Breakthrough in Computational Drug Finding

Crizotinib, a potent, targeted cMet/ALK inhibitor, is a prime model of the use of structure-based design techniques. FDA authorized this in 2011. The hepatocyte growth factor receptor, or c-Met, and its natural ligand HGF are both crucial in controlling several cellular processes. The cMet protein is overexpressed in several malignancies, such as small-cell lung cancer and non-small-cell lung cancer, and there is aberrant cMet signaling activity in many solid and liquid tumors, making c-MET a key target for cancer treatment [24].

9.7.3.1 Computational Advances to Therapeutic Antibody Design and Discovery

Developments in computational methods for antibody analysis have come far enough to be applied more widely in therapeutic pursuits. Antibodies, which are proteins that make up the immune system, play a role in recognizing and eliminating foreign molecules (antigens) or self-antigens from healthy tissue in diseases which are autoimmune. These antibodies possess the ability to bind to a variety of molecular surfaces [25], and this property has been utilized for medicinal use. Now, antibodies are the largest group of biotherapeutics available [26].

9.7.4 Antibody Databases

The availability of appropriate datasets is a prerequisite for computational tools to evaluate and design antibodies. Therapeutic antibody information may be found in resources like TABS and SAbDab-Therapeutic-Antibodies. The majority of other resources may be categorized according to their content's sequence, structure, or experimental data; some databases include all three. The majority of these libraries gather information on both antibodies and nanobodies; however, some, like sdAb-DB, primarily focus on the latter (Table 9.1, Figure 9.6).

Table 9.1: Databases Including Information About Antibody, Nanobody, and Sequence

Database	Links	Description
AbMiner	https://discover.nci.nih.gov/abminer/	Database of available monoclonal antibodies
IgPdb	http://cgi.cse.unsw.edu.au	Database of inferred allelic variants for immunoglobulins
AntigenDB	http://crdd.osdd.net/raghava/	Antigen database
Institute for Analysis and Collection of Nanobodies	http://ican.ils.seu.edu.cn/	Nanobody sequences and structural information
PyIgClassify (license available)	http://dunbrack2.fccc.edu/PyIgClassify/	CDR canonical class database

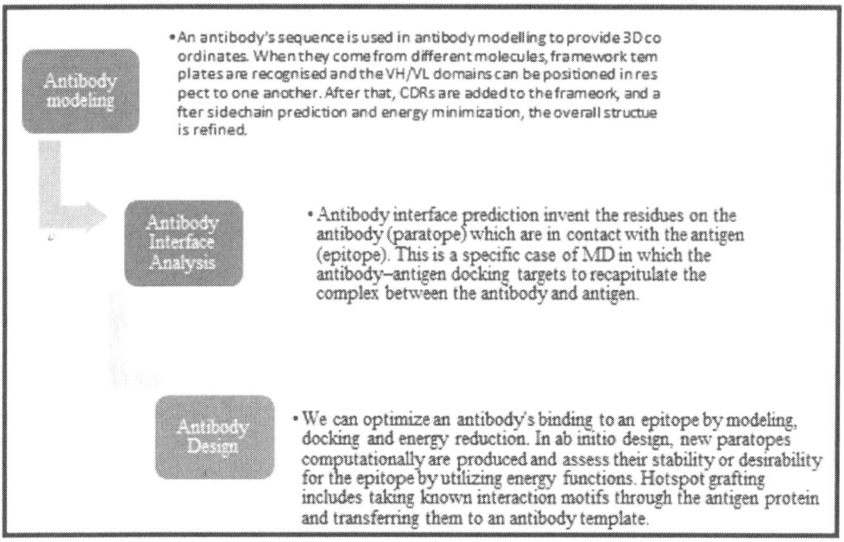

Figure 9.6 Steps for antibody development by computational method.

9.7.5 Unconventional Antibody Layouts: Nanobodies

The first therapeutic approval of caplacizumab in 2018 has considerably increased interest in H chain-only antibodies, or nanobodies. These molecules are found naturally in sharks and camelids (including llamas, alpacas, and camels), and new strategies to take advantage of this have resulted in a variety of databases and analyses tailored specifically to nanobodies.

Only three highly variable loops (CDRH1), (CDRH2), and (CDRH3) are present in the nanobody, and together they create an extensive structural paratope that is situated on a side of bent protein domain. Nanobodies can bind antibody-incomprehensible epitopes in enzyme active locations, viral capsids and G protein coupled receptors because the L chain is absent, causing the CDRs to differ from antibodies in both sequence as well as structure. Recent computational investigations of extensive collections of antibody/nanobody sequences as well as structures show that these molecules have non-trivial systematic dissimilarities. Only nanobodies were shown to have similar degrees of sequence variation to traditional antibodies inside CDRH1 and CDRH2 loops while fewer order and structural variation through their framework regions. In contrast to conventional antibodies, nanobodies, which translate this identical level of sequence variation into augmented structural variety through the CDRH1 and CDRH2 loops but cannot be classified by well-known canonical rules, present additional problems for computational structural modeling tools. In addition, the CDRH3 loop has protrusions that resemble fingers that outspread into epitope holes on antigens due to its average length of three to four residues and significantly higher sequence and structural variability. The average number of additional residues in nanobody paratopes is almost three, which poses a challenge for computational modeling tools. Moreover, aligned sequence positions help to choose paratope than classical antibody VH domain paratopes, which is comparable to the set used by the entire VH-VL paratope. This implies that novel approaches will be needed to apply computational structural modeling methods to nanobody antigen interactions given that the

VL domain exhibits far less structural variability. The modeling difficulty is further increased by the discovery that nanobody paratopes have a substantially wider variety of structural components after extensive analyses of huge sets of nanobody & antigen co-crystal structures. Antigen-contacting residues within nanobodies are dominated by CDRH3 loop residues which are highly variable in both sequence and structure. This suggests that the nanobody–antigen interface will be challenging to model exercising methods that have been developed in the context of classical antibody VH domains. In addition, compared to its antibody equivalent, the nanobody paratope has roughly three additional residues, which makes it more challenging to simulate using standard techniques for modeling antibodies, docking, etc. A rigorous assessment of current antibody computational tools will be helpful to determine the degree to which existing tools become suitable to nanobodies and what improvements would be required for success in computational nanobody creation.

9.7.6 Troubleshooting Computational Techniques in Drug Discovery

9.7.6.1 Homology Modeling

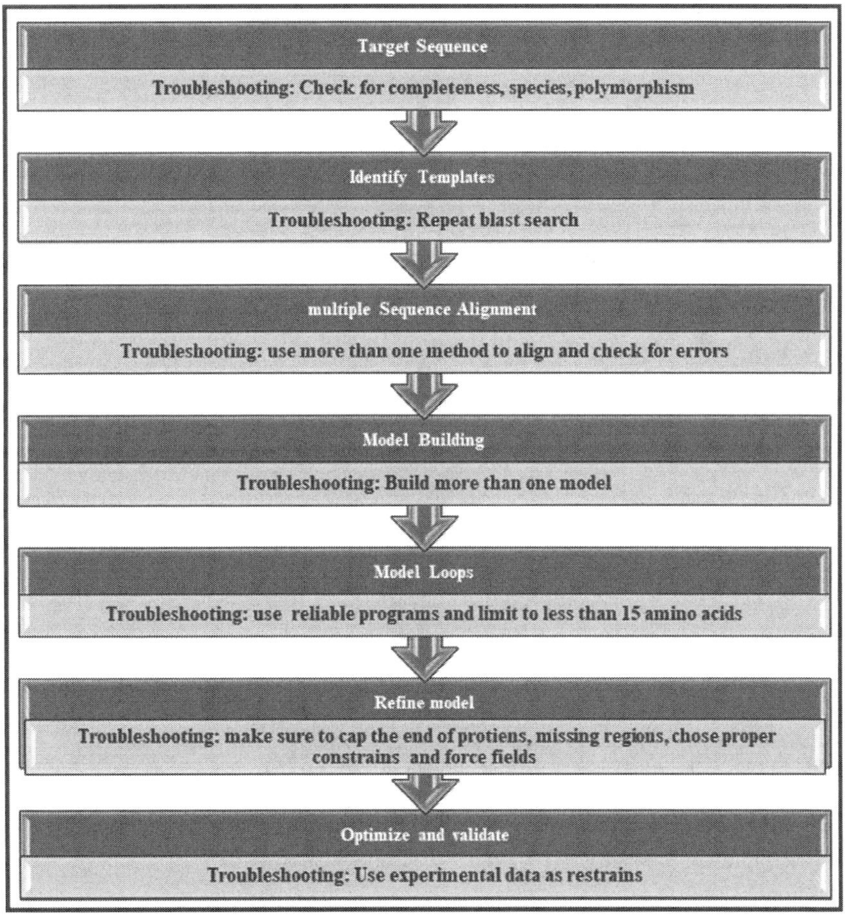

Figure 9.7 Steps for homology modeling.

9.7.6.2 Hybrid Structure-based In-Silico *Screening Method*

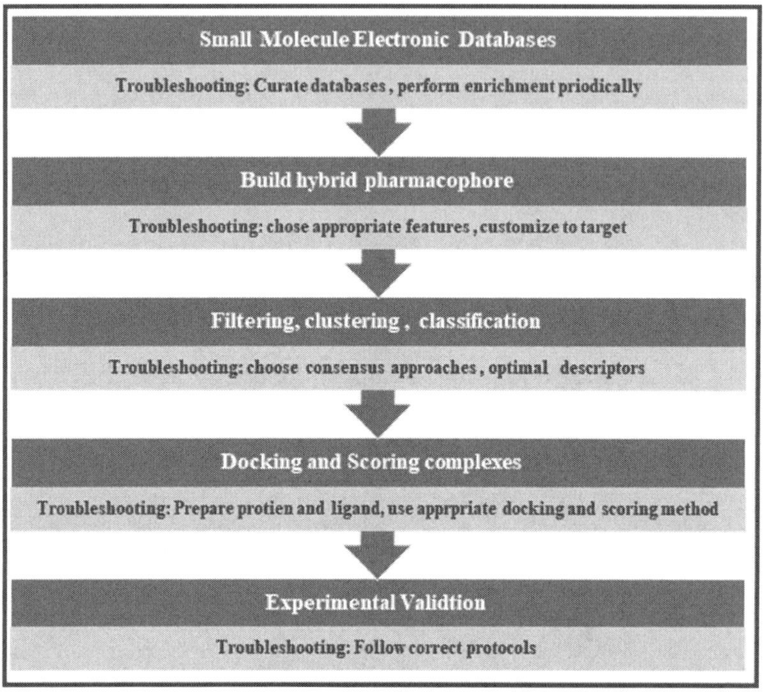

Figure 9.8 Steps for hybrid structure-based in-silico screening method.

9.7.6.3 In-Silico *ADMET Modeling*

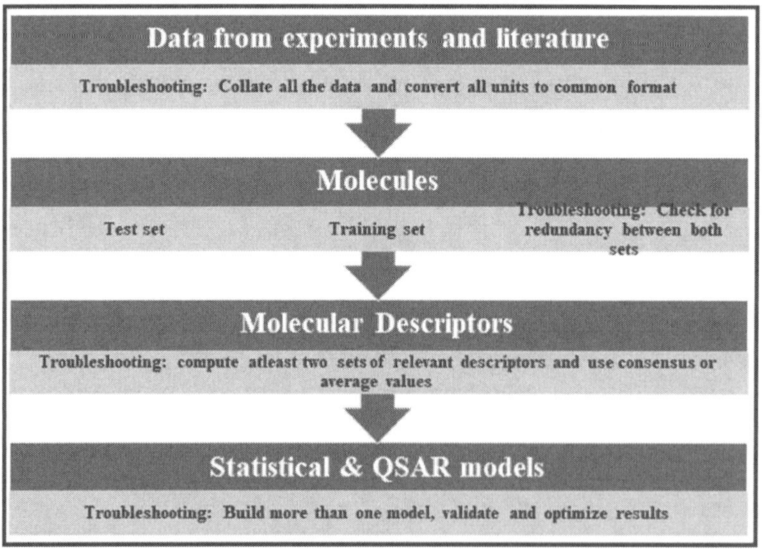

Figure 9.9 Steps for in-silico ADMET modeling.

9.7.6.4 Computational Methods for Innovation of Allosteric Drugs

Drugs targeting allosteric sites have greater promise than those targeting orthosteric sites. Allosteric medicines have a lot of benefits, making them favorable as drug candidates. Spatiotemporal specificity may be seen in allosteric modulators. For instance, they may be restricted in their activity to certain tissues during specific times, which may lessen desensitization, and they may only be active during occurrence of the endogenous ligand. Although there are many potential benefits to allosteric therapeutics, it has proven thought-provoking to predict techniques for discovering allosteric medications.

For past 10 years, new opportunities for creating allosteric medications have emerged thanks to advancements in computational power, prediction algorithms, and extensive structural and biochemical data. Researchers are now able to identify, find, and analyze allosteric sites and effectors to create drugs with greater specificity and selectivity, thanks to cutting-edge computational techniques that can depict and predict allosteric effects at range of scales, from the coordinated atomic movements in one receptor to complex allosteric signaling systems. The combination of:

- Theoretical Markov state model (MSM)-based frameworks and software
- Community MD codes which are able to sample realistic sized systems on single gaming/commodity GPU processors for over 100 nanoseconds per day
- Pocket/druggable site-detection algorithms make it promising for industrial settings with tight deadlines and standards to relate these methods to their drug targets. For example, applying these methods to kinases and GPCRs could prove especially beneficial due to the wealth of existing assays and structure data as well as the issues faced by clinic-based drug candidates [27].

9.7.7 Computational Methods for Experimenting Allostery

9.7.7.1 Protein Sequence Analysis Procedures

An effective method for locating and describing allosteric routes and pockets is protein sequence analysis. We divide sequence-based techniques into two categories: "coupled site" approaches that provide a group list made up of two or more sequence sites that seem to be functionally connected based on their co-evolution. "Single site" methods give individual functional sequence positions lists. All sequence-based analysis techniques face the same difficulties. These difficulties include how to choose and combine clear, pertinent sequences as input; how to understand the results; and how to incorporate sequence analysis findings along with other types of data.

9.7.7.1.1 Single-Site Evolutionary Analysis Procedures

According to meaning single-site evolutionary analysis techniques provide list of anticipated functional sequence locations but are unable provide precise recommendations for links between sites. Single-site approaches for identifying allostery are preferable since they do not have as much noise as correlation analysis frequently does. These studies are extremely interesting due to their simplicity. Typically, fewer parameters need to be adjusted, and the outcomes may be immediately seen by emphasizing important remains on a 3D protein structure.

9.7.7.1.2 Coupled-Site Evolutionary Analysis Procedures

Residue pairings that mutate together more commonly than would be predicted by arbitrary genetic processes are found using second-order sequence analysis.

Because co-evolving residue pairs frequently play crucial roles in allostery and structural integrity, it is hypothesized that they are functionally connected.

A list containing residue pairings along corresponding correlation strengths is the direct result of second-order allostery analysis. The job of combining each of these unique pairwise associations into a comprehensive representation of the protein is distinct. On a fundamental level, the greatest correlations involving a residue or site of interest may indicate thermodynamic connection to further sites, presumably connected to allostery. To investigate these links and find tightly connected networks of co-evolving residues, more intricate techniques include hierarchical clustering or principal component scrutiny.

9.8 FUTURE VISIONS AND PROPOSALS

Although computational approaches are most useful in offering effective and affordable means to direct experimental procedures, they are unlikely to totally replace the discovery process. Insights from structural modeling can be used to identify exposed residues that might be subjected to mutation in order to improve binding, lessen immunogenicity, and reveal hydrophobicity patches that are associated with subpar biophysical characteristics. By creating a foundation for determined surface display libraries to create an antibody binder targeting a particular epitope, forecasting interface information can act as a first step in experimental epitope mapping or as the beginning of a therapeutic campaign. Utilizing the extensive next-generation sequencing (NGS) data will make it simpler to obtain more reliable profiles of "humanness" and "developability," which can guide the development of antibody therapeutics. Because of improvements in MD and VS, CADD has a significant impact on preclinical drug development.

These are still rough approximations, though, and since the early screening tests, MD simulations have gained widespread acceptance. The energies required for binding for the protein–ligand schemes are given by both MD simulations and DFT calculations, but MD simulation, which apply force fields broadly to the entire system, incline to present more favorable binding energy figures that do not continually correspond to the observed values determined experimentally. We support the idea that researchers should use more quantum mechanics (QM) or molecular mechanics (MM) methods, specifically DFT. With the use of this method, the electronic properties of the interacting atoms in system are taken into consideration, and the system may be divided so that the region of significance, such as the active site, can be treated with better reliable and precise QM calculations. In addition, it provides an accurate description of the processes, providing approximations that are more trustworthy and comparable and are in accordance with the findings of in vitro experiments.

Researchers should perform MD simulations, DFT calculations, and experimental assays followed by VS to validate the identified compounds and their target protein interactions.

- It is possible that computational antibody techniques can be used to create alternative immunoglobulin forms, such nanobodies, which have more therapeutic promise.
- The benefits of computational approaches and the situations in which they might be utilized to enhance treatment workflows can be highlighted by systematic benchmarking of the tools.

REFERENCES

1. Sliwoski, G., et al., Computational methods in drug discovery. *Pharmacological Reviews*, 2014. **66**(1): p. 334–395.

2. Sabe, V.T., et al., Current trends in computer aided drug design and a highlight of drugs discovered via computational techniques: A review. *European Journal of Medicinal Chemistry*, 2021. **224**: p. 113705.

3. Yang, L., et al., Development of novel proteasome inhibitors based on phthalazinone scaffold. *Bioorganic & Medicinal Chemistry Letters*, 2016. **26**(12): p. 2801–2805.

4. Leelananda, S.P. and S. Lindert, Computational methods in drug discovery. *Beilstein Journal of Organic Chemistry*, 2016. **12**(1): p. 2694–2718.

5. Hanson, S.M., et al., Capsaicin interaction with TRPV1 channels in a lipid bilayer: Molecular dynamics simulation. *Biophysical Journal*, 2015. **108**(6): p. 1425–1434.

6. Wang, Y., S.A. Shaikh, and E. Tajkhorshid, Exploring transmembrane diffusion pathways with molecular dynamics. *Physiology*, 2010. p. 142–154.

7. Fauman, E.B., B.K. Rai, and E.S. Huang, Structure-based druggability assessment: identifying suitable targets for small molecule therapeutics. *Current Opinion in Chemical Biology*, 2011. **15**(4): p. 463–468.

8. Henrich, S., et al., Computational approaches to identifying and characterizing protein binding sites for ligand design. *Journal of Molecular Recognition: An Interdisciplinary Journal*, 2010. **23**(2): p. 209–219.

9. Halperin, I., et al., Principles of docking: An overview of search algorithms and a guide to scoring functions. *Proteins: Structure, Function, and Bioinformatics*, 2002. **47**(4): p. 409–443.

10. Dias, R., J. de Azevedo, and F. Walter, Molecular docking algorithms. *Current Drug Targets*, 2008. **9**(12): p. 1040–1047.

11. Feher, M., Consensus scoring for protein–ligand interactions. *Drug Discovery Today*, 2006. **11**(9–10): p. 421–428.

12. Voss, N.R. and M. Gerstein, 3V: Cavity, channel and cleft volume calculator and extractor. *Nucleic Acids Research*, 2010. **38**(suppl_2): p. W555–W562.

13. Ghersi, D. and R. Sanchez, Beyond structural genomics: Computational approaches for the identification of ligand binding sites in protein structures. *Journal of Structural and Functional Genomics*, 2011. **12**: p. 109–117.

14. Jorgensen, W.L., Efficient drug lead discovery and optimization. *Accounts of Chemical Research*, 2009. **42**(6): p. 724–733.

15. Durrant, J.D. and J.A. McCammon, Molecular dynamics simulations and drug discovery. *BMC Biology*, 2011. **9**(1): p. 1–9.

16. Fagerberg, J.H., et al., Computational prediction of drug solubility in fasted simulated and aspirated human intestinal fluid. *Pharmaceutical Research*, 2015. **32**: p. 578–589.

17. Huynh, L., et al., Predicting the solubility of the anti-cancer agent docetaxel in small molecule excipients using computational methods. *Pharmaceutical Research*, 2008. **25**: p. 147–157.

18. A. Lipinski, C., F. Lombardo, B.W. Dominy, and P.J. Feeney, Experimental and computational approaches to estimate solubility and permeability in drug discovery and development settings. *Advanced Drug Delivery Reviews*, 2001. **46**(1–3): p. 3–26.

19. Congreve, M., et al., A'rule of three'for fragment-based lead discovery? *Drug Discovery Today*, 2003. **8**(19): p. 876–877.

20. Loew, G.H., H.O. Villar, and I. Alkorta, Strategies for indirect computer-aided drug design. *Pharmaceutical Research*, 1993. **10**: p. 475–486.

21. Wu, F., et al., Computational approaches in preclinical studies on drug discovery and development. *Frontiers in Chemistry*, 2020. **8**: p. 726.

22. Speck-Planche, A. and M. Cordeiro, *Speeding up the virtual design and screening of therapeutic peptides: Simultaneous prediction of anticancer activity and cytotoxicity*, in *Multi-Scale Approaches in Drug Discovery*. 2017. p. 127–147.

23. Lin, X., X. Li, and X. Lin, A review on applications of computational methods in drug screening and design. *Molecules*, 2020. **25**(6): p. 1375.

24. Cui, W., et al., Discovering anti-cancer drugs via computational methods. *Frontiers in Pharmacology*, 2020. **11**: p. 733.

25. Peng, H.-P., et al., Origins of specificity and affinity in antibody–protein interactions. *Proceedings of the National Academy of Sciences*, 2014. **111**(26): p. E2656–E2665.

26. Norman, R. A., et al., Computational approaches to therapeutic antibody design: Established methods and emerging trends. *Briefings in Bioinformatics*, 2020. **21**(5): p. 1549–1567.

27. Wagner, J. R., et al., Emerging computational methods for the rational discovery of allosteric drugs. *Chemical Reviews*, 2016. **116**(11): p. 6370–6390.

10 Natural Products, Nutraceuticals, and Their Chemistry

Hassan Murtaza Naqvi, Khurram Afzal, Shahid Shah, Malik Saadullah, and Samina Afzal*

10.1 INTRODUCTION TO NATURAL PRODUCTS

Natural products are chemical entities obtained from natural sources such as plants, animals, and microorganisms [1]. They have been used for thousands of years in traditional medicine and as a source of food, clothing, and shelter [2]. In recent years, there has been a growing interest in natural products as a source of new drugs, due to their wide range of biological activities and potential therapeutic applications. The use of natural products in drug discovery has been proven to be a successful approach, with hundreds of natural product-derived drugs currently in use for treatment and cure of a lot of ailments [3].

The use of natural products in drug discovery has a long history, being the only productive source of "leads" for the development of drugs [3]. Dating back to the conscious use of the bark of the cinchona tree in the treatment of malaria in the 17th century to the development of anticancer drugs, hundreds of natural products have been discovered to have biological activities and potential to the therapeutic efficacy including the already discussed anticancer, antiviral, antibacterial, antioxidant, cardiotonic, antidiabetic, and anti-inflammatory responses in biological systems.

Natural products have been used as a source of lead compounds for the development of synthetic drugs, as well as a source of inspiration for the development of new drug scaffolds. Natural products are known for their diverse chemical structures and biological activities, making them a valuable source of new drugs [3]. In this era of digitalization where new products are synthesized also by target structure elucidation, plant-based drugs remain not less than 50% of the drug population in the market [2, 4]. The number of natural source-based compounds being evaluated in clinical trials for medicinal use continues to rise along with a similar number of projects investigating natural substances in preclinical trials for drug development. This importance of natural products is credited to the most diverse chemistry of natural substances owing to their evolution in natural environment which made them suitable for use in biological systems [4].

It is estimated that about 0.3 million natural products are biologically active. They have been categorized into five classes depending on their chemistry: (i) terpenoids and steroids; (ii) fatty acids derivatives and polyketides; (iii) shikimic acid derivatives; (iv) non-ribosomal polypeptides; and (v) alkaloids [5].

10.2 SOURCES OF NATURAL PRODUCTS

Natural products can be obtained from more than limited sources that are abundant in nature. The most important sources of natural products are plants, animals, microorganisms, and marine organisms [6].

1. Plants being the most diverse and abundant source of biologically active natural products [2]. It is observed that till 2012, about 70,000 plant species had been assessed for medicinal purposes and of the 252 drugs that are considered important by World Health Organization (WHO), 11% were obtained from the flowering plants only [2].

* Corresponding author

DOI: 10.1201/9781032707273-10

2. The second source of natural products are animals. Billions of dollars are made annually by the use of animal-based drugs today. Of the essential entities proclaimed by WHO, 8.7% of the drugs are obtained from animals [7]. Natural products ranging from medicines to nutraceuticals are included in substances that are obtained from animal sources [7].
3. Microorganism are the third major source of natural products. Of all the naturally occurring biologically active metabolites, 5% are produced by microorganisms [8]. The drugs obtained from microorganisms have often been proved a breakthrough in the field of medicines such as penicillin.
4. Marine environment is the most unexplored world for bioactive compounds. Despite this fact, marine organisms produce a variety of substances that are sources of natural products. In the past five decades, more than 20,000 biologically active substances have been identified from marine sources [9]. Like all other sources the marine environment is still contributing to the field of discovery for natural products.

10.3 CLASSIFICATION OF NATURAL PRODUCTS

On the basis of use, natural products have been classified into following categories [6]:

1. Herbal medicines
2. Essential oils
3. Nutraceuticals
4. Cosmetics

10.3.1 Herbal Medicines

Herbal medicines are the phytochemicals obtained from different types of plants to treat or cure ailments or disorders in human beings. In the modern world, herbal medicines are also called phytopharmaceuticals [10]). In the modern world where computational chemistry is at its peak, herbal medicines of phytopharmaceuticals are still the base of disease treatment of 80% of the world population [10]. The diversity of herbal medicines is so extreme that it includes medicines for strong actions to medicines for very light actions. Herbal medicines are claimed to include treatment for all types of ailments [10]. Medicines such as vincristine, the anticancer drugs; aspirin, the nonsteroidal anti-inflammatory drug; digitalis, the cardioactive drug; and antioxidants increase the importance of herbal medicines to more than health.

10.3.2 Essential Oils

Essential oils are the volatile substances that are present in different parts of the herbs and spices. They are responsible for the odor, taste, and essence of characteristic natural substances. Essential oils have been used from ancient times for the purpose of preservation of food, flavoring, perfuming, and medicines [11].

10.3.3 Nutraceuticals

Nutraceuticals are the food substances which along with providing nutritional benefits also help in the improvement of health or treatment of nutritional deficiency-based disorders or ailments. Nutraceuticals may contain the essential components of nutrition along with vitamins, minerals, and primary metabolites from different natural sources [12].

10.3.4 Cosmetics

There are a lot of natural products that are used as cosmetics and these cosmetics are called herbal cosmetics or natural cosmetics. These cosmetics from natural sources have been used from ancient times although civilizations have debated the use of cosmetics. With the information of few herbs, plants, flora, and their uses, peoples form different ages were able to use natural cosmetics and those uses were even fruitful. In the modern world, the demand for natural cosmetics is again on rise mainly due to their lesser side effects and their natural chemistry to be in compliance with nature. Herbal cosmetics include saffron, sandal, rose water, and aloe vera that are still used today with ease and ability to enhance the aesthetics of human beauty [13].

10.4 NUTRACEUTICALS

10.4.1 Introduction

Nutraceuticals are natural products or substances that are obtained from natural products that are food, ingredients of food, or enhance the benefits of foods for its nutritional value to provide health benefits [12, 14]. The term nutraceuticals originated from two words "nutrition" and "pharmaceutical." It was first used in 1989 by Dr. Stephen L. DeFelice for naturally nutrient-rich food that provides health and medical benefits to aid in ailments [12]. Other names for nutraceuticals are medical foods, dietary and nutritional supplements. Nutraceuticals contain not only the primary metabolites but also the secondary metabolites of nature including protein, lipids, carbohydrates, vitamins, hormones, minerals, and other nutrients. Examples of nutraceuticals are banana, spirulina, soy, moringa, or fish oil containing omega-3 as component [12, 14]. Nutraceuticals include dietary supplements, genetically engineered foods, prebiotics, probiotics, herbal products, isolated nutrients, and processed supplements such as cereals and drinkables (soups and beverages) [12, 15, 16]. Nutraceuticals may also be known as functional foods [17] but according to some there is a little difference between the two. Functional foods are defined as "foods or dietary components that may provide a health benefit beyond basic nutrition" by the International Food Informational Council. Functional foods include prepared meals, already digested or semi-digested food, fortified food for babies, cereals, drinks, breads, and much more [17]. In simple words, nutraceuticals or the functional foods are any eatable substances that may or may not be the combination of one or more natural substances such as orange juice with added calcium, vitamin D, and vitamin C. Other examples may include meshed banana in yogurt as food for children [18], banana shake with milk as instant energy drink in adults [19], or powder of combination of almonds, walnuts, coconut, and raisins as source of vitamins, omega-3, and essential oil [20, 21].

Nutraceuticals have been proved so effective for human health improvement that they must be distinguished from drugs. Until recently no such definition has been developed that may clearly distinguish between the two, but it is established that the drugs are not or never used as food or component of food or supplement of food [15, 17, 22]. The interest in nutraceuticals rather than in drugs is increasing rapidly due to the fact that they are safe as they are already being used in one or other way by us; also nutraceuticals require less time for their research, trials, and approval by the regulatory authorities.

10.4.2 Classification of Nutraceuticals

Nutraceuticals can be classified on the basis of health benefits they provide and their chemistry.

The health benefits provided by the nutraceuticals may be catagorized into two segments.

- General health benefits
- Physiological benefits

10.4.2.1 General Health Benefits

General health benefits are the characteristics of nutraceuticals that can improve the quality of life even in the absence of medical ailments and disorders. These characteristics may aid the normal functions of human body and improve them in one or another way.

Nutraceuticals provide general health benefits by serving as:

- Biochemical reactions substrate
- Enzymatic reaction inhibitors
- Enzymatic reaction co-factors
- Free radical scavengers
- Intestinal absorbents and adsorbents for undesirable substances elimination
- Stability and absorption enhancers for nutrients from gastrointestinal tract (GIT)
- Growth inhibitors of deleterious or harmful intestinal flora
- Growth enhancers for useful and beneficial intestinal flora [12]

The aforementioned functions performed by the nutraceuticals improve the general health of human beings by:

- Improving the value of diet health
- Compensating the nutritional deficiencies in human body
- Improving the cognitive functions of humans
- Increasing the health span of life
- Increasing the time span of life [12].

10.4.2.2 Nutraceutical for Physiological Benefits

Nutraceuticals are also being used for certain medical and physiological disorders. There is a long list of disorders that are being treated or managed by nutraceuticals. Some of the nutraceuticals used for these purposes are:

- Cardiovascular agents
- Anticancer and antitumor agents
- Antiobesity agents
- Antidiabetic agents
- Agents for inflammatory disorders
- Agents for GIT disorders
- Immunity boosters
- Antioxidants

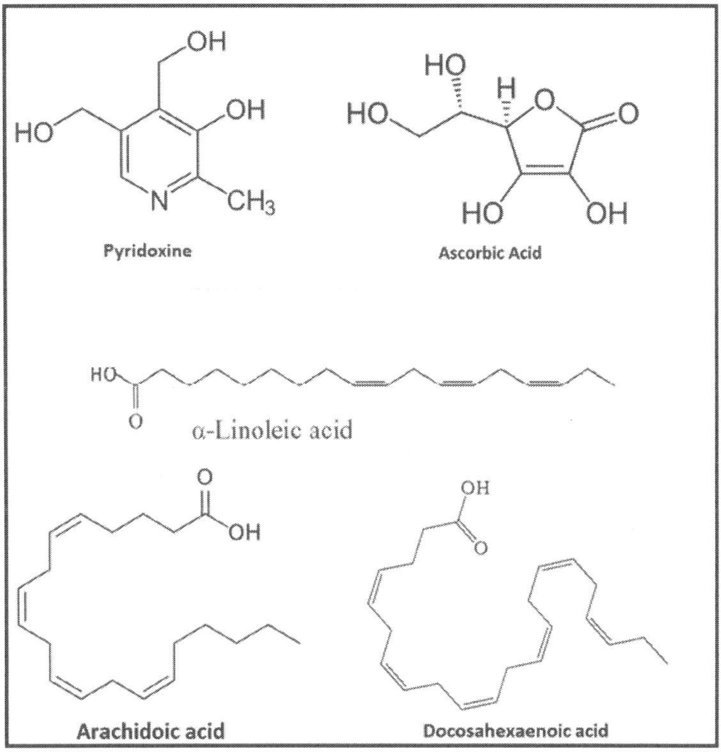

Figure 10.1 Pyridoxine (vitamin B_6), ascorbic acid (vitamin C), essential fatty acids (α-linoleic acid, arachidonic acid, and docosahexaenoic acid).

10.4.2.2.1 Nutraceuticals as Cardiovascular Agents

Disorders of blood vessels and heart are named as cardiovascular disorders. Various kinds of disorders are included in this category which includes all age groups. Nutraceuticals in the form of vitamins, antioxidants, dietary fibers, polyunsaturated fatty acids, and minerals can help to prevent, delay, treat, or manage certain cardiovascular disorders. Magnesium, pyridoxine (vitamin B_6), ascorbic acid (vitamin C), celery, α-linoleic acid (omega-3 fatty acid), and hawthorn serve as blockers of calcium channels (Figure 10.1). Apples, cherries, black grapes, berries, grapefruits, and onions provide flavonoids that serve as angiotensin-converting enzyme (ACE) inhibitors [12, 23]. α-Linoleic acid, arachidonic acid, and docosahexaenoic acid (essential fatty acids) are very important for reducing cholesterol, low-density lipoproteins, and triglycerides to reduce the blood clots in vessels and thus reduce the blood pressure along with soluble, insoluble fibers, and whole grains. Omega-3 fatty acids and other essential fatty acids are very important for the treatment and management of arrhythmias. These essential fatty acids are obtained majorly from milk, egg, and fish. Fibers are obtained from wheat bran, oats, psyllium, and cereals (Figure 10.1) [12].

10.4.2.2.2 Nutraceuticals as Anticancer Agents

Cancer has been a major problem for public health for a long time in progressing countries. Diet and foods containing large quantities of plants are linked with very low incidence of cancers. Green leafy foods, oranges, tomatoes, and lutein-rich foods have been assessed to lower the risk of colon cancer. Ginseng is known to have anticancer properties by preventing inflammation to cancer sequence. Tannins are also known to scavenge free radicals to prevent cancer. Tea, black berries, cranberries, blue berries, and lentils contain such tannins [12]. Cancers that are induced by hormones such as estrogen are prevented efficiently by flavonoids. Phytoestrogens are phytopharmaceuticals that are known to prevent breast and prostate cancers [14]. Ellagic acid, which is a proven and broadly used anticancer agent, is obtained mainly from seeds of red raspberries. Walnuts, strawberries, pomegranates, peacans, and cranberries also serve to provide ellagic acid. A phytochemical class phenolics also demonstrated anticancer activity including gallic acid, ferulic acid, curcumin, and caffeic acid. Curcumin is obtained from turmeric, gallic acid from tea, and caffeic acid from coffee beans mainly. Glucosinolates and the products of glucosinolate that are obtained by hydrolysis, which include isothiocyanates such as sulforaphane [24, 25] and indole derivatives such as indole-3-carbinol [26, 27] are known to lower the risk or fight the onset of colorectal cancer and lung cancer (Figure 10.2). They are also known to detoxify

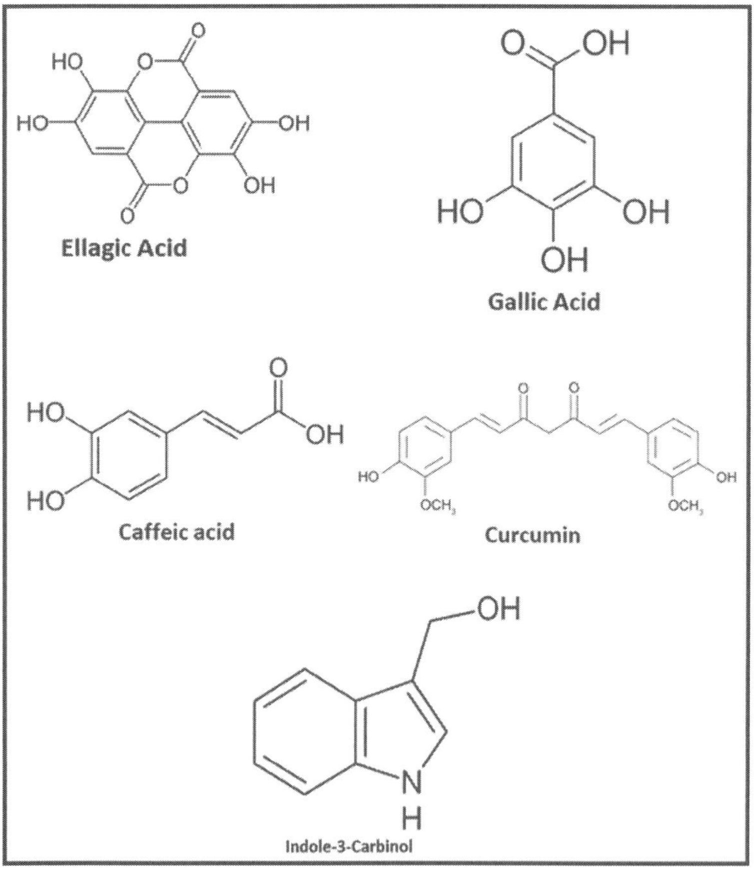

Figure 10.2 Chemical structures of ellagic acid, gallic acid, caffeic acid, curcumin, and indole-3-carbinol.

liver. The glucosinolates and its products are abundantly found in vegetables like mustard, cauliflower, rape seed oil, broccoli, and cabbage [12].

10.4.2.2.3 Nutraceuticals as Anti-obese Agents.

Obese or being fat is a health problem recognized by WHO as global threat. Obesity is a problem leading to a lot of metabolic disorders along with cancers, hypertension, angina pectoris, congestive heart failure, hyperlipidemia, vein thrombosis, respiratory disorders, infertility, and osteoarthritis. There may be many reasons for obesity but top of the list is availability of energy-rich foods such as high-carbohydrate-containing foods, fast foods, snacks and soft drinks. Fat-rich food (vegetable fat) is the second main reason for obesity. Less or no exercise and lazy lifestyle is the third reason in being obese in developing countries. Thus healthy food and balanced lifestyle can play a significant role in the prevention of this problem. Foods that contain less carbohydrates, low vegetable fat, and more fiber are beneficial in maintaining and gaining a normal weight and BMI. Green leafy vegetables, fruits, and red meat are among these foods. Fiber-containing foods include wheat bran, psyllium, and whole grains. Energy stimulants from herbal sources that increase the energy expenditure, such as caffeine, ephedrine, chitosan, and antioxidants, are very effective phytochemicals in avoiding being obese [12]. Green tea, tea, coffee beans, and ephedra are the sources of such energy stimulants. 5-Hydroxytryptophan has been used clinically for the reduction of weight in habitual obese patients as nutraceuticals [28]. Vitamins also play a main role in losing weight and gaining healthy body index (Figure 10.3). Vitamin D along with little exercise has been

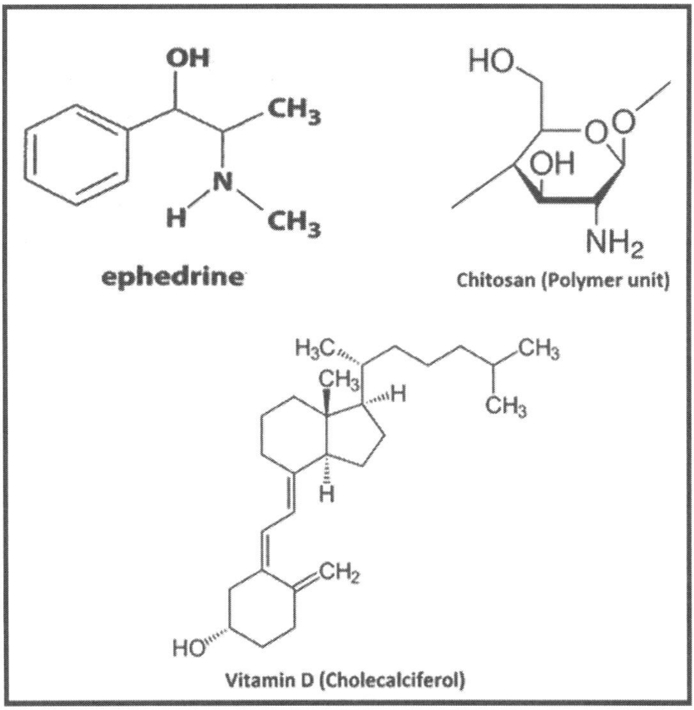

Figure 10.3 Chemical structure of ephedrine, chitosan, and vitamin D.

known to reduce or maintain healthy weight in individuals [29]. Same is the case with vitamin C. When taken with exercise, it helps to burn fat more [30]. Vitamin B complex has also been known for energy regulation and consumption in the body such as vitamin B_6 is known to regulate gluconeogenesis and glucogenolysis to provide energy to the body [31].

10.4.2.2.4 Nutraceuticals as Anti-Diabetic Agents

Diabetes mellitus is becoming the most growing disorder in the world identified by health authorities. It is of two types mainly: type I (insulin dependent) and type II (insulin independent). Very high levels of glucose not maintained or regulated in the blood cause this disease. Type I diabetes (occurs due to autoimmune disorder) prevails in 5% and type II (occurs mainly due to obesity) prevails in 95% of the total diabetic patients. There is another type of diabetes that occurs during pregnancy and is it is known as gestational diabetes. Diabetes of any type in the modern day is very effectively managed by diet and food even with high blood random sugar. Healthy lifestyle and proper diet is very effective in avoiding and preventing diabetes, especially type II [32, 33]. Taking more amount of isoflavones found in leguminous plants has proved most effective in preventing and reducing the occurrence of type II diabetes [34]. Omega-3-fatty acids such as α-linoleic acid have been used traditionally to reduce glucose tolerance in diabetic patients. Insulin is necessary in the body to form omega-3 fatty acids, which in the case of diabetes may be depleted in the body [12]. Insulin resistance is modulated efficiently by docosahexaenoic acid (DHA). DHA also

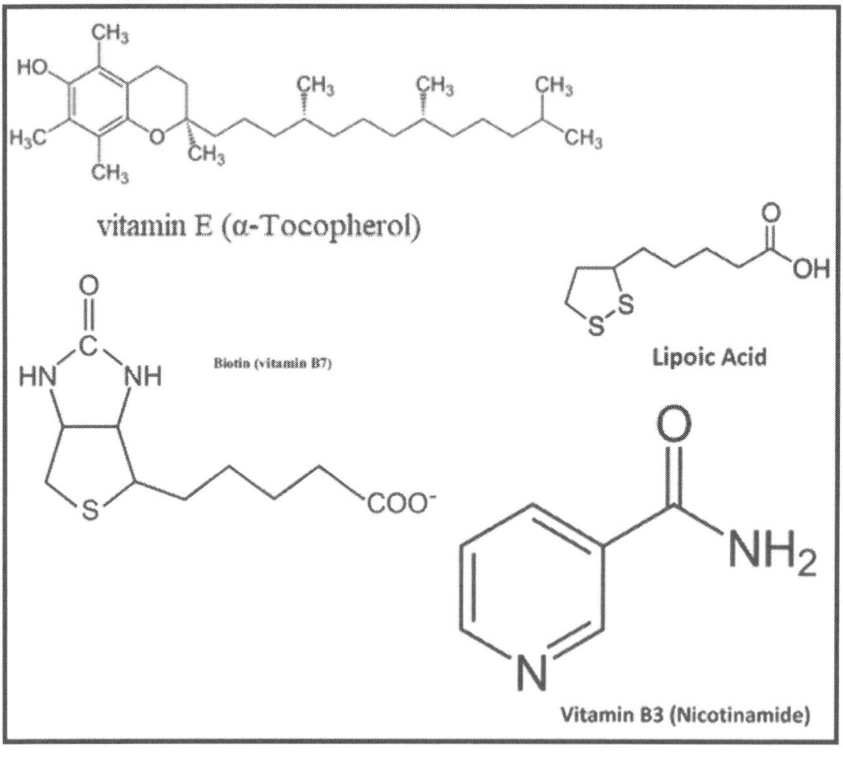

Figure 10.4 Chemical structures of vitamin E, lipoic acid, biotin, and vitamin B_3.

plays a vital role in the development of neuronal health in diabetic patients. DHA is necessary for gestational diabetes to maintain normal sugar levels during pregnancy by increasing insulin sensitivity. Magnesium, calcium, chromium, and vitamin D are reported to improve sugar controls in diabetic patients. Sugar levels also have been controlled in patients with fibers from psyllium husk. Traditionally, the extracts from cinnamon and bitter melon have been used for long to prevent and treat patients with diabetes [12]. Antioxidants such as lipoic acid that are used traditionally and clinically have proved to be reducing pathologies associated with diabetes. Vitamin B1, B3, B6, B7 B_{12} and B_9 are clinically used to treat ailments associated with diabetes such as neuropathy, nephropathy, cardiomyopathy, and retinopathy [14, 35, 36]. Vitamin C and E (α-tocopherol) and antioxidant by different mechanisms, also have been reported to aid in treating associated ailments during diabetes by protecting lipid membranes against peroxidation [37]. The potential source of most of these vitamins and minerals is *Moringa oleifera* which not only serves as the source of dietary nutrients but also its use in hyperglycemia has been reported. Aloe vera not only gives a large amount of vitamin E but it also provide anthraquinones and lectin which are known to be having antidiabetic properties. There are other micronutrients that are helpful in diabetes including coenzyme Q10, amino acids (glutamic acid and cysteine), L-carnitine, and inositol [38]. Figure 10.4 mentions the chemical structures of vitamin E, lipoic acid, biotin, and vitamin B3.

10.4.2.2.5 Nutraceuticals as Anti-inflammatory Agents

Inflammation is the defense response of the body against injury, irritation, and any other foreign stimuli to repair by healing. Inflammation is characterized by redness, swelling, and pain. There are many inflammatory disorders, including polymayalgia rheumatica, rheumatoid arthritis, bursitis, gouty arthritis, and shoulder tendinitis. Nutrients like soybean, unsaponifiables, sadinosylmethionine, avocado, capsicum, ginger, chondroitin, and glucosamine (Figure 10.5) have been reported to have valuable

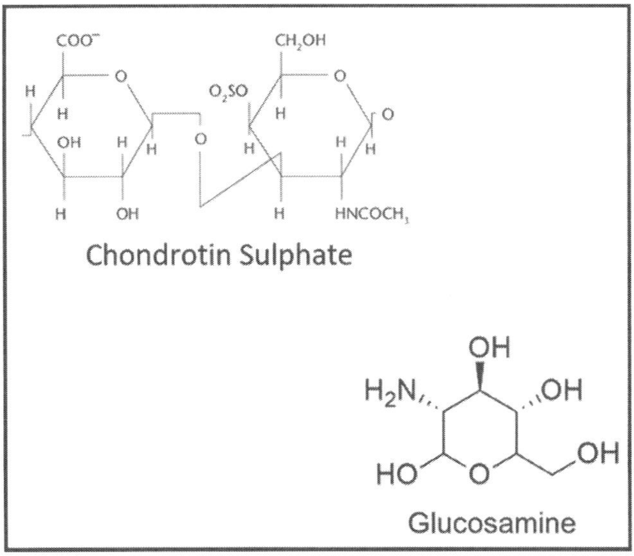

Figure 10.5 Chemical structures of chondroitin sulfate and glucosamine.

anti-inflammatory affects. Use of vitamin C and D has also proven beneficial. Anti-inflammatory effects of *Uncaria guianensis* and *U. tomentosa* have been used clinically for the treatment of wounds and inflammation. The aqueous extract of these two species is known to contain strong anti-inflammation and antioxidant response [12]. Also, the hot water extract of Buckwheat showed considerable anti-inflammatory and antioxidant effects and its use during inflammation is promoted [39]. Heteropolysaccharides from mushroom are also known to have antioxidant and anti-inflammatory properties such as *Lentinula edodes* [40].

10.4.2.2.6 Nutraceuticals as Immunity Boosters

Immune system of the human body is responsible for fighting against a lot of diseases and disorder causing agents. Optimal immune system requires balanced and proper diet with all nutrients. There are such nutrients that are most suitable to provide boost to our immune system. *M. oleifera* is among the plants that are particularly used clinically to enhance the immune system [41]. The powder of dry leaves and stalks of moringa plant is taken directly or with eatables to harness its full advantage. The extracts of coneflowers, such as *angustifolia* and *perpurea* species, produce unsaturated alkyl ketones that are helpful in immunity [42]. The herbs of genus *Sambuca*, such as goldenseal and elderberries extract produce immune boosting effect by antibiotic effects. Herbs of *Astralagus* species such as *mongolicus* and *membranaceous* are effective immune boosters. Astralagus provides nutrients to promote the stem cell production in lymph tissues and bone marrow to promote immune response. Several probiotics used for intestinal health and immunity have been promoted in the clinical setting. The strains of *Bifidobacterium*, *Streptococcus*, and *Lactobacillus* are used to improve intestinal health, functions and immunity [12]. *Curcuma longa* has been also known to stimulate cellular immunity interfering inflammation, contributing to overall immune boosting. There are other plants such as *Panax ginseng* and *Withania somnifera* which enhance immunity by fighting stress [42]. Minerals and vitamins have an established role in the immune system promotion. Vitamins C, D, and E are well known for their immunomodulatory effects. Zinc and selenium have been clinically used for boosting the immune system when taken regularly [43, 44].

10.4.2.2.7 Nutraceutical as Antioxidants

Dietary antioxidants are defined as "a dietary antioxidant is a substance in food that significantly decreases the adverse effects of reactive oxygen species (ROS), reactive nitrogen species or both on normal physiological function in humans" [45]. Normal physiological processes generate energy and metabolites to fight against invasions, along with these, ROS and nitrogen species are produced. If these species are not dealt with by the body with antioxidants, then these start to produce the adverse effects also known as oxidative stress. Many disorders arise in the body due to long oxidative stress effects resulting in cancer, cardiovascular disorders, inflammation, infection, low immunity, endocrinological and/or neurological disorders. Therefore, there is a continuous need of the body for antioxidant supply [45]. There are a lot of substances that act as antioxidants in the food and diet that we take routinely. Taking fruits, vegetables, tea, and olive oil provides enough antioxidants to the body which include vitamin E, vitamin C, and carotenoids. Lycopene obtained from tomatoes also provides considerable amounts of antioxidants. Flavonoids not only prevent damage to lipid membrane by peroxidation and atherosclerosis but also promote vascular relaxation as scavengers of free radicals. Coenzyme Q10 obtained from fish oil, animal liver, or whole grains is also a strong antioxidant [46].

10.4.2.2.8 Nutraceuticals as Energy Boosters

To live a healthy lifestyle it is very important to be active, and to remain active one needs proper metabolism and energy throughout time. There are such foods and nutrients which ensure proper metabolism and enough energy production in the body to remain active and healthy. Beetroot extract is used traditionally to boost energy and has been claimed by many researchers [47]. Human metabolism is also boosted by taking proper quantity of vitamins such as B complex as nearly all vitamin B are necessary for food metabolism that generates energy [14]. Certain antioxidants also boosts the metabolism of body, such as astaxanthin, glutathione, catechin, anthocyanin, quercetin and omega-3 fatty acids [48, 49].

10.4.3 Chemical Classification of Nutraceuticals

Plants, animals, and microorganisms give us different types of compounds. These compounds are classified on the basis of their chemical characteristics. Nutraceuticals can also be defined on the basis of their chemistry. The chemical classification of nutraceuticals is as follows:

1. Carotenoids
2. Collagen hydrolysates
3. Dietary fibers
4. Essential fatty acids
5. Flavonoids
6. Phenols
7. Plant sterols
8. Prebiotics and probiotics
9. Saponins
10. Soy proteins
11. Phytoestrogens
12. Thiols/Sulfides
13. Tannins
14. Polyols

Classification of natural substances or nutraceuticals into some categories is not possible on the basis of chemistry. Some of the major classes along with their major sources and benefits that are mostly discussed are presented in Table 10.1. More work is being done on the classifications and more effort is required upon their research (Tables 10.2 and 10.3).

Table 10.1: Some Common Nutraceuticals, Their Sources and Benefits [14, 50]

Chemical Class	Natural Source	Uses/Benefits
Carotenoids		
α-Carotene	Carrots, sweet potato	Scavenge free radicals
β-Carotene	Vegetables and fruits	Scavenge free radicals
L-lutein	Green leafy vegetables	Vision health
Tocotrienol	Palm oil, grains	Cardiovascular disease (CVD) risk reduction, anticancer effects
Lycopene extract	Tomatoes	Antioxidants, reduce prostate cancer risk
Zeaxanthein	Corn, citrus fruits, egg	Vision health
Collagen Hydrolysate		
Collagen hydrolysate	Gelatin source	Osteoarthritis associated issues
Dietary Fiber		
Fiber insoluble	Wheat bran	Colon and breast cancer risk reduction
β-Glucan	Oats	CVD risk reduction
Fiber soluble	Psyllium husk	CVD risk reduction
Whole grains	Cereals	CVD risk reduction
Essential Fatty Acids		
Omega-3-fatty acids	Tuna, marine oils, fish	Enhances visual, mental function, CVD risk reduction
Docosahexaenoic acid	Tuna, marine oils, fish	Enhances visual and mental function, CVD risk reduction
CLA	Meat, milk, cheese	Enhances composition of body, cancer risk reduction
Monosaturated fatty acids	Tree nuts	Coronary heart disease risk reduction
Flavonoids		
Anthocyanidins	Fruits, green vegetables	Scavenge free radicals, cancer risk reduction
Catechins	Tea, coffee	Scavenge free radicals, cancer risk reduction
Flavonones	Citrus fruits	Scavenge free radicals, cancer risk reduction
Flavones	Green leafy vegetables, fruits	Scavenge free radicals, cancer risk reduction
Curcumin	Turmeric rhizome	Anti-inflammatory, antioxidant, blood thinner
Glucosinolates, Isocyanates, Indoles		
Sulforaphane	Vegetables (broccoli), horse raddish	Stimulates detoxification enzymes, cancer risk reduction
Phenols		
Caffeic acids	Citrus fruits, vegetables, and fruits	Effective like antioxidant, degenerative risk reduction
Plant Sterols		
Stanol ester	Soy, wheat, corn, wood oil	Cholesterol absorption inhibition to lower cholesterol levels

Table 10.1 *(Continued)*: Some Common Nutraceuticals, Their Sources and Benefits

Chemical Class	Natural Source	Uses/Benefits
Pro/Prebiotics		
Fructo-oligosaccharides	Onion powder, Jerusalem artichokes, shallots	Improve gastrointestinal tract (GIT) health
Lactobacillus	Dairy products, yogurt	Improve GIT health
Saponins		
Saponin	Soy foods, soybeans, foods containing soy protein	Anticancer, LDL reduction, cholesterol reduction
Soy Proteins		
Soy proteins	Soy-based diet, soybeans	Heart disease risk reduction
Phytoestrogens		
Isoflavones	Soybeans, legumes	Lower LDL and cholesterol, antioxidant, cancer risk reduction
Lignans	Flaxseed, rye	Breast and colon cancer risk reduction
Ganistein, Daidzein	Soy-based diet, soybeans	Menopause symptoms reduction
Thiols/Sulfides		
Diallyl sulfide	Garlic, scallions, onions, leeks	Cholesterol reduction, LDL reduction, boosts immune system
Dithiol thiones and allyl methyl trisulfide	Green leafy vegetables	Cholesterol reduction, LDL reduction, boosts immune system
Tannins		
Proanthocyanidins	Cocoa, chocolate, cranberries	CVD risk reduction, effective in UTI
Polyols		
Polyols, sugar alcohols, xylitol, sorbitol	Present in sweet foods	Improve oral hygiene, dental hygiene

Table 10.2: Customary Nutrients and Their Clinically Proved Health Benefits

Customary Nutrients	Clinical Health Benefits
Vitamins	
Vitamin **A** (retinol and related compounds)	Antioxidant, necessary for growth, development and health, vision health, protects and maintains mucous membranes and skin health, cancer risk reduction
Vitamin **D** (cholecalciferol and related compounds)	Essential for immune system. Teeth and bone health, stimulates the calcium absorption throughout body
Vitamin **E** (tocopherol and related compounds)	Anti-inflammatory, antioxidant, boost immunity, essential for blood, nerve tissues, lungs and skin
Vitamin **K** (minaquinone and related compounds)	Stimulate and maintain blood clotting
Vitamin **C** (ascorbic acid and related compounds)	Antioxidant. Essential for healthy skin, teeth, gums, and bones, aid in healing, improve immunity, resist against common cold

(Continued)

Table 10.2 *(Continued)*: Customary Nutrients and Their Clinically Proved Health Benefits

Customary Nutrients	Clinical Health Benefits
Vitamin $\mathbf{B_1}$ (thiamine and related compounds)	Necessary for food metabolism and neurological functions, aid in energy generation from food
Vitamin $\mathbf{B_2}$ (riboflavin and related compounds)	Necessary for food metabolism and neurological function, aid in energy production from food, essential for healthy skin and vision
Vitamin $\mathbf{B_3}$ (niacin and related compounds)	Necessary for food metabolism and neurological function, aid in energy production from food
Vitamin $\mathbf{B_5}$ (pantothenic acid and related compounds)	Essential for the cholesterol, fatty acids, steroids and intra-neuronal acetylcholine synthesis
Vitamin $\mathbf{B_6}$ (pyridoxine)	Essential for protein metabolism, aid in energy production from protein food
Vitamin $\mathbf{B_7}$ (biotin and related products)[51]	Necessary for food metabolism and neurological function, strengthen hairs and nails [51]
Vitamin $\mathbf{B_9}$ (folic acid and related compounds)	Essential for genetic materials, important in first trimester of pregnancy for embryo, blood cell formation, and heart health
Vitamin $\mathbf{B_{12}}$ (cobalamin and related compounds)	Necessary for food metabolism, essential for genetic materials, important for blood cells production and health of neurons
Minerals	
Calcium salts	Responsible for building and maintenance of teeth, gums and bones, essential in muscular, glandular, and nerve functions
Iron salts	Essential for red blood cells, hemoglobin, carry and transfer oxygen to cellular levels. Aid in energy production
Magnesium salts	Important for muscular and nerve function, teeth and bone health
Phosphorous salts	Responsible for building and maintenance of teeth, gums and bones, essential for genetic materials, energy production, storage and consumption
Potassium salts	Essential for balance of pH and water in body, nerve conduction, maintaining blood pressure and muscular function [52]
Chloride salts	Maintain balance of fluids, constituents of digestive juices and other bodily fluids [53]
Chromium salts	Essential for food metabolism by aid from insulin, antioxidant
Cobalt salts	Vitamin B_{12} component
Copper	Important in collagen, hemoglobin and energy formation, required for heart health and iron absorption
Iodine	Important constituent of hormones (tri- and tetra-iodothyronine)
Selenium	Antioxidant, required for healthy heart function
Zinc	Boosts immune system, required for cell reproduction, growth, development and heath maintenance in kids, aids in healing, sperm reproduction, and increases testosterone level
Manganese	Important part of many enzymes, required for metabolism, energy production and neurological functions [54]
Molybdenum	Development and maintenance of nervous system, involved in energy production and prevention of disorders like anemia and sexual impotence [55]

Table 10.2 *(Continued)*: Customary Nutrients and Their Clinically Proved Health Benefits

Customary Nutrients	Clinical Health Benefits
Miscellaneous nutrients	
Levo Carnitine	Fatty acids oxidation, enhances oxidative phosphorylation
Choline	Manage fat metabolism and fatty liver
Vitamin **F**	Required for membrane development, synthesis of hydroxyfatty acids, prostaglandins and leukotrienes
Inositol	Lipotropic, required for transportation of amino acids, sodium, and potassium
Taurine	Important for photoreceptor activity of retina, leucocyte antioxidation, conjugation of bile acid, improves cardiac function, enhances sperm motility and growth, enhances insulin activity

Table 10.3 Nutraceutical Brands of the United States

Company	Product	Source
Cedar's Mediterranean Foods	Probiotics	*Bacillus coagulans*
Irish Pride Bakeries	Kids bread	Omega-3
McNeil Nutrients	Splenda essentials	Wheat bread
Amazing Grass	Antioxidants	Raspberry, rice
Gaspari Nutrition	Phenorex	Bitter orange
Assure Food and Beverage Company	Cardiotonic	Kiwi, strawberry, tangerine, melon
Europharma	Vectomega	Salmon
Multifunction Supplements	Brain Shiner	Ginseng, *Ginkgo biloba*
Whitestone	Fuze	Fruits
Resvitale	Collagen enhanced chews	Red wine
Forgiven Bottling Group	Alcohol burner	*Rhodiola rosea* extract
Jamieson Laboratories	Glucosamine–collagen complex	Chicken cartilage
Pacific Health Inc.	Fish oil	Salmon
Next foods	Probiotics	Coconut water
Xigo Health	Daytime restore & Nighttime repose	*G. biloba*, ginseng
Metagenics	Cognisure	Proline-rich complex
Whitewave food company	Milk proteins + DHA	Algal oil
Jarrow Formulas	Rice protein	Brown rice
Arizona Beverage company	Rescue Water	Lemon, berry, orange
Shamrock Farms	Calcium Plus Milks	Milk supplement
Cadbury North America	Trident Vitality	Mint and white tea
Vitapill LLC	Bite amin	B vitamins, *Echinacea*, vitamin C, garlic, capsaicin, curcumin, quercetin

REFERENCES

1. Enodiana OI. The chemistry of natural product: Plant secondary metabolites. *International Journal of Technology Enhancements and Emerging Engineering Research*. 2016;4:1–9.

2. Veeresham C. Natural products derived from plants as a source of drugs. *Journal of Advanced Pharmaceutical Technology & Research*. 2012;3:200–1.

3. Atanasov AG, Zotchev SB, Dirsch VM, Orhan IE, Banach M, Rollinger JM, et al. Natural products in drug discovery: Advances and opportunities. *Nature Reviews Drug Discovery*. 2021;20(3):200–16.

4. Newman DJ, Cragg GM. Natural products as sources of new drugs over the last 25 years. *Journal of Natural Products*. 2007;70(3):461–77.

5. Abdel-Razek AS, El-Naggar ME, Allam A, Morsy OM, Othman SI. Microbial natural products in drug discovery. *Processes [Internet]*. 2020;8(4) 470.

6. Khaleel FD. Natural products and its scope and applications. 2018.

7. Costa-Neto EM. Animal-based medicines: Biological prospection and the sustainable use of zootherapeutic resources. *Anais da Academia Brasileira de ciências*. 2005;77:33–43.

8. Demain AL, Sanchez S. Microbial drug discovery: 80 years of progress. *The Journal of Antibiotics*. 2009;62(1):5–16.

9. Gerwick WH, Fenner AM. Drug discovery from marine microbes. *Microbial Ecology*. 2013;65(4):800–6.

10. Kamboj VP. Herbal medicine. *Current Science*. 2000;78(1):35–9.

11. Fometu S, Shittu S, Herman R, Ayepa E. Essential oils and their applications: A mini review. *Advances in Nutrition and Food science*. 2019;4:1–13.

12. Pulipati S, Srinivasa B, Sree N, Kumar U, Shaheela S, Krishna J, et al. Nutraceuticals. *The Indian Pharmacist*. 2016;13(10):21–30.

13. Kumar D, Rajora G, Parkash O, Antil M, Kumar V. Herbal cosmetics: An overview. *International Journal of Advanced Scientific Research*. 2016;1(4):36–41.

14. Verma G, Mishra MK. A review on nutraceuticals: Classification and its role in various diseases. *International Journal of Pharmacy & Therapeutics*. 2016;7(4):152–60.

15. Dudeja P, Gupta RK. Nutraceuticals. In: Gupta RK, Dudeja, Singh M, editors. *Food Safety in the 21st Century*. 2017. p. 491–6.

16. Daliu P, Santini A, Novellino E. From pharmaceuticals to nutraceuticals: Bridging disease prevention and management. *Expert Review of Clinical Pharmacology*. 2019;12(1):1–7.

17. Wildman REC. *Handbook of nutraceuticals and functional foods*. 2016; CRC Press.

18. Roy DKD, Saha T, Akter M, Hosain M, Khatun H, Roy MC. Quality evaluation of yogurt supplemented with fruit pulp (banana, papaya, and water melon). *International Journal of Nutrition and Food Sciences*. 2015;4(6):695–9.

19. Harrop GA. A milk and banana diet for the treatment of obesity. *Journal of the American Medical Association*. 1934;102(24):2003–5.

20. Pan F, Wang X, Wen B, Wang C, Xu Y, Dang W, et al. Development of walnut oil and almond oil blends for improvements in nutritional and oxidative stability. *Grasas y Aceites*. 2020;71(4):e381.

21. Pradhan C, Peter N, Dileep N. Nuts as dietary source of fatty acids and micro nutrients in human health. In: *Nuts and nut products in human health and nutrition*. 2021; Intech.

22. Santini A, Cammarata SM, Capone G, Ianaro A, Tenore GC, Pani L, et al. Nutraceuticals: Opening the debate for a regulatory framework. *British Journal of Clinical Pharmacology*. 2018;84(4):659–72.

23. Temple NJ, Gladwin KK. Fruit, vegetables, and the prevention of cancer: Research challenges. *Nutrition*. 2003;19(5):467–70.

24. Prieto MA, López CJ, Simal-Gandara J. Glucosinolates: Molecular structure, breakdown, genetic, bioavailability, properties and healthy and adverse effects. *Advances in Food and Nutrition Research*. 2019;90:305–50.

25. Connolly EL, Sim M, Travica N, Marx W, Beasy G, Lynch GS, et al. Glucosinolates from cruciferous vegetables and their potential role in chronic disease: Investigating the preclinical and clinical evidence. *Frontiers in Pharmacology*. 2021;12:767975.

26. Popolo A, Pinto A, Daglia M, Nabavi SF, Farooqi AA, Rastrelli L. Two likely targets for the anti-cancer effect of indole derivatives from cruciferous vegetables: PI3K/Akt/mTOR signalling pathway and the aryl hydrocarbon receptor. *Seminars in Cancer Biology*. 2017;46:132–7.

27. Megna BW, Carney PR, Nukaya M, Geiger P, Kennedy GD. Indole-3-carbinol induces tumor cell death: Function follows form. *Journal of Surgical Research*. 2016;204(1):47–54.

28. Cangiano C, Ceci F, Cascino A, Del Ben M, Laviano A, Muscaritoli M, et al. Eating behavior and adherence to dietary prescriptions in obese adult subjects treated with 5-hydroxytryptophan. *The American Journal of Clinical Nutrition*. 1992;56(5):863–7.

29. Slusher AL, McAllister MJ, Huang C-J. A therapeutic role for vitamin D on obesity-associated inflammation and weight-loss intervention. *Inflammation Research*. 2015;64:565–75.

30. Johnston CS. Strategies for healthy weight loss: From vitamin C to the glycemic response. *Journal of the American College of Nutrition.* 2005;24(3):158–65.

31. Manore MM. Vitamin B6 and exercise. *International Journal of Sport Nutrition and Exercise Metabolism.* 1994;4(2):89–103.

32. Astrup A. Healthy lifestyles in Europe: Prevention of obesity and type II diabetes by diet and physical activity. *Public Health Nutrition.* 2001;4(2b):499–515.

33. Sami W, Ansari T, Butt NS, Ab Hamid MR. Effect of diet on type 2 diabetes mellitus: A review. *International Journal of Health Sciences.* 2017;11(2):65.

34. Duru K, Kovaleva E, Danilova I, Van der Bijl P, Belousova A. The potential beneficial role of isoflavones in type 2 diabetes mellitus. *Nutrition Research.* 2018;59:1–15.

35. Ko S-H, Ko S-H, Ahn Y-B, Song K-H, Han K-D, Park Y-M, et al. Association of vitamin B12 deficiency and metformin use in patients with type 2 diabetes. *Journal of Korean Medical Science.* 2014;29(7):965–72.

36. Deshmukh SV, Prabhakar B, Kulkarni YA. Water soluble vitamins and their role in diabetes and its complications. *Current Diabetes Reviews.* 2020;16(7):649–56.

37. Davì G, Santilli F, Patrono C. Nutraceuticals in diabetes and metabolic syndrome. *Cardiovascular Therapeutics.* 2010;28(4):216–26.

38. Sharma R, Amin H, Prajapati P. Plant kingdom nutraceuticals for diabetes. *Journal of Ayurvedic and Herbal Medicine.* 2016;2(6):224–8.

39. Kang HW. Antioxidant and anti-inflammation effects of water extract from buckwheat. *Culinary Science and Hospitality Research.* 2014;20(6):190–9.

40. Lee C-B. Anti-inflammation activity of water extracts from *Hericium erinacium* among medicinal mushrooms. *Culinary Science and Hospitality Research.* 2012;18(4):233–42.

41. Fajri M, editor. The potential of *Moringa oleifera* as immune booster against COVID 19. IOP Conference Series: Earth and Environmental Science; 2021: IOP Publishing.

42. Di Sotto A, Vitalone A, Di Giacomo S. Plant-derived nutraceuticals and immune system modulation: An evidence-based overview. *Vaccines.* 2020;8(3):468.

43. Shakoor H, Feehan J, Al Dhaheri AS, Ali HI, Platat C, Ismail LC, et al. Immune-boosting role of vitamins D, C, E, zinc, selenium and omega-3 fatty acids: Could they help against COVID-19? *Maturitas.* 2021;143:1–9.

44. Korf H, Decallonne B, Mathieu C. Vitamin D for infections. *Current Opinion in Endocrinology, Diabetes and Obesity.* 2014;21(6):431–6.

45. Cornelli U. Antioxidant use in nutraceuticals. *Clinics in Dermatology.* 2009;27(2):175–94.

46. Giugliano D. Dietary antioxidants for cardiovascular prevention. *Nutrition, Metabolism, and Cardiovascular Diseases: NMCD.* 2000;10(1):38–44.

47. Ashraf S, Sayeed SA, Ali R, Vohra F, Ahmed N, Alam MK. Assessment of potential benefits of functional food characteristics of beetroot energy drink and flavored milk. *BioMed Research International.* 2022:1971018.

48. Takami M, Aoi W, Terajima H, Tanimura Y, Wada S, Higashi A. Effect of dietary antioxidant-rich foods combined with aerobic training on energy metabolism in healthy young men. *Journal of Clinical Biochemistry and Nutrition.* 2019;64(1):79–85.

49. Amos D, Cook C, Santanam N. Omega 3 rich diet modulates energy metabolism via GPR120-Nrf2 crosstalk in a novel antioxidant mouse model. *Biochimica et Biophysica Acta (BBA) – Molecular and Cell Biology of Lipids.* 2019;1864(4):466–88.

50. Agarwal SS, Paridhavi M. Nutraceuticals: A Modern Approach. Herbal Drug Technology. 2nd ed. 2007 p. 711–4.

51. Bistas KG, Tadi P. Biotin. StatPearls [Internet]. 2021.

52. Weaver CM. Potassium and health. *Advances in Nutrition.* 2013;4(3):368S–77S.

53. Tkacheva N, Eliseeva T. Chlorine (Cl)–importance for the body and health, where it is contained. *Journal of Healthy Nutrition and Dietetics.* 2022;2(20):28–34.

54. Watts DL. The nutritional relationships of manganese. *Journal of Orthomolecular Medicine.* 1990;5(4):219–22.

55. Atwal PS, Scaglia F. Molybdenum cofactor deficiency. *Molecular Genetics and Metabolism.* 2016;117(1):1–4.

11 Natural Ingredients and Cosmetics

*Malik Saadullah**, Ayesha Rafiq, Asra Shanzeh Shabbir, and Ghulam Abbas*

11.1 INTRODUCTION

Natural ingredients are chemical substances obtained from naturally occurring living things that have the potential to greatly improve human health or treat a variety of ailments. The primary sources of natural goods are land and marine species, specifically plants, animals, and microorganisms. Organosulfur, phytosterols, polyphenols (e.g., stilbene, simple phenolic acids, curcuminoids, flavonoids, and isoflavonoids), and terpenoids (e.g., carotenoids) are examples of natural products that also play a significant pathological role in the prevention of aging, cancer, diabetes and its complications, hepatotoxicity, oxidative stress, and obesity. According to World Health Organization (WHO), 80% of the populations in Africa and Asia rely on natural medicines for their primary healthcare [1].

Plant-based substances have been used since the existence of mankind to enhance their appearance. The effect of the internet and other social media has increased knowledge of both the health advantages provided by the flora and other forms of nature as well as the concerns associated with the use of various chemicals in cosmetics. As an outcome, the use of natural ingredients is receiving more attention in the cosmetic industry [1].

Around 3000 BC, Egyptians were fond of using cosmetics which were mainly derived from plant sources. Egyptian women bathed in milk that was sour, containing lactic acid and α-hydroxy acid. Today these are found in many peeling solutions as these acids are known to exfoliate the skin and make it look brighter and smoother. A mixture of crocodile dung with herbs was used on the face as a mask. For the treatment of dryness, eczema, acne, antiaging, and hair care, a wide variety of natural sources are employed in skin care products. Plant essential oils have been used to create skin emollients, hair care products, and perfumes. Coconut, olive, and jojoba oils are few examples. In addition, Indian women continue to use natural skin and hair care products like *Emblica officinalis* (amla), *Acacia concinna* (shikakai), and *Callicarpa macrophylla* (priyangu). Women in Indonesia use powdered coffee seeds as facial exfoliants because they contain caffeic acid, an antioxidant that tightens collagen fibers and encourages the synthesis of new collagen. Women of the Middle East used the urine of camel to beautify their hair [1].

11.2 PHYSIOLOGY OF THE SKIN

The largest organ of the body, the skin, is responsible for maintaining bodily homeostasis and controlling the loss of fluid, electrolytes, and biomolecules (Figure 11.1). It also keeps the body active and protects it from external forces (biological, chemical, and physical). The epidermis, dermis, and hypodermis are the three layers of the skin. The stratum corneum, stratum lucidum, stratum granulosum, stratum spinosum, and stratum basale are the five distinct sublayers that make up the epidermis, which is the skin's outermost layer. Melanocytes are found in the stratum basale, the epidermis' base layer that connects to the dermis. Melanin is transferred to keratinocytes by the melanocytes' dendrites. Darkening of the eyes, hair, and skin is caused by the creation of melanin, which shields against ultraviolet (UV) light damage [2].

*Corresponding author

 DOI: 10.1201/9781032707273-11

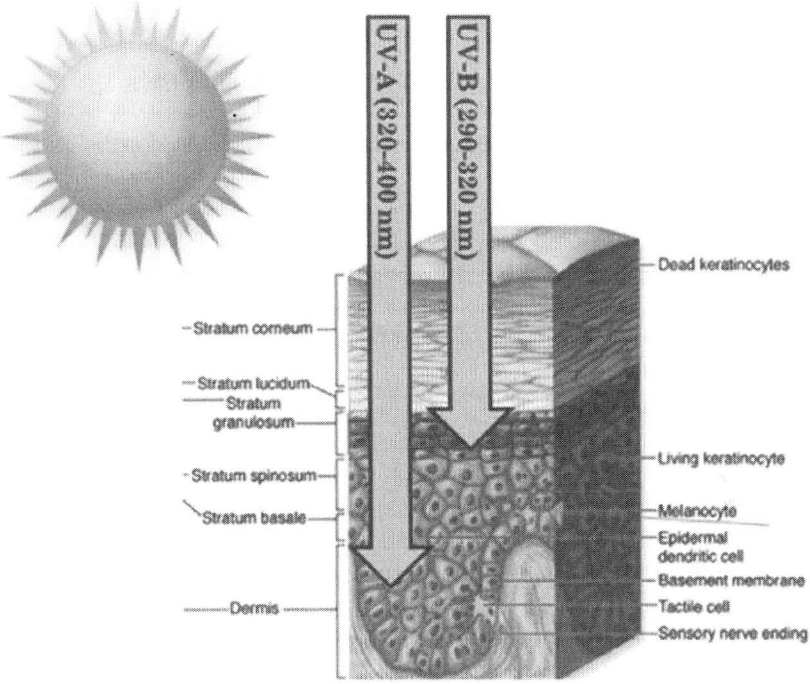

Figure 11.1 Structure of skin.

11.3 HUMAN AGING AND THEIR TYPES

Human aging can be detected by looking at the skin. Free radicals and reactive oxygen species (ROS), which cause oxidative stress in skin cells and serious damage to DNA and biomembranes, are the most significant variables linked to skin aging. The two types of skin aging are photoaging and chronologic aging. Photoaging is caused by environmental factors including prolonged exposure to sunlight (UV radiation), which degrades the structure and function of the skin.

Natural cosmetics, also known as organic cosmetics, have increased the growth of the entire cosmetics industry. This is due to the growing correlation among beauty and health and well-being, particularly in recent years and globally. Furthermore, the abundance of bioactive chemicals found in the plant and marine kingdoms that may cure, protect, and maintain the attractiveness of human skin has long been known. However, it is important to know that natural products are not always secure and efficient. To provide the ideal circumstances for their use, it is crucial to assess both their safety and effectiveness [2]

11.4 MODEL SYSTEM FOR EVALUATION IN COSMETICS

11.4.1 Antioxidant Activity

The body's cells and tissues go through a variety of biochemical processes that are necessary for the organism to function properly. Free radicals like ROS are frequently produced as a result of these processes. The body typically has defenses against ROS production through its natural supply of antioxidants; however, in most circumstances, these natural antioxidants can be reduced owing to ROS overproduction, putting the body cells at risk of oxidative stress.

Endogenous antioxidants are typically needed by the body to meet its needs. ROS are produced when the skin is exposed to too much UV light, and this can cause photo aging of the skin. The significance of antioxidants for maintaining healthy skin has prompted an ongoing search for substances that can scavenging reactive oxygen species [3].

11.4.2 Anti-inflammatory Activity

A physiological reaction to injury, inflammation is typically characterized by pain, diminished functioning, redness, fever, and the swelling. Inflammatory disease may be exacerbated by the overproduction of inflammatory mediators such NF-B, TNF-, IL-1, IL-6, IL-8 (interleukins), ICAM-1 (intercellular adhesive molecule-1), activated COX-2, PGE2, 5-LOX, and iNOS (inducible nitric oxide (NO) synthase). Atopic dermatitis is a chronic, inflammatory skin condition brought on by a variety of physiological and environmental causes that is typically characterized by rash, redness, and intense itching. Natural extracts or items with anti-inflammatory properties are frequently employed in cosmetic formulations [3].

11.4.3 Anti-hyaluronidase Activity

Enzymes called hyaluronidases break down hyaluronic acid (HA), a crucial part of the extracellular matrix. They can be found in a wide variety of groups of creatures, including fish, snakes, creatures such as insects, and vertebrates. They are widespread in nature. Six distinct hyaluronidases, including HYAL1-4, PH-20, and HYAL-P1, have recently been identified in humans. The extracellular matrix contains short-chain HA that is produced when long-chain HA is degraded by the enzyme HYAL-2. Short-chain HA acid binds to the CD44 receptor and is then taken inside the cell. The cell contains HYAL-1 which is in charge of destroying short-chain HA.

With time, the amount and quality of HA in the skin decreases, making it more difficult for the skin to regenerate, repair, and retain moisture. Wrinkles develop as a result of hyaluronidase's overactive action. The safeguarding of the connective tissues of the skin depends on the inhibition of HA degradation. These enzymes are crucial targets for the creation of anti-aging cosmetic products because they stimulate various pathways of signaling that hydrolyze ECM. Finding specific hyaluronidase inhibitors is essential because different hyaluronidase isoforms regulate various physiological processes, the identification of effective inhibitors that specifically target a specific kind of hyaluronidase despite affecting another is crucial [3].

11.4.4 Anti-Tyrosinase Activity

There are two types of melanin: red/yellow pheomelanin and brown/black eumelanin that are produced as a result of melanogenesis, which is caused by UV radiation. L-tyrosine or L-dihydroxyphenylalanine (L-DOPA), which is oxidized to dopaquinone to produce eumelanin and pheomelanin, are precursors for melanin formation. Tyrosinase (glycoprotein) an enzyme, catalyses the first step of this process, which converts L-tyrosine to dopaquinone. Pheomelanin and eumelanin are generated by the end product quinone. So, the most effective melanogenesis target is direct inhibition of tyrosinase.

By controlling the messenger RNA expression of tyrosinase and tyrosinase-related the proteins (TRP1 and TRP2), the signal transduction pathways increase the production of melanin. As a result, in addition to inhibitors

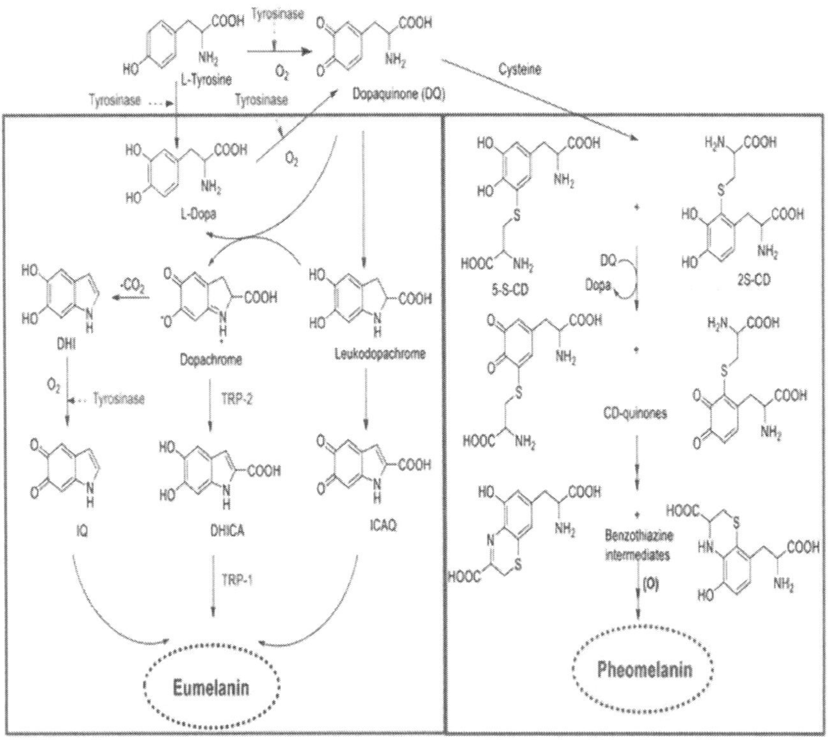

Figure 11.2 The following compounds are involved in the formation of melanin: 5-S-cysteinyldopa (5-S-CD), 2-S-cysteinyldopa (2-S-CD), 5,6-dihydroxyindol (DHI), indole-5,6-quinone (IQ), 5,6-dihydroindole-2 (DHICA), indole-2-carboxylic acid-5,6-quinone (ICAQ); tyrosinase-related protein 1(TRP-1); tyrosinase-related protein 2 (TRP-2).

of these enzymes, natural whitening agents also include the tyrosinase regulators TRP1 and TRP2 [223]. MITFs are transcription factors related with microphthalmia that control the intricate process of melanogenesis. Therefore, substances that prevent MITF expression will also prevent the whole melanogenesis process (Figure 11.2) [3].

11.4.5 Anti-elastase and Anti-collagenase Activity

Elastin and collagen, two structural proteins found in the extracellular matrix (ECM), are produced by dermal fibroblasts. In the skin, these two proteins have a wide range of protective functions. A metalloprotease called elastase breaks down elastin. Continuous exposure to elastase damages elastic fibers, resulting in a loss of skin's elasticity and the appearance of wrinkles. In order to find novel active cosmetics to delay wrinkles and skin aging, it is important to screen natural elastase inhibitors because it has been discovered that elastase activity increases dramatically with age.

Collagen is the most important part of the outermost layer of cells of skin and provides the basis for restoring skin strength and suppleness; as a result, the aging process is brought on by the destruction of collagen in reaction to UV light. Elastin, collagen, and hyaluronic acid levels in the extracellular matrix of the skin decline with age, causing flexibility loss and strength that can result in wrinkles.

The extracellular matrix is degraded by matrix metalloproteinases (MMPs), zinc-dependent endopeptidases linked to a number of physiological and pathological disorders. A membrane-anchored metalloproteinase called MMP-1 breaks down collagen. Frequently, prolonged exposure to UV radiation, which generates a lot of ROS, causes skin aging. Because of the ROS that are generated, AP-1 (activated protein factor 1) is stimulated, which causes collagen to break down and wrinkles to appear on the skin. Consequently, the natural AP-1 and MMP inhibitors can be utilized as cosmetics ingredients [3].

11.4.6 Antimicrobial Activity

Nonpathogenic microorganisms like fungus, *Staphylococcus aureus*, and streptococci regularly colonize the skin. Age or environmental variables including humidity, temperature, and the production of sebum affect density and dispersion of microflora of skin. Microorganisms can cause inflammatory skin conditions such atopic dermatitis, folliculitis, and others. It is widely recognized that an increase in the colonization of skin by bacteria like *S. aureus* is linked to atopic dermatitis. Rather than synthetic antimicrobial agents, the cosmetics industry is always looking for bioactive components from natural sources. It is even possible that these germs are resistant to popular topical antimicrobial drugs [3].

11.5 COSMETICS

The word "cosmetic," which is derived from the Greek word "kosm tikos," means proficient in decoration—"kosmein" meaning artistic, and "kosmos" meaning balance [3].

11.5.1 Classification of Cosmetics [3]

A "cosmetic material" is a material with a gentle impact on many different areas of the human body, such as hair, skin, feet, and hand nails, and is used to beautify, clean, promote visually appealing products, silky smooth safeguard, and keep the human body in excellent shape. Toilet soaps, shower gels, creams and lotions, powdered cosmetics, deodorants, perfumes, depilatories, and antiperspirants are just a few examples. Simple chemicals including emollients, thickeners, emulsifiers, preservatives, stabilizers, and neutralizers as well as physiologically active agents like antioxidants, anti-acne ingredients, perfumes, colors, and sunscreens make up the basic constituents of cosmetic formulations [4]. Table 11.1 provides a classification of cosmetics.

Table 11.1: Classification of Cosmetics

Cosmetics	Examples
Embellishment	Lip colors, perfumes
Personal cleansing	Shampoos, deodorants, and soaps
Maintenance	Moisturizers, shaving creams
Skin and hair care	Topical care products and toothpaste
Active	Antiseptics
Protective	Sunscreen and sag-reducing products
Corrective	Face masks and hair dyes

11.5.2 Cosmeceutical

Contrarily, the term "cosmeceutical" refers to a mix of medications and cosmetics. Thus, it refers to products with bio actives that are applied topically to aesthetically influence the biological function of the skin by supplying the nutrients required for healthy skin to improve its appearance, anti-aging activity, texture, and radiance. These products have benefits similar to those of medical drugs. With a strong focus on well-being, contemporary skincare products offer both a quick response (like a cosmetic) and a lasting effect (like a medication). The many improvements in dermatological goods have led to cosmeceuticals being the fastest-growing market and, as a result, the next generation of skincare products [5].

11.5.3 Nutricosmetics

In recent years, the term "nutricosmetics," which combines cosmetics with food and pharmaceuticals, has emerged as one of the newest trends in skincare. In other words, it refers to the utilization of edible components that have high-quality and potent bio-actives for skin restoration. It represents the improvement of skin health and aesthetics through the use of functional foods. The development of nutriacosmetics with high concentrations of numerous bio-actives, for instance ceramide, collagen, elastin, and hyaluronic acid and the nutrients that can improve or preserve the skin's structure and function, is on the rise in the cosmetics industry. These bio-actives have also been associated to considerable inhibition of enzymes like collagenase, elastase, hyaluronidase, and tyrosinase, retaining the elasticity, moisture level, and stability of the structure, and limiting the formation of free radicals and blocking melanogenesis (Table 11.2) [6].

Topical creams and lotions have long used active substances derived from natural sources (e.g., plants, animals, and microbes). The widespread use of natural components in a variety of industries (such as the pharmaceutical, nutraceutical, and cosmeceutical) is a result of consumer concerns and the natural compounds' enormous potential. Polyphenols, terpenes, fatty acids, and vitamins that support various bioactivities can be linked to cosmetic items found in natural sources. For instance, natural skin care products with antioxidant, hyperpigmentation-inhibiting, and antibacterial characteristics are advantageous for enhancing the tone, texture, and look of the skin [12].

Table 11.2: Illustrations of Some Natural Substances and Their Potential for Nutrition-Based Cosmetics

Natural Products	Nutricosmetics Potential	References
Edible mushroom	Anti-tyrosinase and antioxidant	[7]
Coca pod husks	Anti-elastase, anti-tyrosinase, and UVB sunscreen properties	[8]
Safranal *Crocus sativus* Linn	Actions that inhibit hyaluronidase, collagenase, elastase, and protect against the sun	[9]
Black seed (*Nigella sativa* L.)	Oil for the treatment in skin conditions	[10]
Ocimum sanctum Linn.	Anti-aging skin materials	[11]

11.6 VARIOUS FACTORS CAUSING SKIN PIGMENTATION

Skin pigmentation is enhanced by various factors such as hormone levels, inflammation, age, endocrine disorders and environmental exposures to UV rays etc. The production of melanin in epidermis can increase as a result of exposure to UV radiation, which is produced by the melanocytes located between the dermis and the epidermis. Increased melanin synthesis and deposition in the epidermis also causes the skin pigmentation.

Tyrosinase is essential for the production of melanin, hence melanogenesis inhibitors' most popular and effective target is the direct suppression of tyrosinase catalytic activity. Tyrosinase inhibitors make up the majority of cosmetic or skin-whitening products [13].

11.7 SKIN WHITENING AGENTS FROM NATURAL PRODUCTS

11.7.1 Well-Known Tyrosinase Inhibitors

The well-known tyrosinase inhibitors are mentioned in Table 11.3 [13].

11.7.2 Phenolic Compounds and Stilbenes

Oleuropein can be obtained from olive leaves. This particular compound possesses both antioxidant and free radical scavenging properties. The protective effect is noticed when gel and emulsion containing oleuropein are applied topically. The phenylethanoid salidroside is derived from *Rhodiola rosea*. Due to its favorable effects on numerous disease processes, used in Chinese medicine as traditional for a long time; is a promising ingredient for treatments intended to brighten and even out skin tone as well as to reduce redness due to its capacity to prevent melanin synthesis and reduce inflammation. In addition, the skin may benefit from its antioxidant activity in preventing damage from environmental stressors like UV radiation and pollution. Resveratrol is a polyphenol phytoalexin. It is natural compound that can be found in a variety of sources including some Polygonum plants red grapes, berries, and peanuts. It is known for its antioxidant and anti-inflammatory properties, which make it a popular ingredient in cosmetics. Some cosmetics products contain up to 5% pure Resveratrol in order to take advantage of its potential benefits for the skin. Resveratrol protects skin from photo aging. Oxyreveratrol and chlorophorin

Table 11.3: Tyrosinase Inhibitors

Active Ingredients	Source	Applications
Hydroquinone	*Castoreum (fungus)*	Treatment of hyperpigmentation Recent Kligman formulations (hydroquinone, retinoic acid, and corticosteroids)
β-Arbutin	*Serratulae quinquefoliae*	Skin-whitening creams, antioxidants A gel formulation used as hyperpigmentation
Aloesin	Aloe vera	Topical cosmetics
L-ascorbic acid	Vitamin C	Photo-protective and antioxidant
Kojic acid	*Aspergillus* and *Penicillium*	Antioxidant, photodamage, hyperpigmented scars
Azelaic acid (dicarboxylic acid)	*Pityrosporum ovale* Wheat, barley, and rye	Reduced pigmentation, acne, and inflammation
Ferulic acid (cinnamic acid derivative)	Wheat, oats, parsley, apples, grapes, and spinach	Skin-whitening formulations and antioxidant

have shown tyrosinase inhibitory activity with oxyreveratrol; being inhibitor with an activity 32 times better than koji acid. Piceatannol is another compound with antioxidant properties that is present in the seeds of passion fruit (*Passifora edulis*). These compounds have potential to be used in cosmetics and other skincare products to improve skin tone and reduce hyperpigmentation [14].

11.7.3 Flavonoids

The most well-known classes of flavonoids, which are derived from chalcone, are the cyclic molecules. Glycosides, which are the form of flavonoids, are present in nature. Flavonoids' ability to act as antioxidants on the skin is perhaps its most well-known function. Thereby stopping the production of ROS and skin aging.

Apigenin It is a particular class of flavonoid that is frequently found in many different kinds of plants, including aromatic herbs, grains, and fruits and are renowned for their anti-inflammatory and antioxidant activities. On skin cells exposed to UVB radiation, apigenin helps to protect DNA in skin cells from damage brought on by UVB radiation-induced oxidative stress, which is a significant contributor to the development of skin cancer. *Catechin, Kaempferol, Morin and Quercetin* they all shows tyrosinase inhibitory activities. *Phyla Nodiflora* a popular herbal tea in China that researchers used to find the key ingredient that might have lightening effects on the skin. The plant's above-ground components were extracted with methanol before being subjected to partition chromatography to separate the pure flavonoid compound eupafolin, which may have skin-lightening effects because it significantly decreased cellular melanin production by modulating this signaling pathway. *Steppogenin*, a compound isolated from the branches of the Cudrania tricuspidata plant, has been found to have potent tyrosinase inhibitory effects. It has been shown to have ten times more potent tyrosinase inhibitory activity than kojic acid. It is possible to treat disorders like melanoma and hyperpigmentation by reducing tyrosinase activity. It has been claimed that the isoflavone component *Genistein*, which is present in soybeans, possesses anti-photo aging and anti-photo carcinogenic properties. Due to its anti-inflammatory and antioxidant qualities, lessen harmful effects caused by oxidative stress and inflammation brought on by UV radiation, which in turn may help to prevent skin cancer. *Liquirtin* is a whitening agent for the skin. Treatment of melasma is beneficial with 20% Liquirtin ointment [15].

11.7.4 Chalcons

Chalcones are known to be the building blocks for the flavonoids and isoflavonoids that having pharmacological effects, such as antioxidant, anti-inflammatory, antibacterial, or anticancer capabilities. Their capacity to inhibit the enzyme tyrosinase is one of their intriguing biological functions. The tyrosinase inhibitory actions of *Kuraridin* and *Kuraridinol* are 34 and 18 times stronger than those of kojic acid, respectively. A chalcone derivative known as *Licochalcone A* was isolated from roots of the licorice plant *Glycyrrhiza*. *Morachalcone A* is a chalcone derivative that is extracted from jackfruit, and one of its intriguing properties is its capacity to block tyrosinase (an enzyme involved in the formation of melanin in the skin) is the capacity to stop melanin synthesis. It has been discovered that the catechol on the A ring of the molecule serves as a metal chelating agent in the presence of copper ions, which can limit the activity of tyrosinase and act as a competitive inhibitor of the enzyme. However, the catechol on the B ring of the molecule is converted to o-quinone in the presence of both copper ions and tyrosinase, which can form a covalent bond with the enzyme and irreversibly inhibit its activity. The hop plant, *Humulus lupulus L.*, contains the highest concentration of the prenylated chalcone chemical known as *Xanthohumol*. Tyrosinase-related protein-1 genes, which are important for the production of melanin, are most affected by this. This causes keratinocytes to produce less melanin and human melanocytes to produce less melanin. The prenyl side chain is crucial for xanthohumol's biological activity because it is thought to help it interact with and modify the activity of proteins in the cell [16].

11.7.5 Phenylpropanoids and Coumarins

Both *Caffeine* and *p-coumaric acid* are naturally occurring byproducts of the phenylpropanoid pathway and have far stronger inhibitory effects on tyrosinase activity than kojic acid. These substances have a structural resemblance to L-tyrosine, the natural substrate of tyrosinase, and they can attach to site of the enzyme. By doing so, they block L-tyrosine from binding to the enzyme and obstruct the hydroxylation and oxidation reactions that follow. The capacity of caffeic acid and p-coumaric acid to chelate the copper ions in the active site of tyrosinase, which are essential for the enzyme's catalytic activity, is assumed to be the cause of their inhibitory effect. Caffeic acid and p-coumaric acid interfere with tyrosinase's regular catalytic cycle by attaching to the copper ions, which inhibits the enzyme's activity. It has been demonstrated that the polyphenolic chemical *Rosmarinic acid*, which is present in many plants like rosemary, basil, and mint, can prevent UVB-induced cellular damage. *Verbascoside* may be a viable element in the creation of a sunscreen composition because of its photoprotective qualities, high SPF value, antioxidant properties, and wound-healing capabilities. It was made from *Buddleja scordioides* extract [17].

11.7.6 Alkaloids

For facial melasma, the use of a *Diacetyl boldine*-containing night cream at night and a TGF-1 biomimetic oligopeptide-68/sunscreen cream during the day proved effective and safe. Lumiskin with diacetyl boldine, a component of Revitol's "Skin Brightener" that affects tyrosinase expression, is present in that product. Histidine is an amino acid that is predominantly obtained by animals through dietary sources, and its derivative *L-ergothioneine* contains sulphur. L-*ergothioneine* is a key component in anti-aging cosmetics due to its outstanding ROS scavenging and anti-inflammatory properties. It is a tripeptide, *Glutathione*. It is produced intracellularly and is made up of the amino acids glycine, glutamic acid, and L-cysteine. Its ability to lighten skin was discovered by accident, and since topical applications have started to appear on the market, its popularity has grown quickly. Tyrosinase inhibition, which causes a change in color from eumelanin to pheomelanin, and antioxidant activity are the suggested mechanisms for skin whitening. Vitamin B3/niacin is *Niacinamide* due to its antioxidant properties, it prevents melanosomes from moving from melanocytes to keratinocytes and from interacting with one other. Niacinamide as a cosmeceutical component could aid patients with diminished NAD+ pools in the skin as a result of external or internal stress or the elderly minimize skin aging and hyperpigmentation. From sea cucumbers, *Porphyra-334* was obtained. Porphyra-334 liposome-containing sunscreen formulations can lessen skin lipid oxidation and enhance skin aging metrics including wrinkle depth, low elasticity, and roughness. Natural UV-A radiation (315–400 nm) and UV-B radiation (280–315 nm) have been exposed to cyanobacteria, which have an impact on cellular biomolecules. Different UV radiation defence mechanisms were created by these photoautotrophs. These defence systems include DNA repair, avoiding UV rays, and shielding cells with photoactive substances including carotenoids and mycosporine-like amino acids (MAAs). MAAs are extremely stable over an extensive range of pH and temperature values. These outstanding cosmeceutical chemicals, whose market has been growing dramatically over time, stand out for their photoprotective ability to block the damaging effects of UV B and UVA radiation as well as their antioxidant capacities in hydrophilic and lipid solutions. *Shinorin*, a naturally occurring antioxidant and UV absorption molecule [18].

11.7.7 Terpenoids: Sesquiterpenoids, Diterpenoids, and Triterpenoids

Cedrol and *Widdrol*, two anti-melanogenic sesquiterpenes derived from the *Juniperus chinensis* plant (Cupressaceae), have showed remarkable antioxidant and anti-tyrosinase activity. A highly oxygenated diterpene called *Gagunin D* was discovered in the marine sponge *Phorbas sp*. By reducing expression of tyrosinase and boosting its breakdown, it was discovered to have anti-melanogenic properties. Additionally, it prevents the expression of proteins linked to the metastasis of melanosomes. Gagunin D and its analogues are prospective candidates for skin-lightening cosmeceuticals due to their multifunctional qualities. An important class of natural compounds are terpenoids. Some of these triterpenoids, such as *Arjunilic acid* from *Rhododendron collettianum* have tyrosinase inhibitory activity. *Ginsenoside Rd, Ginsenoside Re* and *Floralginsenoside* were extracted from the *Panax ginseng* berry, similarly reduced the formation of melanin. Due to the numerous health advantages, it offers, including toughness, skin flexibility, and the management of pimples, scars, and stretch marks, essential oils like orange are rapidly gaining popularity. Due to their antibacterial qualities, orange and lemon oil derivatives are excellent ingredients for skincare and hair care [19].

11.7.8 Fatty Acids

Seed or nut oil provides the fatty acids utilized in cosmetics and personal care products. They are separated into two groups: saturated (such as lauric, palmitic, and stearic) and unsaturated (such as linoleic, linoleic, and oleic). The most often used fatty acid in cosmetics is linoleic acid, which is used to cure acne and moisturize the skin. The linoleic acid treatment almost completely eliminated pigmentation, restoring the skin's natural color. In comparison to long carbon-chain fatty acids, short carbon-chain fatty acids (C-12 to 16), such as palmitic acid, were more effective at enhancing skin permeation. Fatty acids, particularly *Omega-3* (linolenic acid) and *Omega-6* (linoleic acid), play an aesthetic role in hair. Fatty acids are an efficient way to stop hair loss and used in the treatment of hyperpigmentation of skin. Due to its moisturizing and smoothing qualities, omega-3 from rubber seeds may be used as a component in lip balm [20].

11.7.9 Carotenoids

The best-known carotenoids that have historically been used in a variety of industries as antioxidants are ß-carotene, lycopene, lutein, astaxanthin, and zeaxanthin. Due to their oxidative impact, carotenoids serve as a photoprotective against UV radiation. Both lycopene and ß-carotene serve as organic sun protection for the skin and found in Carrots, tomatoes, and other yellow vegetables. The carotenoids were applied topically or orally, and administration of combined oral and topical lutein and zeaxanthin applications demonstrated the best antioxidant protection. Individual compounds, however, also show a sizable impact of skin photo aging and skin elasticity. One benefit that carotenoids give is skin whitening. In addition to whitening the skin and preventing hair problems, the use of carotenoids-rich dietary supplements as an adjunct to oral application of commercial lightening cream for the treatment of melasma is helpful [21].

11.7.10 Vitamins

Vitamins and antioxidants topically through cosmetics can help prevent free radical damage and possibly even repair it. In addition, some vitamins may be good for the skin because of anti-inflammatory and anti-pigmentation properties, as well as their effects on collagen production, keratinization, and bruising.

The Food and Drug Administration (FDA) granted vitamin A the first approval as an anti-wrinkle agent, improving the appearance of the skin's surface and having antiaging effects. The retinoid family includes fat-soluble substances like vitamin A. Among the best anti-aging treatments are vitamin A and its derivatives. Retinoid control cell proliferation, differentiation, and death. The anti-wrinkle effects of retinoids increase keratinocyte proliferation, fortify the epidermis's barrier function, reduce water loss, stop collagen deterioration, and block metalloproteinase activity.

It was found that Vitamin A had photoprotective properties because it decreased the number of lipids peroxyl radicals in skin exposed to UV light. Among the most powerful anti-aging treatments are vitamin A and its analogues, especially retinol. The stratum corneum (outermost layer of skin) and to some extent; the dermis is both penetrated by fat-soluble retinol. Retinol penetrates the interior of the keratinocyte after entering it and binds to the proper receptor. The cytosolic retinol-binding protein receptors have a high affinity for retinol. In the epidermis, retinoids may affect growth factor secretion and transcription. They are in charge of the growth of the epidermis' living layer, the reinforcement of the epidermis' barrier function.

Retinoids are frequently employed in numerous cosmetic compositions for topically applied usage. Table 11.4 lists some instances of vitamin A and its equivalents being utilized in cosmetics [22].

11.7.10.1 Vitamin B, C E, and K

A water-soluble component called vitamin B is present in many foods, but it is especially abundant in whole grain foods and leafy green veggies. Pantothenic acid, or vitamin B5, is called pathenol. Because it functions as a humectant, escalating the water concentration and enhancing the elasticity of hair, it has long been a component of hair care products. Because it can draw moisture

Table 11.4: Vitamin A, Its Analogues and Its Applications

Vitamin A and Its Analogues	Applications in Cosmetics
Retinol (a precursor of retinaldehyde)	Used in depigmentation, dryness, and anti-wrinkle treatment
Retinoic acid	Used to treat hair psoriasis and persistent inflammation of the scalp
Retinaldehyde or retinal, a precursor of retinoic acid	Used in the treatment of wrinkle as stabilizer
Retinyl acetate and palmitate	Antioxidants and having stabilizing properties in wrinkle treatment
Tazarotene	Provides photoprotection from sunlight and used in the treatment of psoriasis and acne

into the skin's outermost layer and moisturize the skin. Panthenol is an efficient moisturizer in cosmetics. Ascorbic acid (AA), a hydrophilic molecule that is a form of vitamin C, is a potent antioxidant that can counteract oxidative stress through an electron transport process. Vitamin C can lessen the levels of unstable radicals of oxygen, nitrogen, and sulfur radicals. Lipid peroxidation by peroxide radicals can be stopped using vitamin C. In addition, vitamin C encourages the absorption of folic acid, calcium, and iron, preventing allergic reactions. Vitamin C can reduce ROS produced by the sun's rays and other environmental variables like smoke and pollution when applied topically. For the treatment of freckles, melasma, and sunspots, vitamin C has been shown to be successful. Soy, almonds, whole-wheat flour, oils, and other foods are particularly high in vitamin E, a lipid-soluble vitamin. It has many health advantages for the eyes and circulatory system because of its capacity to minimize lipid peroxidation. Vitamin E used topically has been shown to have many cutaneous advantages. The powerful antioxidant activity of vitamin E is by far its most significant quality. Because vitamin E and its analogues have the capacity to squelch free radicals, specifically lipid peroxyl radicals. The most active form of vitamin E is tocopherol, although vitamin E esters administered topically have also been demonstrated to penetrate the epidermis. The hepatic synthesis of many clotting factors depends on phytonadione (vitamin K). There is a reason for utilizing topically vitamin K to repair and avoid some of the vascular symptoms of aging. Vitamin K is predominantly found in green leafy vegetables along with gut bacteria. It was shown that topically 1% vitamin K given twice daily was useful for both hastening the healing of existing bruises and preventing new ones [23].

11.8 SKIN ANTI-AGING AGENTS FROM NATURAL PRODUCTS

All animals, including humans, have a protective layer on their bodies called the skin. Changes in the skin are impacted by both internal and external influences. One of the most visible indications of aging is the skin's alterations, which include loosening of the skin, lines and wrinkles a lack of moisture, and age spots along with fat loss and a loss of the skin's natural smoothness. The epidermis gradually thins with age, and the skin's natural capacity to heal itself declines. Additionally, as people age, their skin grows thinner, paler, clearer, and has larger pigmented spots due to a decrease in melanocytes. As we age, our bodies generate less both elastin and collagen or lose their elasticity, which points to the necessity for anti-aging treatments. Collagen production can be increased by using anti-aging products or treatments. In order to decrease wrinkles, acne, and fine lines, therapies for aging are also required. They also assist to tighten up the skin [23]. Various factors involved in human aging are shown in Figure 11.3.

Skin damage induced by aging has oxidative stress as one of its main contributors. A characteristic of aging skin, increased matrix metalloproteinase and heightened collagen breakdown are brought on by the buildup of hydrogen peroxide-related ROS. Natural elements for skin and hair care, including antioxidants, have been developed into a variety of potential cosmetic products. By scavenging free radicals or stimulating their disintegration and suppressing diseases, natural antioxidants are efficient at avoiding their development. Some substances prevent or repair the oxidative damage caused by oxygen-induced chain reactions that start or spread throughout cells in the body [23].

Antioxidants that work as metal chelators, oxidative inhibitors of enzymes, or enzyme cofactors donate an electron to oxidizing species, stopping the radical chain reaction and subsequent production of reactive oxidants. Antioxidants

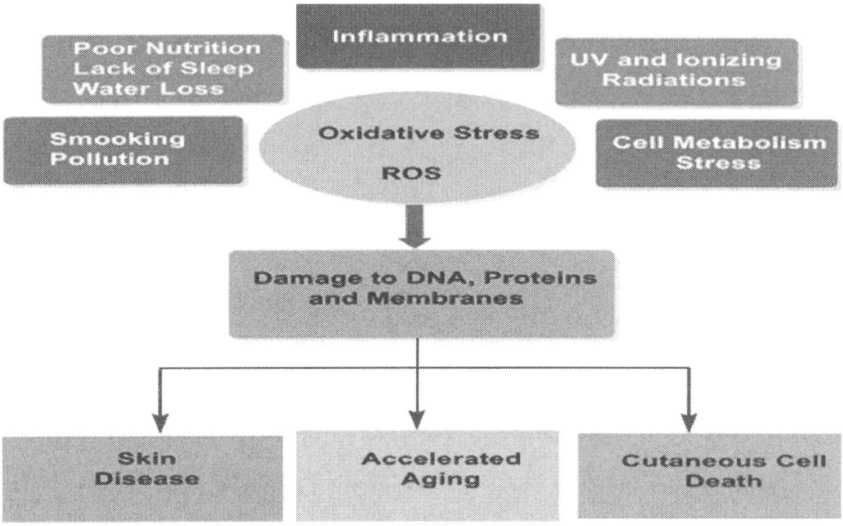

Figure 11.3 Various factors involved in human aging.

are additionally utilized as stabilizers to stop the oxidation of lipids. Antioxidants may therefore perform a number of tasks when present in a product. Some naturally derived antioxidants and their uses in cosmetics are shown in Table 11.5 [23].

Transcommunic acid from Japanese red pine (*Pinus densiflora*) is obtained. A naturally occurring diterpenoid having known to inhibit-microbial, anti-tumor, and antioxidant properties is trans-communic acid. Due to its UVB protection, it is an interesting component for cosmeceutical therapies. The perennial herb *Glycyrrhiza glabra,* often referred to as Licorice has a sweet flavor. One of its most well-known medications in ancient times was this one. Skin eruptions such dermatitis, eczema, pruritus, and cysts can benefit from licorice and also having antimicrobial, antiseptic, and anti-inflammatory qualities.

Table 11.5: Natural Antioxidants, Their Sources and Applications

Natural Antioxidants	Sources	Cosmetics Applications
Apple	Phenolic compounds	DNA protection against UV radiation
Basil leaves	Phenolic compounds	Anti-acne, anti-aging, remove dead skin cell
Chest nut	Polyphenols	Moisturizer
Ginkgo biloba leaves	Flavonoids	Anti-inflammatory and antioxidant and prevent UVB-induced photoaging
Grape	Anthocyanins and phenolic compounds	UV radiation defense, antioxidant and antiaging properties, skin lightening, anti-inflammatory properties, and wound healing
Mango	Carotenoids	Prevents skin aging, antioxidant and wound healing agent
Tomato	Flavonoids and lycopene	Provide wound healing and UV protection, antioxidant and protection from cell damage

Glycyrrhizin, glycyrrhetinic acid, flavonoids, also asparagine, isoflavonoids, and chalcones are all present in licorice. The primary component of licorice extract's hydrophobic fraction, glabridin, has been demonstrated to suppress tyrosinase activity. The efficient inhibition of diacetyl production by *G. glabra* extract of root without adverse bacterial effects is another intriguing result. Diacetyl is a major component of the foul smells that come from the axillae, foot, and head areas and is created by local skin microbes that can aid in the creation of efficient deodorant compounds. Treatment with piceatannol decreased UVB-induced skin stress indicating that it might be employed in preparations for photoprotective cosmetics. A diterpene called sclareol was discovered from the *Salvia officinalis* plant, regarded as a source of the natural scent and efficient cosmetic ingredient that prevents UVB-induced photoaging [24].

11.9 MOISTURIZERS FROM NATURAL PRODUCTS

The outermost stratum corneum (SC) of our skin serves as a strong barrier to keep noxious substances or allergens out of the body, but it also serves the vital purpose of binding water molecules. The skin's ability to retain moisture is thus ensured by its adherence of water molecules from the environment and the deeper skin tissues. Skin moisture is crucial for preserving the epidermis' ability to operate as a barrier, as well as the operation of several enzymes and regular cellular functions. The NMF (natural moisturizing factor), a group of hygroscopic substances found in the SC, prevents TEWL (trans-epidermal water loss). Moisturizers increase the hydration of the skin in two ways:

a. By forming an occlusive layer on the skin's surface that prevents water molecules from evaporating

b. By supplying hygroscopic chemicals to the SC that attach and retain water.

Moisturizers can be effectively delivered into the skin using colloidal delivery systems such micro- and nano-emulsions, lipid- and polymer-based particle systems, and others.

In adult skin cells (corneocytes), NMF (natural moisturizing factor) is a combination of hygroscopic components made primarily of amino acids and derivatives of them, like carboxylic acids and pyrrolidone. Electrolytes, urea, and lactic acid are additional components. Due to their capacity to efficiently bind molecules of water from external sources, these hygroscopic substances function as effective humectants, aiding in the maintenance of the corneocytes' moisture levels. Additionally, the hydrated NMF's strong ionic interaction with keratin reduces water's mobility, lowering the interactions of molecules between keratin fibers and enhancing skin elasticity. Aquaporins (AQPs) are membrane proteins that are essential for the formation of water channels in biomembranes. Water, glycerol, and various other tiny-molecule solutes are primarily transported through biomembranes by AQPs. In addition, AQPs regulate hydration in the innermost layers of the epidermis and speed up water molecule movement within the various layers of skin tissue. In terms of subtypes, AQP3 is the one that is most prevalent in skin. An anionic polysaccharide that belongs to an anionic is Hyaluronic acid. Because of its hydrophilic nature, it may bind a lot of water. It is one of the dermis' primary constituents. By connecting the SC's hydrophobic zone to its lipids, this humectant completely hydrating the SC and modifies its physical structure. Molecular weights for HA range widely, from 2×10^5 to 10^7 Da. The physicochemical characteristics of HA can be affected by its average molecular weight. The molecular weight of HA affects both its biological function and penetration in the skin. HAs with different molecular weights have various impacts on the skin; those with a high molecular weight hydrate, those with a medium molecular weight promote cell repair and wound recovery, and those with a low molecular weight fight aging. In recent years, there has been an increase in demand for HA products. Today, HA is undoubtedly one of the most commonly utilized active ingredients in cosmetic formulations. From 0.2 to 1% concentrations are used in cosmetic formulations. Based on particular characteristics, HA has a wide range of applications: high moisture absorption, the elasticity and biologic compatibility [25].

11.10 EMOLLIENTS

Due to its emollient and occlusive qualities, lipids and oils are commonly utilized in skin moisturizing formulations. They improve skin hydration by minimizing TEWL and forming a thin layer on the skin's surface. One of the most utilized occlusive ingredients in skin-moisturizing compositions is petroleum jelly. The majority of it is made up of long-chain hydrocarbons that are aliphatic. Petrolatum significantly reduced skin water loss (> 98%) and improved skin hydration. Petrolatum increases the skin's barrier function and promotes lipid production in addition to its effect on water retention inside the SC. Fatty acids such as lanolin, beeswax, mono-, di-, and triglycerides are additional emollients and occlusive [26].

11.10.1 Humectants

Hyaluronic acid, glycerol, panthenol, sorbitol, serine, and as well as pyrrolidone are carboxylic are actually examples of hygroscopic compounds with a high or low molecular weight that have been employed as humectants to increase skin moisture. Their primary function is to bind and retain molecules of water within the SC, which provides moisture to the epidermal tissue and increases skin moisture [26].

11.10.2 Exfoliants

Exfoliants' primary purpose is to remove the outermost layer of cells that are dead, giving the skin a smooth and radiant appearance. Several exfoliants, including α-hydroxyl acids (AHA), retinol, salicylic acid (SA), and the enzymes (such as bromelain, papain, and the proteases from Bacillus), have been used in skin formulations to encourage exfoliation by interfering with corneocyte cohesiveness, enhancing the development of the skin barrier, and maintaining a youthful appearance [26].

11.11 MARINE-DERIVED COMPOUNDS USED IN COSMETICS

The seas' diverse habitats and environmental settings support a vast diversity of microbial life. A wide range of bioactive compounds has been produced as a result of various biological adaptations triggered by the distinctive features of various marine habitats. The marine species that have the most potential for creating bioactive substances which can be used for a variety of commercial applications, notably cosmetic and cosmeceutical ones, include fungi and bacteria. In comparison to other natural molecules used in photoprotective, anti-aging, and skin-whitening products for face, body, and hair care, mycosporines and mycosporine-like amino acids, carotenoids, exopolysaccharides, fatty acids, and chitosan may represent a sustainable, affordable, and quick-production alternative. In addition, due to consumer demand for natural products being driven by health concerns and fashion trends, research is being forced to find novel compounds from the plentiful as well as alternative sources represented by marine organisms [27]. Major types of cosmetics made with bioactive

Table 11.6: Major Types of Cosmetics Made with Bioactive Substances Derived from Bacteria, Fungi, and Thraustochytrids

Sources of Microorganisms	Natural Ingredient	Bioactive Compound	Applications	References
		Photoprotective compounds		
Marine fungi *Trimmatostroma salinum*, *Cryptococcus liquefaciens*	Mycosporine	Mycosporine–glutaminol–glucoside	UV screening agents	[27]
marine sponge *Halichondria panicea*	Benzodiazepine alkaloids	Circumdatins I, C, G	UV-A protecting agents	
		Anti-aging compounds		
Marine fungi and bacteria such as *Agrobacterium* sp., *Bacillus* sp., *Zymomonas mobilis*	Polysaccharides	Exopolysaccharides (EPS)	Used as anti-wrinkles agents	
		Skin whitening products		
Marine bacteria genus *Pseudomonas*	Alkyl halides	Methylene chloride	Inhibition of tyrosinase	
Marine fungus, *Microsporum* sp	Anthraquinones	Chrysophanol	Tyrosinase inhibition	
		Moisturizing agents		
Marine fungi and bacteria such as *Actinobacter*, *Streptomyces*, *Bacillus*, *Azotobacter*, *Corynebacterium*	Protein polysaccharide complexes, glycolipids, lipopeptides		Skin moisturizing agents	

substances derived from bacteria, fungi, and thraustochytrids (decomposers) are presented in Table 11.6 [27].

Cosmetic products contain preservatives to stop deterioration and shield against microbial contamination. Parabens, also known as 4-hydroxybenzoate alkyl esters, are widely employed as preservatives in the food and cosmetics sectors. It was discovered that the ascidian-isolated marine bacteria produced 4-hydroxybenzoate alkyl and its esters. These substances were successful in stopping the development of yeast, mold, and gram-positive bacteria [27].

11.11.1 Current Applications of Algae-Derived Metabolites in Cosmeceutical

Phlorotannin, a bioactive substance found in brown algae, has various cosmeceutical properties, including being an anti-melanogenic, antioxidant, known to inhibit inflammation, and anti-aging properties. Similarly, fucoidan as a sulfated polysaccharide derived from brown algae, aids in the prevention of cancer, melanogenesis, and inflammation. Fucoxanthin, a carotenoid found in brown, red, green, and microalgae, has antioxidant, anti-melanogenic, and anti-aging properties. Microalgae and red and green seaweeds include mycosporine-like amino acids (MAAs), that also have antioxidant, anti-inflammatory, and anti-aging properties. In the food, pharmaceutical, agricultural, and cosmetic industries, marine algae are utilized for a variety of applications. It's interesting to note that the FDA has approved phycocyanin, a natural color addition used in food and cosmetics because it is non-toxic, natural, and biodegradable. It is typically found in red algae or cyanobacteria. Astaxanthin, a carotenoid used in cosmeceutical products by companies like Unilever, L'Oreal, and Henkel, is vital in eliminating free radicals that exist in the human body. Furthermore, because of their stabilizing, thickening, and gelling properties, red algae of the genus *Gracilaria* have been extensively utilized in cosmetics. There are several commercially available *Gracilaria* species products, such as Thalasso®'s moisturizing lotion, Sea Laria®'s hydrogel soap, and Balinique®'s facial mask. Algenist, an anti-aging moisturizer with microalgae oil and alguronic acid coming from algae, Protulines®, an anti-aging agent made from a protein-rich extract of Arthrospira, and Dermochlorella, an anti-wrinkling agent made from *Chlorella vulgaris* extract are just a few of the skincare products made using algae as a base. As a result, bioactive substances obtained from algae may one day be used as cosmeceutical skincare products [28].

11.12 ADVANCED TECHNOLOGIES FOR SKINCARE

Growth factors, stem cells, neural rejuvenation, DNA-customized formulas, youth-promoting nanotechnologies, mitochondria, enzyme-boosting treatments, plasma therapies, and hormone replacement therapies are being used more and more to regenerate aging skin as a result of the development of advanced technology and biological treatments [29].

11.13 VARIOUS DELIVERY SYSTEMS IN COSMECEUTICALS

Numerous types of nanoparticles, including dendrimers, liposomes, niosomes, polymeric nanoparticles, nanosuspensions, and nano-emulsions, have dominated the cosmetics industry. The usage of these formulations helped to reduce adverse effects such as alopecia, nausea, vomiting, diarrhea, exhaustion, and skin rash as well as poor bioavailability [29].

The topical application of natural ingredients in cosmeceuticals, nutricosmetics, and cosmetics is generally known to improve skin protection and may even be able to counteract the effects of free radicals, in enhancing keratinization, reducing skin pigmentation and bruising, and promoting collagen formation. However, issues with permeation, its bioavailability low water solubility, stability, and excretion make it difficult to use plant-based goods and their bio-actives. The beneficial qualities of these bioactive are typically altered by exposure to poor pH and temperature conditions, enzymatic activity, light, metal ions, oxygen, and water. Hence, various delivery systems and advanced preparation methods and techniques are used for the incorporation of natural products which are discussed in subsequent subsections [29].

11.13.1 Liposome

A lipid-based delivery method for natural compounds is the liposome. In addition, it often employed carriers for encapsulating bioactive in both cosmetics and different pharmaceutical applications. It is a cell-like lipid-based colloidal delivery system, a promising method for encasing and protecting both hydrophilic and lipophilic compounds due to its biodegradability, safety, and enhancement of transcellular transport through transient disruption of cellular lipophilic bilayers. Other advantages are strengthening solubility enhancing the therapeutic index, lowering toxicity, overcoming drug resistance, and sustained and focused drug delivery [29].

11.13.2 Micro- and Nano-encapsulation

The incorporation of natural products or purified plant extract into these delivery systems has been developed to increase the bioavailability, stability, and solubility, as well as to control and protect the release of natural ingredients within the GI tract. The advantages of nanocapsulations are creating stable capsules, shielding the substance from air, moisture, and UV rays, controlling release, reducing agglomeration of fine powders, enhancing the sticky materials' handling, and avoiding chemical interactions with outside influences [29].

11.13.3 Nano-emulsion, Phytosomes, Nanoliposomes, and Niosomes

Improved bioavailability and cell membrane penetration of the active substances; encapsulation of both hydrophilic and hydrophobic medicines, reducing the harmful and adverse effects of medications, stabilizing sensitive drugs, slowing down the pace of drug breakdown, and increasing effectiveness.

11.13.4 Polymeric and Inorganic Nanoparticles

Low toxicity, improved medication bioactivity, immune system evasion ability, and improved drug penetration into and through the skin [29].

Some of the Nano formulations used in cosmetic applications are summarized in Table 11.7 [30].

11.13.5 Nanotechnology

Recent advances in nanotechnology have made it possible for skin moisturizing solutions to outperform more traditional cosmetic formulations. Furthermore, penetration improvement would be more significant than topical application. The use of nano- and micro-vectors for the trapping of bio-actives, among them polysaccharide-based polyelectrolyte complexes (PECs) and in specific chitosan-based polyelectrolyte complex nanoparticles, is expanding due to their noteworthy antibacterial properties, biocompatibility, biodegradability, excellent loading efficacy, and hydrophilicity [31].

Multiple saccharides combine to form polysaccharides, which can be a long chain of branched or unbranched molecules. They are considered to be polymers. These naturally occurring polymers are hydrated, forming a gel structure known as a hydrogel or hydrocolloid and having a moisturizing

Table 11.7: Nano Formulations Used in Cosmetic Applications Are Summarized

Plants	Active Ingredient	Delivery System	Uses
White tea	Phenolic compounds	Polymeric nanoparticle	protects, boosts bioactivity, and increases the bioavailability
Aloe vera	Phenolic compounds	Liposomes	Increases collagen synthesis and bioavailability
Achyrocline satureioides	Flavonoid compound (quercetin)	Nano emulsion	Increase absorption of the drug on the skin
Centella asiatica	Asiaticoside Madecassoside Asiatic acid Madecassic acid	Nano encapsulation	Skin protection enhancement
Curcuma longa	Curcumin	Liposome, ethosome, transferosome	Improved hydration protection and absorption into the skin

Table 11.8: Polysaccharides Are Categorized in Medicinal and Cosmetic Products

Natural Ingredient	Source	Applications
Cellulose (anionic water-insoluble polymer)	Plants with hemicelluloses, lignin, and microorganisms	Anionic water-insoluble polymer plants with hemicelluloses, lignin, and other extractives.
Gum (anionic or nonionic polysaccharides)	Acacia (*Astragalus gummifer*, *Anogeissus latifolia*)	Emollient, hydrating effect and used in skin nourishing cosmetics.
Chitosan (deacetylation of chitin derivative)	Arthropod exoskeleton or the cell walls of fungus and yeast	Potential ingredients in skin- and hair-care products, primarily in body creams, mascaras, hair shampoos and conditioners hair foams, and moisturizing agents for UV protection, skin cleansing, and skin protection [8].
Okra Polysaccharide (high content of fruit mucilage)	*Abelmoschus esculentus*	Moisturizing alcohol-based hand rub product-maintained skin from hydration
Aloe vera	Aloe	Skin nourishing effect
Chia seeds	Aztec chia	Polysaccharide gel has a moisturizing effect
Tea (polysaccharide)	*Camellia sinensis*	Good moisturizer
Papaya fruit	*Carica papaya*	Skin hydrating polysaccharides
Tamarind seed	*Tamarindus indica*	Tamarind polysaccharide skin moisturizing effects
Mycelium	*Aspergillus niger*	Improved skin hydration by decreasing skin roughness
β-Glucan	*Aureobasidium pullulans*	Used in skin moisturizing cosmetics

effect. The moist gels are classified as biopolymers since they are extremely compatible with biological tissues and are biodegradable in addition to it they are widely available from natural sources, which highlights their significance in applications for health benefits, including cosmetics [31].

Due to their ability to hold water in their polymerized networks via swelling, polysaccharides are used in cosmetic products as thickeners, emulsifiers, gelling agents, conditioners, healing and suspending agents, moisturizers, hydrators. As a result, functional polysaccharides are typically classified as follows with different charges (cationic, nonionic, anionic, or amphoteric) polymers. The use of these biopolymers as active ingredients in cosmetics is very common [31].

Based on their ability to moisturize skin, polysaccharides are categorized in medicinal and cosmetic products as mentioned in Table 11.8 [31].

Several other factors must be considered when conducting specific studies and developing products for cosmetics. The method of skin whitening has undergone significant expansion in recent years. The use of complex mixes aimed at several processes, such as tyrosinase expression, melanosome transfer, antioxidant, and anti-inflammatory effects, has frequently replaced single medications that block tyrosinase. Encouragement of substances that are safe and biological activity, including specific mechanism and unique function responsible for these activities, have been demonstrated in cell culture is crucial. In addition, it is still advised to use in vitro assays for screening natural substances in cell culture before evaluating them on human skin models.

Moreover, the use of nanotechnology in cosmetics for beauty will only increase, and compositions will be improved to allow for better absorption of active chemicals. As the human genome is being fully decoded, genes involved in aging and skin are being revealed one by one. Future product releases will be increasingly specialized and advanced [3].

The management and prevention of skin aging illnesses associated with inflammation offer significant promise for herbal remedies customized as cosmetics, cosmeceuticals, and nutricosmetics. Additionally, using various delivery systems would greatly improve the bioavailability, stability, solubility, and monitored profile of the bioactive, supporting the growing interest in and need for safe and natural alternative skincare regimens in the multibillion-dollar cosmetic industry [2].

REFERENCES

1. Mohd-Setapar HM. Natural ingredients in cosmetics from Malaysian plants: A review. *Sains Malaysiana*. 2018; 47(5):951–9.

2. Ahmed IA, Mikail MA, Zamakshshari NH, Mustafa MR, Hashim NM, Othman R. Trends and challenges in phytotherapy and phytocosmetics for skin aging. *Saudi Journal of Biological Sciences*. 2022 Jun 23; 29(8):103363.

3. Liu JK. Natural products in cosmetics. *Natural Products and Bioprospecting*. 2022 Dec; 12(1):40.

4. Costa R, Santos L. Delivery systems for cosmetics - From manufacturing to the skin of natural antioxidants. *Powder Technology*. 2017; 322:402–16.

5. Wanjari N, Waghmare J. A review on latest trend of cosmetics-cosmeceuticals. *International Journal of Pharma Research & Review*. 2015; 4:45–51.

6. Dini I, Laneri S. Nutricosmetics: A brief overview. *Phytotherapy Research*. 2019; 33(12):3054–63.

7. Abd Razak DL, Fadzil NH, Jamaluddin A, Abd Rashid NY, Sani NA, Manan MA. Effects of different extracting conditions on anti-tyrosinase and antioxidant activities of *Schizophyllum* commune fruit bodies. *Biocatalysis and Agricultural Biotechnology*. 2019 May 1; 19:101116.

8. Campos-Vega R, Nieto-Figueroa KH, Oomah BD. Cocoa (*Theobroma cacao* L.) pod husk: Renewable source of bioactive compounds. *Trends in Food Science and Technology*. 2018; 81:172–84.

9. Madan K, Nanda S. In-vitro evaluation of antioxidant, anti-elastase, anti-collagenase, anti-hyaluronidase activities of safranal and determination of its sun protection factor in skin photoaging. *Bioorganic Chemistry*. 2018; 77:159–67.

10. Hossain MS, Sharfaraz A, Dutta A, Ahsan A, Masud MA, Ahmed IA, Goh BH, Urbi Z, Sarker MMR, Ming LC. A review of ethnobotany, phytochemistry, antimicrobial pharmacology and toxicology of *Nigella sativa* L. *Biomedicine & Pharmacotherapy*. 2021; 143:112182.

11. Chaiyana W, Anuchapreeda S, Punyoyai C, Neimkhum W, Lee K-H, Lin W-C, Lue S-C, Viernstein H, Mueller M. *Ocimum sanctum* Linn. As a natural source of skin anti-ageing compounds. *Industrial Crops and Products*. 2019; 127:217–24.

12. Emerald M, Emerald A, Emerald L, Kumar V. Perspective of natural products in skincare. *Pharmacy & Pharmacology International Journal*. 2016; 4(3):72.

13. Pillaiyar T, Manickam M, Namasivayam V. Skin whitening agents: Medicinal chemistry perspective of tyrosinase inhibitors. *Journal of Enzyme Inhibition and Medicinal Chemistry*. 2017; 32:403.

14. Lam RYY, Lin ZX, Sviderskaya E, Cheng CHK. Application of a combined sulphorhodamine B and melanin assay to the evaluation of Chinese medicines and their constituent compounds for hyperpigmentation treatment. *Journal of Ethnopharmacology*. 2010; 132:274.

15. Amer M, Metwalli M. Topical liquiritin improves melasma. *International Journal of Dermatology*. 2000 Apr; 39(4):299–301.

16. Philips N, Samuel M, Arena R, Chen YJ, Conte J, Natarajan P, Haas G, Gonzalez S. Direct inhibition of elastase and matrixmetalloproteinases and stimulation of biosynthesis of fibrillar collagens, elastin, and fibrillins by xanthohumol. *Journal of Cosmetic Science*. 2010 Mar 1; 61(2):125–32.

17. Acevedo JA, Castañeda CM, Benitez FJ, Durán DA, Barroso VR, Martínez CG, Muñoz LJ, Martínez CA, de Vivar AR. Photoprotective activity of *Buddleja scordioides*. *Fitoterapia*. 2005 Jun 1; 76(3–4):301–9.

18. De la Coba F, Aguilera J, Korbee N, de Gálvez MV, Herrera-Ceballos E, Álvarez-Gómez F, Figueroa FL. UVA and UVB photoprotective capabilities of topical formulations containing mycosporine-like amino acids (MAAs) through different biological effective protection factors (BEPFs). *Marine Drugs*. 2019 Jan 14; 17(1):55.

19. Lee DY, Lee J, Jeong YT, Byun GH, Kim JH. Melanogenesis inhibition activity of floralginsenoside A from *Panax ginseng* berry. *Journal of Ginseng Research*. 2017 Oct 1; 41(4):602–7.

20. Lee N-Y, Wong LP, Siti Hamidah M-S, Zuhaili I, Mohd Azizi CY, Ida-Idayu M. Application of rubber (*Hevea brasiliensis*) seeds oil extracted using supercritical carbon dioxide in cosmetics. *Journal Technology*. 2014; 69(4):55–59.

21. Teo WL, Gan E, Jinghan A, Chuah SY, Alain K. Double blind placebo controlled trial to evaluate of the effectiveness of a dietary supplement rich in carotenoids as adjunct to topical lightening cream for the treatment of melasma: A pilot study. *Pigmentary Disorders*. 2015; 2(164):1–6.

22. Rossetti D, Kielmanowicz MG, Vigodman S, Hu YP, Chen N, Nkengne A, Oddos T, Fischer D, Seiberg M, Lin CB. A novel anti-ageing mechanism for retinol: Induction of dermal elastin synthesis and elastin fibre formation. *International Journal of Cosmetic Science*. 2011; 33:62–69.

23. Hoang HT, Moon JY, Lee YC. Natural antioxidants from plant extracts in skincare cosmetics: Recent applications, challenges and perspectives. *Cosmetics*. 2021 Nov 10; 8(4):106.

24. Park JE, Lee KE, Jung E, Kang S, Kim YJ. Sclareol isolated from *Salvia officinalis* improves facial wrinkles via an antiphotoaging mechanism. *Journal of Cosmetic Dermatology*. 2016 Dec; 15(4):475–83.

25. Juncan AM, Moisă DG, Santini A, et al. Advantages of hyaluronic acid and its combination with other bioactive ingredients in cosmeceuticals. *Molecules*. 2021; 26:4429.

26. Hara-Chikuma M, Verkman AS. Physiological roles of glycerol-transporting aquaporins: The aquaglyceroporins. *Cellular and Molecular Life Sciences CMLS*. 2006 Jun; 63:1386–92.

27. Corinaldesi C, Barone G, Marcellini F, Dell'Anno A, Danovaro R. Marine microbial-derived molecules and their potential use in cosmeceutical and cosmetic products. *Marine Drugs*. 2017 Apr 12; 15(4):118.

28. Thiyagarasaiyar K, Goh BH, Jeon YJ, Yow YY. Algae metabolites in cosmeceutical: An overview of current applications and challenges. *Marine Drugs*. 2020 Jun 19; 18(6):323.

29. Ribeiro AM, Estevinho BN, Rocha F. Microencapsulation of polyphenols-The specific case of the microencapsulation of *Sambucus Nigra* L. extracts-A review. *Trends in Food Science & Technology*. 2020 Nov 1; 105:454–67.

30. Gibis M, Zeeb B, Weiss J. Formation, characterization, and stability of encapsulated hibiscus extract in multilayered liposomes. *Food Hydrocolloids*. 2014 Jul 1; 38:28–39.

31. Kanlayavattanakul M, Lourith N. Biopolysaccharides for skin hydrating cosmetics. *Polysaccharides: Bioactivity and Biotechnology*. 2015:1867–1892.

12 Toxicology of Natural Substances

Malik Saadullah, Muhammad Usama Munir, Muhammad Fakhar-e-Alam, Shahab Nasir, and Muhammad Hanif*

12.1 INTRODUCTION

Biological poisons, whether found in plants, microorganisms, and animal tissues, are the most potent poisons, contrary to the assumption that natural foods and medications are superior than synthetic ones. These organic toxins may take the form of benzodiazepines, tannins, glycosides, imidazole, polypeptides, enzymes, viruses, and a variety of other stereochemical types. There is no doubt that such chemicals may be responsible for several ailments that are currently classified as "idiopathic."

Generally speaking, there are two groups of toxic substances: ones that can be generated synthetically and those that are found in nature. Nevertheless, the difference between these classes is frequently only a matter of dose. An acutely poisonous substance is one that either kills within a few hours or several days of intake or causes just a brief disturbance before full recovery. They include the majority of known alkaloids, cardiac glycosides, and bacterial toxins. As the causal relationship between the two events (effect and cause) can be easily shown, their hazardous qualities have been known for a very long time. The majority of the immediate effects induced by such drugs are investigated using conventional pharmacological techniques. It has not been sufficiently demonstrated if all of their effects are truly reversible or whether any of these hazardous substances might cause persistent lesions [1].

Furthermore, toxicology refers to the investigation of how chemicals interact with biological systems and how those interactions affect living things negatively. Another one is that these actions need physical connection with the substances in the complex molecules or their metabolic byproducts. The degree of a harmful impact is proportional to the chemical dose, or, in other words, the chemical potential must be enough to cause the system to deviate from its usual range of function, and the degree of deviance is determined by the law of mass action. This is probably what Paracelsus said in the third place. A portion of this statement may be regarded to constitute allergic mechanisms. There is a dosage response following the initial contact of a person who subsequently develops allergic sensitization; this hypersensitive individual then transforms into a modified individual who subsequently reacts according to dosage but at a new dose level. About the following, it goes without saying that the contact and its harmful outcome will be in accordance with all chemistry and physics rules. Finally, if tested under identical circumstances, the chemical's harmful impact will be repeatable [2].

It is sometimes believed that because plants have been since the beginning of human development and synthetic chemicals are more recent, the defenses animals have developed to deal with the toxicity of organic products will not shield us from manufactured chemicals. In spite of this, a variety of natural poisons, some of which have existed throughout the evolutionary history of vertebrates, induce cancer in vertebrates. For instance, mold aflatoxins have been linked to the development of cancer in fish, rats, mice, monkeys, and perhaps humans. Since at least five hundred million years ago, plants have been developing and perfecting their chemical weapons, manufacturing these

*Corresponding author

 DOI: 10.1201/9781032707273-12

poisons at great fitness cost. Plants wouldn't have been organically chosen to generate these compounds if they were ineffective at keeping predators away [3].

According to Paracelsus, toxicity is a characteristic of all substances, not only a small subset known as poisons. Seen from a different perspective, this generic statement has a contradictory implication: whether or not anything is a "poison," the manifestation of toxicity is dependent on variables unrelated to the substance's fundamental properties, such as the dosage to which a person has been exposed. As a result, labeling chemical compounds as dangerous or non-toxic makes no sense, according to both Paracelsus and contemporary toxicologists. Only in respect to the circumstances of an individual's exposure to a substance do toxicity and non-toxicity exist. As a result, toxicology is engaged with the whole chemical component universe, and its goal, may be regarded as being focused on all aspects that determine harmful events rather than just processes [4].

Due to their anti-inflammatory, immunomodulatory, antioxidant, and antibacterial properties, a variety of plant extracts and other natural substances (PENS) have been used to improve animal health, growth, and performance for a long time [5]. Humans are typically exposed to several components (such as food, medications, consumer goods, and drinkable water) at varying exposure levels and via a variety of exposure pathways concurrently or sequentially. On the basis of in animal toxicity data, which are produced from exposure to specific compounds at fixed dosages through a specific exposure route and medium, human health risk is, however, primarily addressed. Moreover, the potential of simultaneous action or interactions is typically not considered when uncertainty factors are applied to account for various species and interpatient variance. Thus, new research initiatives in the arena of toxicology are aimed at:

- Determining the likelihood or absence of occurrence in people of such possible combination or interaction harmful effects from exposure to mixtures of chemicals.
- Examining the best approaches to study these complex interactions together [5].

Many toxicants need to consider any potential interactions between them since failure to do so might result in inaccurate results. As a consequence, interaction of different compounds in toxicology are studied for a variety of purposes, including (a) figuring out the causes of poisoning (i.e., toxicodynamic and/or toxicokinetic), (b) to develop effective biological and pharmacological treatment regimens and antidotes against potentially dangerous toxins, (c) evaluating the safety of different chemical mixtures under various exposure scenarios for both humans and animals. Moreover, dietary components may cause different chemical reactions with various substances to which a person may be exposed. For instance, it is known that the tannins found in several plant-derived extracts hinder the iron's absorption from the stomach. Similar to this, grapefruit juice inhibits the microsomal cytochrome P450 enzyme, also known as the isoform CYP3A4, which increases the plasma half-lives of a variety of substances. It might be appropriate to say that toxicologists currently face a significant challenge that calls for proper attention when assessing different risks of health related to mixing of interacting chemicals found in botanical medicine combinations, in nonnutritive plant-derived toxins in worth eating foods, in dietary supplements, and in the environment. Toxicologists have conducted a great deal of research to comprehend chemical-chemical interactions, and they have been effective in comprehending the fundamental processes of interactions. The chemical reaction between two substances may be toxicodynamic or toxicokinetic in nature, depending on the mechanism of interactions. The reactivity of tissue to some other chemical is changed in a toxicodynamic kind of interaction. The interactions at receptor sites are mostly to blame for this. Contrarily, in toxicokinetic reactions, one chemical may change how another chemical is absorbed, distributed, metabolized, and excreted, affecting the quantity and durability of the toxin at receptor sites [6].

12.2 TOXICOLOGY BY DIFFERENT NATURAL COMPOUNDS

12.2.1 Lathyrism

There has been a lot of new information revealed on the causes of lathyrism in both humans and animals. The sweet pea, *L. odoratus*, is to blame for this ailment. The components of L. odoratus, -aminopropionitrile and its N-7-L-glutamyl derivative, have been found to be the cause of the bone abnormalities seen in animals consuming the seeds. Similar bone abnormalities are brought on by aminoacetonitrile, bis(β-cyanoethyl) amine, and -mercaptoethylamine (Figure 12.1). When *L. odoratus* seeds were given to pregnant rats, particularly between days 13 and 22, the offspring experienced aortic aneurysms and skeletal alterations, but their mothers did not experience any negative consequences. The development of collagen appears to be connected to the biochemical defects. Neurotoxicity is caused by other poisonous nitriles found in some species of *Lathyrus* and *Vicia*, including cyanoalanine and the dipeptide 7-glutamyl-β cyanoalanine [1].

12.2.2 Embryopathic and Cytotoxic Effects

Investigations into this subject are crucial and may be applicable to humans as embryopathies in animals are frequently brought on by the fodder consumed by the pregnant animal. Remarkable abnormalities have been brought about in sheep fetus by feeding them Veratrum californicum around days 10–15 of pregnancy. Death happened at delivery because the gestation period was typically prolonged, the infant weight heavier, and they had lesioned such hydrocephalus, proboscis, cyclopia in one eye, and blockage of the respiratory passages. When administered to sheep on days 10–15 of gestation, some of the components of "wild corn" from the veratrum steroidal alkaloids (whose precise structures have not yet been determined) recreate this state, but not at any other point in pregnancy.

Natural substances may include oestrogenic chemicals that might pose specific risks, such as miroestrol (Figure 12.2). This substance was extracted from the tuberous roots of the *Pueraria mirifica* (Leguminosae) plant, which is native to North Thailand and has historically been used as a renewing medicine for elderly people. It is not advised for use by children. Even when administered orally, this drug has Oestrogenic action on level with that of natural oestrogens.

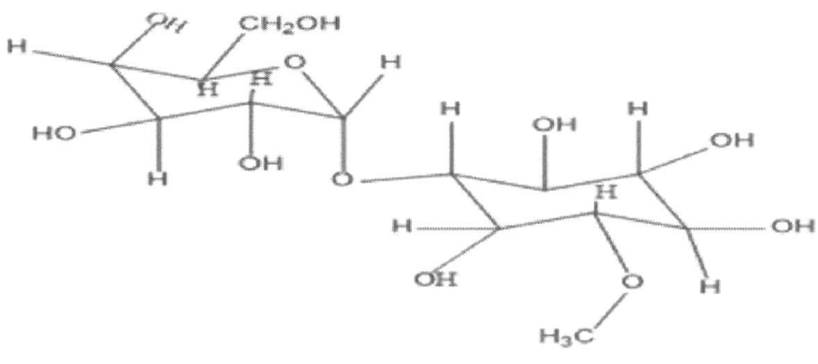

Figure 12.1 Molecular structure of lathyritol derived from the seed of *Lathyrus odorus.*

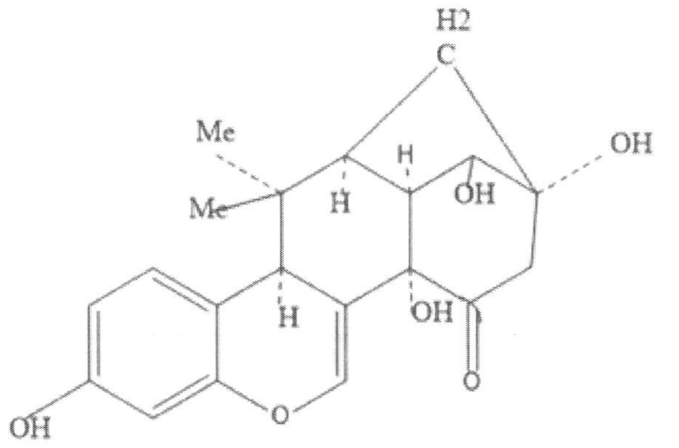

Figure 12.2 Chemical structure of miroestrol.

Recently, cardiac glycosides from *Ornithogallum umbillatum* bulbs, *Apocynum cannabinum* L., and a variety of comparable synthetic, steroidal chemicals have been found to exhibit cytotoxic effects in tissue cultures. At the μg level or even below, several of the chemicals tested were quite potent. Testing of these substances for carcinogenic activity is clearly necessary, as it provided the frequently noted link between cytotoxic and malignant activity. Apocynaceae plant extracts have been employed historically in arrow poisons; several species of Ornithogallum have been recognized to be harmful to sheep as well as other animals. Convallaria majalis L, often known as lily of the valley, has a compound called convallatoxigenin (Figure 12.3) that has a steroidal structure and an, -unsaturated lactone ring. Even in Britain, where herbalists sell it for free, it continues to be used therapeutically.

12.2.3 Neoplastic Diseases

Kaposi sarcoma is an example of multi pigmented sarcoma that has been reported in Africans. It also happens to Italians, although American Blacks are less likely to experience it. Its cause is unidentified. Onyalai is a disorder that causes purpura, bleeding from several orifices, and blood vesicles to appear on the cheeks and hard palate. Esophageal carcinoma in Transkei has been reported. Although the cause has not been determined, it is more common in some families and may be related to food.

The first-time pyrrolizidine alkaloids were proposed as potential "natural" etiological variables in the high prevalence of liver illnesses (liver cirrhosis, and primary liver malignancies) among Africans and in several other tropical

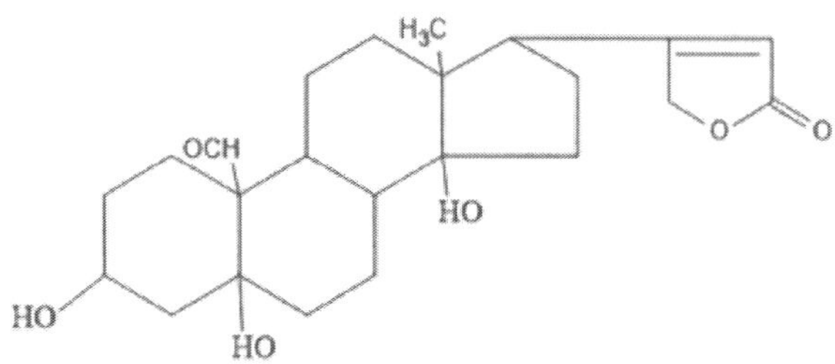

Figure 12.3 Chemical structure of convallatoxigenin.

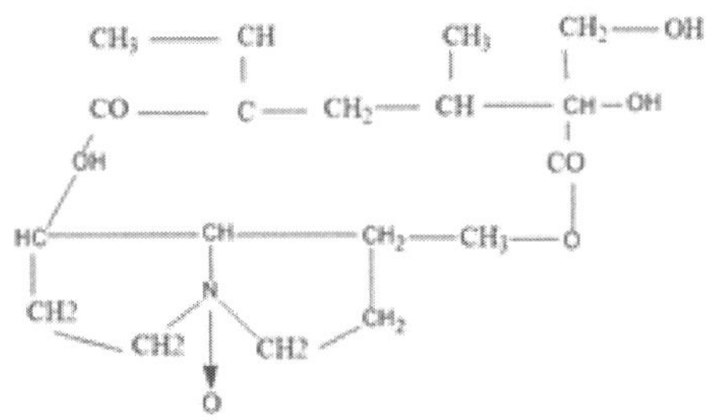

Figure 12.4 Chemical structure of isatidine (N-Oxide of retrorsine).

and subtropical regions around 15 years ago. This assertion was supported by rat trials using histidine, which is retrorsine's N-oxide. Since then, a number of other natural hepatotoxic and/or hepatocarcinogenic compounds have been identified. These include fungus-derived compounds like luteoskyrin, a chlorine-containing cyclic peptide, griseofulvum, Sopp., an acid from *P. rubrum* and other chemicals from different Penicillium species. Some funguses are similarly poisonous (Figure 12.4).

The discovery of the carcinogenic properties of cycasin, a glycoside of methylazoxymethanol found in the nuts and leaves of the tree *Cycas circinalis* L., which can be found throughout the tropics and subtropics, was of particular interest. After proper preparation, which includes soaking in water, the nuts—a source of starch—are frequently eaten as food. Aqueous emulsions of fresh cycad nuts have been applied as dressings for skin ulcers and wounds and seem to hasten their healing. Encephalartos and Zamias are two other Cycadaceae species; some are poisonous and, like Macrozamia communis, may contain macrozamin (also known as cycasin). The unstable aglycone methylazoxymethanol that was produced in the free form by hydrolyzing cycasin with emulsion, the-glucosidase of almonds, is the active ingredient in both drugs. Both oral and intraperitoneal administration of methylazoxymethanol (Figure 12.5) causes cancer. While cycasin is only active when taken orally, its more stable derivative from acetylation is similarly active by both routes. Evidently, the carcinogen is created as a result of the glycoside's breakdown in the digestive system. In addition to mesenchymal tumors, bile duct and colon tumors may also be created with just one intravenous injection of methylazoxymethanol.

This natural substance shares strong molecular similarities with the man-made hepatocarcinogen dimethyl nitrosamine; both likely function through diazomethane produced during metabolic breakdown. Elaiomycin has been discovered, a related glycoside, from Streptomyces hepaticus. This substance results in cirrhosis of the liver in guinea pigs and possesses antibacterial activity against Mycobacterium tuberculosis, it might also result in liver tumors.

Crotalaria aridicola is thought to be the cause of Chillagoe illness, which results in esophageal tumours in horses. Possible causes for esophageal cancer in livestock in Kenya include comparable factors. It would be important to find

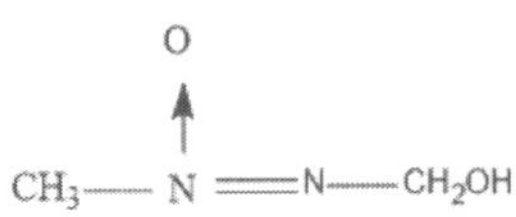

Figure 12.5 Chemical structure of methylazoxymethanol.

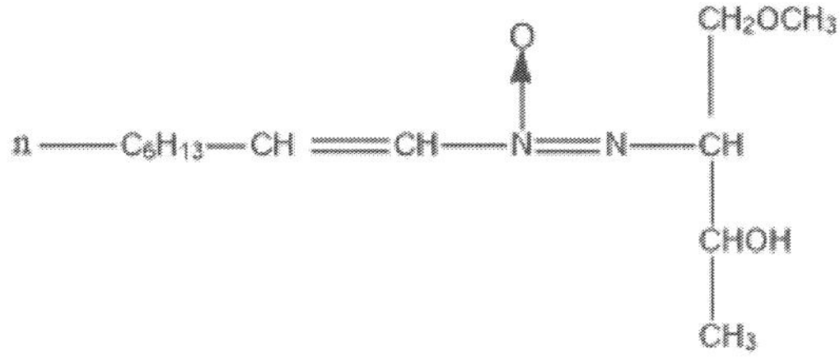

Figure 12.6 Chemical structure of elaiomycin.

out if the high rate of esophageal tumors among Bantu people in Transkei was maybe caused by their use of certain plant ingredients. They might theoretically produce alkyl nitroso compounds during fermentation if they are high in nitrates. Experimental research has shown that certain alkyl nitroso compounds can cause rat esophageal cancers (Figure 12.6).

Dimethyl nitrosamine and other unexplained nitroso compounds have recently been found in some batches of herring meal that has been preserved with sodium nitrite. When fed to sheep, a species that is especially susceptible to the hepatotoxic effects of dimethyl nitrosamine, these herring meals were hepatotoxic. Different strained or cqfl-unsaturated lactones are abundantly found in natural goods, the discovery of their carcinogenic qualities' points to the necessity for more research on this kind of chemical.

12.2.4 Other Diseases

In South Africa, the disease known as geeldikkop in sheep may also be caused by chronic selenium poisoning. It has been shown that eating the plant, *Lippia rehmanni* Pears, from which different scientists isolated a chemical known as "icterogenin," caused the photosensitization disease with jaundice. The structures of icterogenin and similar pentacyclic triterpenes, which are also found in *Lippia pretorensis* pears and *Lantana camara* L. In anaesthetized rabbits with carmulated bile ducts, other few chemists in 1964 investigated the effects of these compounds and several of their derivatives. The fl-22-angeloyloxy group, which is present in icterogenin, and the fl-equatorially directed hydroxyl group at C-3 or at C-24, both significantly increased the icterogenic activity. The second one seems to be five times as poisonous as icterogenin (Figure 12.7).

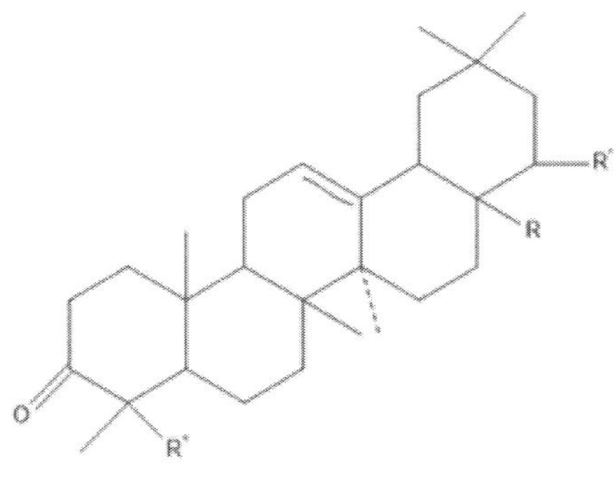

Figure 12.7 Chemical structure of icterogenin.

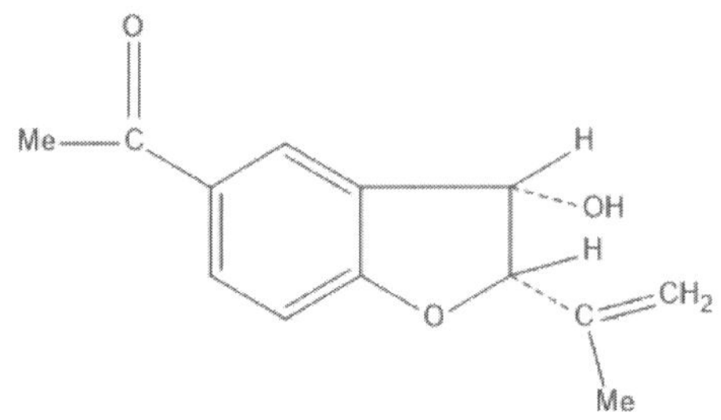

Figure 12.8 Chemical structure of toxol.

Nancy Hanks Lincoln, the mother of Abraham Lincoln, died of milk poisoning, which is another illustration of harmful substances being removed in milk. The hazardous plant in question was the Jimmy weed. The compound toxol, which is (-)-2-isopropenyl-3-hydroxy-5-acetyl-2,3-dihydrobenzofuran, is most likely to blame for the plant's toxicity. Tremetone appears to be less hazardous than toxol. The plant *Eupatorium urticaefolium* also exhibits it (Figure 12.8).

12.3 TOXICOLOGY OF SIMPLE AND COMPLEX MIXTURES

Understanding the fundamentals of chemical interaction and combined action, as well as the differences between component-interaction analysis and whole-mixture analysis, are essential for studying the toxicology of mixtures of chemicals and accurately assessing their potential health risks. Additionally, it is crucial to distinguish between Feronetal's simple and complicated combinations. A simple mixture is one that has a known composition on both a qualitative and quantitative level and is composed of only a small number of compounds (for example, ten or less). A complicated mixture is one that contains tens, hundreds, or even thousands of compounds, the qualitative and quantitative makeup of which is unknown. Pesticide mixtures, drug combinations, or a collection of allyl alcohol esters are examples of the former; working environments, water sources, welding fumes, or natural flavoring compounds are examples of the latter category.

The toxicokinetic phase, which includes processes of absorption, distribution, metabolism, and excretion, and the toxicodynamic phase, which includes the impact of toxins on the receptor, cell target, or organ, are both potential sites for interactions. Since metabolism is a key factor in cytotoxicity, enzyme induction or inhibition, or both, are the most apparent instances for connections in the toxicokinetic phase. The toxicity of other compounds can be significantly influenced by substances that affect the quantity of biotransformation enzymes. Uptake and excretion may also be ongoing activities. Along with competition between stimuli for the biological transforming enzymes themselves, there is also competition amongst substrates for the same pumps (Figure 12.9).

A detailed characterization of the situation of no or zero communication is necessary when testing for a potential relationship between multiple agents. There have been many different empirical techniques put forth, but there is still no universally accepted definition of zero interaction 7–10. The three fundamental ideas known as impact or reaction additivity, independence, and dosage additivity (Loewe additivity) may be linked to practically all of these techniques. With the exception of few niche areas like the hazardous consequences of combined irradiation, the independence criteria do not appear to be applied much in toxicology.

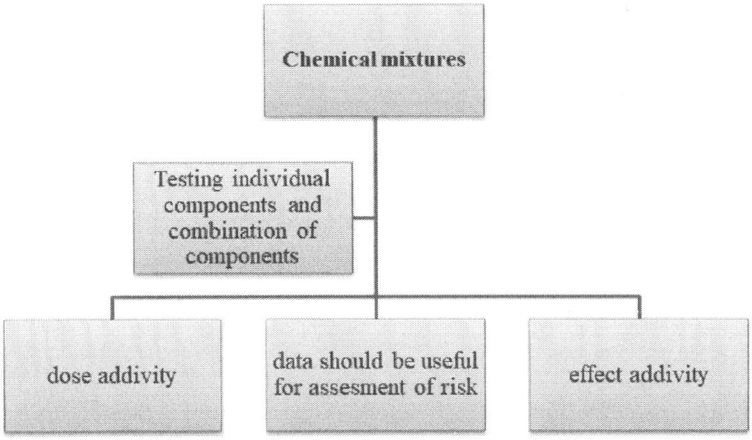

Figure 12.9 Fundamental ideas for reaction additivity.

12.4 CHALLENGES TO DEAL WITH SIMPLE AND COMPLEX MIXTURES

The challenging nature of a single chemical research and a combination design investigation varies significantly. The amount of substances in a mixture and if it is important to evaluate potential existing chemical interactions will both have a significant impact on the experimental strategy of a mixed design research. Mixture design studies still require substantial additional evaluation and cross validation regardless their potential and existing application. It is better to evaluate a combination in these investigations at high (effective) doses as well as low (realistic) values.

An integrated toxicological testing approach must now include in in vitro assays. Additionally, as a result of advancements in science in the disciplines of molecular and cell biology, it is now possible to analyze the mechanism of toxicity of mixes using in vitro test systems in high-throughput test scenarios. Toxicology will now be able to use the latest developments in the fields of molecular science and chip engineering to identify the toxic effects of combinations at a molecular level via effectiveness and safety assessment in pharmaceutical research. In this approach, proteins that have been affected by a toxin may be recognized, and information about altered metabolic pathways can then be retrieved. The initial step towards revealing the topological arrangement and dynamics of metabolic, signaling, and regulatory networks is the accumulation of such proteomic signatures in databases. Proteomics and toxicogenomic advancements will considerably expand our knowledge of toxicity pathways, and this, in turn, may be anticipated to have a considerable influence on our understanding of the methods for combined-action evaluation [5].

12.4.1 Toxicology of Nutmeg Abuse

The nutmeg tree, *Myristica fragrans*, produces seeds that are used to make the culinary spice known as nutmeg. From a toxicological point of view, it is best characterized as an addictive replacement for psychotomimetic drugs. In the past, nutmeg intoxication was on the rise at the beginning of the century, but cases declined until a rebound in the middle of the 1960s.1 Since then, nutmeg was reduced to the status of a lawful or natural high that teenagers occasionally

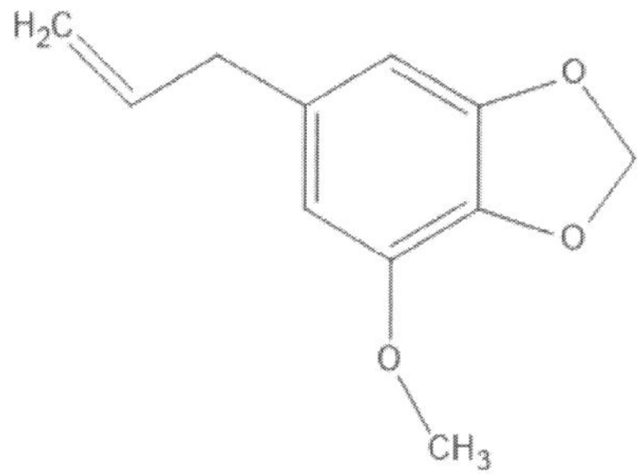

Figure 12.10 Chemical structure of myristicin.

rediscover. Because the adverse effects are often unpleasant and scary and may cause self-referral for medical care, widespread and persistent nutmeg misuse is not a prevalent issue (Figure 12.10) [7].

12.4.2 Toxicology of Herbs

The study and use of botanicals that have been discovered to have therapeutic characteristics is known as herbal medicine, plant medicine, or phytotherapy. Different methods towards herbal therapy have emerged in many civilizations. Western herbology emphasizes plants that can be found in North America and Europe, Chinese herbology depends on the customs and methods of traditional Chinese medicine, and Ayurvedic medicine is based on the basis of Indian herbology.

Animals with particular health issues and illness situations may get nutritional supplements as part of dietary medicine, often known as therapeutic nutrition. Nutraceuticals, a term recently coined to describe substances reported to have properties of both nutrients and pharmaceuticals, may include vital nutrients (amino acids), not vital nutrients (omega-3 fatty acids), herbal remedies, organ extracts, or enzymes used for treating or avoiding disease [8].

12.5 ADVERSE OR TOXIC REACTIONS

An unfavorable, dangerous, destructive, or fatal effect is referred to as a negative or toxic response to a drug. Only two of the other subclassifications of adverse responses will be discussed here. An unfavorable or hazardous reaction that is dose-dependent may be brought on by a substance's pharmacology or inherent chemical features. An illustration would be diarrhea following aloe consumption. A genetically determined idiosyncratic or dosage-independent response may show excessive reactivity to a small amount or excessive insensitivity to a high dose, among other symptoms (Figure 12.11) [8].

12.5.1 Drug and Herb Interactions

Drug interactions with natural items are becoming more and more common. In certain situations, the herb or vitamin boosts the effects of the medication, necessitating a decrease in the medication's dosage. In other instances, the natural substance may change how a medicine is metabolized or absorbed. Two natural compounds may occasionally interact, boosting or opposing the effects of either of them. For instance, potassium losses are increased by both aloe and licorice. On the basis of an understanding of the mechanism of action, interactions can be anticipated to some extent. However, it becomes challenging to forecast when the treatment mechanism is unclear [8].

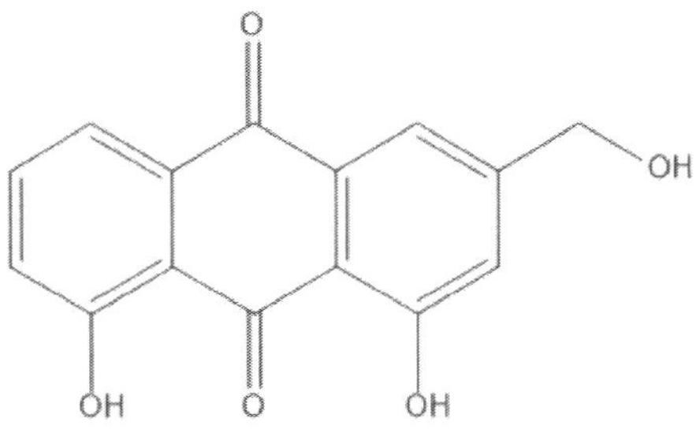

Figure 12.11 Chemical structure of Aloe emodin.

12.5.1.1 Toxicology of Emerging Capsinoids

Red peppers, or capsicum fruits, are used as seasonings to enhance tasting and improve the appeal of bland dishes all over the world. Red pepper's main pungent ingredient, capsaicin, has been shown to elevate energy expenditure, induce the release of catecholamines, and reduce body fat storage in test

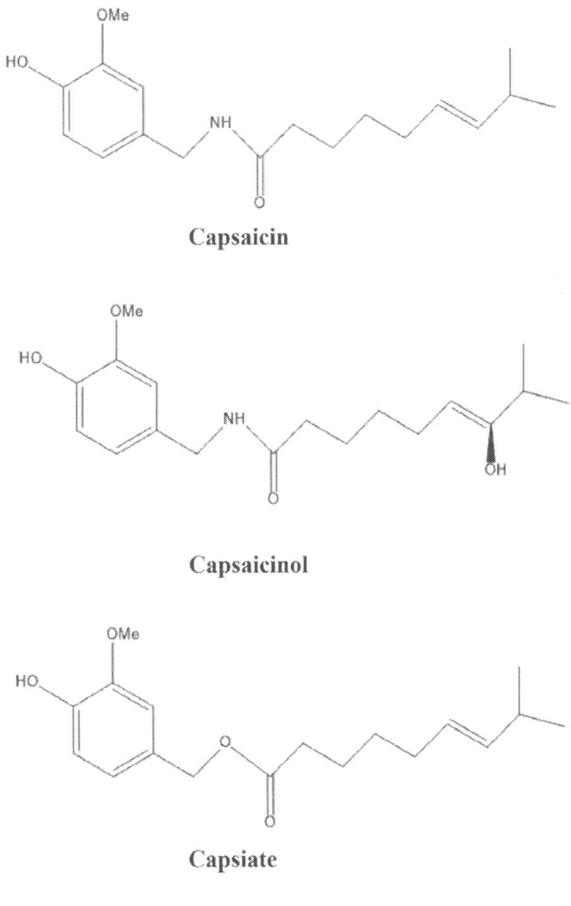

Figure 12.12 Chemical structure of capsaicin, capsaicinol, and capsiate.

animals. As a result, capsaicin is thought to be a possible element of dietetic treatment for obese and diabetic patients. The consumption of it to humans is mainly prohibited by its powerful odor and poisonous nature (Figure 12.12) [9].

12.5.1.1.1 Physiological and Pharmacological Effects of Capsinoids

Similar to a mouse dosage of hot capsaicin, only one oral treatment of capsinoids raises temperature in the body, utilization of oxygen, and blood adrenaline levels. According to research, capsinoids stimulate sympathetic nerve activity and improve the breakdown of energy in both lab animals and humans. Along with pharmacological tests, the researchers Kawabata et al. discovered indicating capsinoids enhance the use of energy in wild-type mice but not in TRPV1 knockout mice, demonstrating that capsinoids enhance energy expenditure via TRPV1 [9].

Uncoupling protein 1, a mitochondrial inner-membrane molecule that can disperse a proton gradient before it is used to generate ATP for oxidative phosphorylation and play a role in heat production, is also said to be increased by Capsiate. The effect of Capsiate on fat storage in human research are given in addition to research animal studies. Human body fat is said to be reduced with capsinoids therapy [9].

12.5.1.1.2 Toxicity of Capsinoids

Acute toxicity After the administration of lethal dose of capsinoids in both male and female mice, toxicity was observed. According to the results the no deaths occurred in 10 male and 10 female mice. After 6 hours of treatment, several behavioral changes—including a staggered walk, drowsiness, tremor, and gasping—were gone [9].

Chronic toxicity A 26-day study of administration of extract was done on both male and female mice. With the exception of male rats receiving the maximum dosage, no changes in physical condition, including weight and food/water intake, were noticed. Salivation was observed in all mice. The number of segmented neutrophils, aspartate aminotransferase (AST), alanine aminotransferase (ALT), lactate dehydrogenase, and liver cell focal necrosis all increased in high-dose males. The liver weight of the same animals showed small increases as well [9].

12.6 EFFECT OF PENS AGAINST MEMBRANE DAMAGE INDUCED BY ENTEROTOXIGENIC ESCHERICHIA COLI K88 IN PIG INTESTINAL CELLS

Enterotoxigenic *Escherichia coli* (ETEC) variant K88 is the most frequent research projects cause of piglet enteric illnesses and diarrhea. Infectious disorders are an important cause of death in weanling pigs. Since many years ago, medicines have been used to lower piglet infection levels caused by pathogens, however numerous bacteria are now developing antibiotic resistance. Finding replacements to in-feed antibiotics as preventative antibacterial and growth regulators is consequently urgently needed, especially in light of the recently complete restriction on the use of antibiotics as growth regulators. Because of their anti-inflammatory properties, immunomodulatory, antioxidant, and antibacterial effects, a variety of plant preparations and other naturally occurring compounds have been recommended for use in enhancing animal health, growth, and performance. In order to enhance the health condition of piglets by preventing ETEC infection, we have thus hypothesized that PENS may serve as alternatives to in-feed antibiotics. An in vitro model using ETEC K88-infected pig intestinal epithelial IPEC-1 cells and several PENS with varying activity has been employed to test this concept. We have already demonstrated that ETEC K88 may cause membrane rupture in human intestinal Caco-2 cells, which is followed by inflammatory responses. Even though they haven't been fully studied, IPEC-1 cells can develop into mature ones and show enterocytic characteristics such polarization, microvilli, and tight connections. Prior to using the PENS to assess any potential protective benefits against ETEC-induced membrane damage, it was first evaluated for potential harmful effects on membrane permeability [10].

12.7 METHODS APPLIED TO THE IN VITRO PRIMARY TOXICOLOGY

The study of all-natural products, unprocessed extracts, and isolated chemicals using in vitro techniques is becoming an accepted method for both screening and mechanistic research. There is still much room for study and discovery because the majority of the products that have been identified so far have only been examined on a small number of the numerous tissue and cell models that are accessible. The first phase of in vitro experimentation

Figure 12.13 Method to evaluate the types and mechanisms of cell death.

should be to identify the amounts that are hazardous to the model, whether studying cytotoxic, detrimental (toxic), or beneficial effects. Cellular toxicity may or may not be a goal on its own depending on the research's objectives and expected outcomes (Figure 12.13) [11].

12.7.1 Apoptosis/Necrosis Differentiation

The assay is based on the detection of phosphatidylserine translocation, a mechanism that takes place in the first stages of apoptosis, from the interior to the exterior side of the plasma membrane. In a Ca^{2+}-dependent manner, the serum protein annexin V, which has an unclear function, has a great affinity for phospholipids like PS. In the year 1995, Vermes et al. created an assay to identify PS externalisation using an Anexin V that was FITC-labeled. The approach was combined with PI labelling to enable the recognition of later stages of necrosis and apoptosis during which the integrity of the plasma membrane is disturbed. These two indicators enable distinguishing between apoptosis and necrosis and may be simultaneously identified by flow cytometry. Depending on the stimuli and the cell type, apoptosis is expected to continue for between 12 and 24 hours; hence, annexin V/PI interaction only offers information on a certain time-point, neglecting the overall number of cells that have gone through apoptosis. The test has a high level of stability since PS exposure alone causes apoptosis. Working with herbal products that include saponins, however, requires caution since they have the potential to cause membrane integrity disturbance and allow PI to enter the cytoplasm [11].

12.7.2 Assays Based on the Assessment of DNA Damages

Cells may initiate death programs depending on the level of DNA damage, particularly double strand breaks. Investigating this potential cause of mortality is made feasible by the detection of harm to DNA at concentrations sensible lower than the IC50 [11].

12.7.2.1 Comet Assay

The single-cell gel electrophoresis assay, also known as the comet test, is a short-term genotoxicity test that is often used to quantify (i) DNA strand breakdowns, crosslinks, and alkali-labile sites caused by a variety of physical or chemical agents, as well as (ii) incorrect cutting repair events in individual eukaryotic cells. Lysed and electrophoresed individual cells embedded in agarose. When used in combination with PI staining, fluorescence microscopy makes it possible to see damaged DNA fragments that are moving outside of the cell's nucleus following electrophoresis. The result is a picture of a "comet" with a unique head made up of solid DNA and a tail made up of flexible DNA loops or fragments of broken DNA. The classification of comets can be done visually or using morphological criteria derived from image analysis and intensity profile integration [11].

12.7.2.2 Ames Test

An in vitro approach for evaluating genotoxicity is the Ames technique. It is a bacterial reverse mutation experiment that uses *E. coli* or *Salmonella typhimurium* histidine-dependent or tryptophan-dependent auxotrophic mutants. Competitive pressure from a medium deficient in the necessary amino acid causes reverse alterations and the development of clusters that are measured in the presence of a carcinogenic product. Salmonella strains need to be diversified since each one may test for a different kind of mutagenic agent on its own [11].

12.7.2.3 Micronucleus Test

An acentric chromosomal segment left over after mitotic cell division is known as a micronucleus and can be seen in the cytoplasm of the interphase phase cells as a little extra nucleus. Agents that primarily cause chromosomal growth or loss or chromosomal breakdowns can also create micronuclei. Fluorescent DNA labelling and microscope analysis make it simple to find an index of chromosomal breakage and loss. By using FISH analysis or by categorising the size of micronuclei, clastogens and aneugens are distinguished. The test is regarded by the OECD as one of most effective and trustworthy for detecting genotoxicity. The exam is straightforward to administer, easy to score, and sensitive, but it is challenging to improve in a high-throughput form. However, automated tests using image analysis to size-classify micronuclei and distinguish aneugens from clastogens have just recently been created [11].

12.7.2.4 UDS Assay

The UDS test quantifies chemically stimulated DNA excision repairing by spotting the incorporation of labelled thymidine (3H TdR). It is assumed that DNA damage came first before the activation of DNA repair mechanisms, demonstrating a chemical's capacity to cause DNA damage. The UDS assay's failure to determine if an infectious agent is mutagenic, is one of its main drawbacks; in fact, it offers no information on the accuracy of DNA repair and fails to recognise DNA lesions treated by processes except for excision repair [11].

12.8 COMPARATIVE TOXICOLOGY OF NATURE'S AND SYNTHETIC CHEMICALS

It is frequently believed that because plants have been used since the beginning of human development whereas artificial substances are more recent, the defenses animals have developed to deal with the cytotoxicity of natural chemicals would not shield us from artificial chemicals.We believe this premise to be false because a number of reasons [3].

- Since there are so many natural substances that might have hazardous effects, it makes sense that the majority of animal defenses have evolved to be of a broad kind. Humans are well-equipped against toxins because of general defenses that provide defense not just against natural substances but also against artificial substances. Included in these defenses are the following. The surface layers of the mouth, oesophagus, stomach, intestine, colon, skin, and lungs disappear every few days. (a) The constant shedding of cells exposed to poisons. (b) The induction of numerous general detoxifying mechanisms, such as antioxidant defenses or glutathione transferases for detoxifying alkylating agents: Human cells exposed to low dosages of an oxidant, such as radiation or hydrogen peroxide, induce antioxidant defenses and become more resistant to higher doses. These defenses are efficient against both natural and manmade oxidants, such as the herbicide paraquat, and can be triggered by both. (c) The active expulsion of synthetic or natural planar hydrophobic compounds from intestinal and hepatic cells. (d) DNA repair: This can be induced in response to DNA damage and is effective against DNA adducts produced from both artificial and natural substances. (e) Animals' smell and gustatory perception of chemicals that are sour, sharp, stringent in nature, and pungent warn against a variety of toxins and may be more effective in doing so for a few natural toxins that have played a significant role in the evolution of food toxicity than for some synthetic toxins. Nevertheless, it is probable that these signals are also broad defenses that are keeping an eye on specific toxicity-related components, as several synthetic dangerous substances have astringent, acrid, or pungent flavors. Some of these characteristics are present in mustard, pepper, garlic, onions, and other foods, yet people frequently disregard the warnings [3].

- Despite the fact that several natural poisons, some of which have existed throughout the evolutionary history of vertebrates, induce cancer in vertebrates. For instance, mould aflatoxins have been linked to cancer in fish, rats, mice, monkeys, and perhaps even people. Eleven mould poisons have been demonstrated to cause cancer, while 19 have been found to cause clastogenic effects. Despite being present throughout evolution, several common elements, such as lead, cadmium, beryllium, nickel, chromium, selenium, and arsenic salts, are carcinogenic or clastogenic at high levels. In animal nutrition, selenium and chromium are crucial trace nutrients [3].

REFERENCES

1. Barceloux DG. Medical toxicology of natural substances: Foods, fungi, medicinal herbs, plants, and venomous animals. 3rd ed. 2008.

2. Waddell WJ. The science of toxicology and its relevance to MCS. *Regulatory Toxicology and Pharmacology*. 1993;18(1):13–22.

3. Ames BN, Profet M, Gold LS. Nature's chemicals and synthetic chemicals: Comparative toxicology. *Proceedings of the National Academy of Sciences*. 1990;87(19):7782–6.

4. Paumgartten FJR. Risk assessment for chemical substances: The link between toxicology and public health. *Cadernos de Saúde Pública*. 1993;9:439–47.

5. Groten JP, Feron VJ, Sühnel J. Toxicology of simple and complex mixtures. *Trends in Pharmacological Sciences*. 2001;22(6):316–22.

6. Bhat AS, Ahangar AA. Methods for detecting chemical–chemical interaction in toxicology. *Toxicology Mechanisms and Methods*. 2007;17(8):441–50.

7. Sangalli BC, Sangalli B, Chiang W. Toxicology of nutmeg abuse. *Journal of Toxicology: Clinical Toxicology*. 2000;38(6):671–8.

8. Harman J. The toxicology of herbs in equine practice. *Clinical Techniques in Equine Practice*. 2002;1(2):74–80.

9. Watanabe T, Ohnuki K, Kobata K. Studies on the metabolism and toxicology of emerging capsinoids. *Expert Opinion on Drug Metabolism & Toxicology*. 2011;7(5):533–42.

10. Roselli M, Britti M, Le Huërou-Luron I, Marfaing H, Zhu W, Mengheri E. Effect of different plant extracts and natural substances (PENS) against membrane damage induced by enterotoxigenic *Escherichia coli* K88 in pig intestinal cells. *Toxicology in Vitro*. 2007;21(2):224–9.

11. Bunel V, Ouedraogo M, Nguyen AT, Stévigny C, Duez P. Methods applied to the in vitro primary toxicology testing of natural products: State of the art, strengths, and limits. *Planta Medica*. 2014;80(14):1210–26.

Index

Printed in the United States
by Baker & Taylor Publisher Services